State-Of-The-Art Sensors Technology in France 2016

Special Issue Editors

Nicole Jaffrezic-Renault
Gaelle Lissorgues

MDPI • Basel • Beijing • Wuhan • Barcelona • Belgrade

MDPI

Special Issue Editors
Nicole Jaffrezic-Renault
Institute of Analytical Sciences
University of Lyon
France

Gaelle Lissorgues
ESIEE Noisy-le-Grand
France

Editorial Office
MDPI AG
St. Alban-Anlage 66
Basel, Switzerland

This edition is a reprint of the Special Issue published online in the open access journal *Sensors* (ISSN 2072-6651) in 2017 (available at: http://www.mdpi.com/journal/sensors/special_issues/ technology_france_2016).

For citation purposes, cite each article independently as indicated on the article page online and as indicated below:

Lastname, F.M.; Lastname, F.M. Article title. *Journal Name*. **Year**. *Article number*, page range.

First Edition 2018

ISBN 978-3-03842-652-3 (Pbk)
ISBN 978-3-03842-653-0 (PDF)

Table of Contents

About the Special Issue Editors

Nicole Jaffrezic-Renault received her engineering degree from the Ecole Nationale Suprieure de Chimie, Paris, in 1971 and the Doctorat dEtat s Sciences Physiques from the University of Paris in 1976. As Director of Research at the Centre National de la Recherche Scientifique, past president of the chemical micro sensor club (CMC2), president of the Analytical Division of the French Chemical Society, her research activities in the Institute of Analytical Sciences, include conception and design of (bio)chemical sensors and their integration in microsystems. She coordinates several European and national projects for the development of microsystems for biomedical and environmental monitoring and for food safety. She published more than 500 papers with more than 8500 citations (H index: 45).

Gaelle Lissorgues graduated from Ecole normale Suprieure de Cachan (1993), obtained her PhD in applied physics (1997) at University Paris 6, and her research habilitation in electronics and physics at University Paris Est (2006). Since September 1997, she is working at ESIEE PARIS engineering school, first as Assistant professor, then associate professor in 2002, and finally Full Professor since 2007. She is also the Head of the Health, Energy and Environment Department at ESIEE PARIS since 2012. Her research activity is performed within the ESYCOM research lab at University Paris Est on microsensors and microtechnologies for biomedical applications. She is focusing on microelectrode arrays for Neurosciences applications like cortical or retinal implants, on volatile organic compound detection based on artificial nose concepts, or on wearable medical sensors for health diagnostic or monitoring (intra ocular pressure sensor, heart pulse wave detection). She published more than 30 journals, 70 conference proceedings, and 11 patents.

Preface to "State-Of-The-Art Sensors Technology in France 2016"

The french community of sensors and microsystems is structured around three organizations:

- Chemical Microsensor Club (CMC2): chemical sensors and biosensors
- French Group of Bioelectrochemistry (GFB): biosensors and biofuel cells
- Electronics, Electrical Engineering, Automatic (EEA) club : physical sensors and systems

These organizations are active in the animation of the french communities: annual meeting, thematic meetings and information letters and websites. In 2017, CMC2 organized the European conferences Bioelectrochemistry 2017 (3–7 July 2017 in Lyon) and Eurosensors 2017 (3–6 September 2017 in Paris). Sensors was then the editor of the Proceedings of Eurosensors 2017.

In order to be able to reflect the research activity in France, in 2016, in the field of sensors and microsystems, the guest editors contacted the researchers through the cited three organizations. 11 contributions were then published in this book. The book includes research articles that consolidate our understanding of the state-of-the-art in this area and also four reviews on hot fields in sensor technology (nanomaterials, electronic tongue and optical fibre networks).

We thank all the contributors, CMC2, GFB, EEA club and Sensors' staff for giving us the opportunity to spread throughout the world, the research activity in France, in 2016, in the field of sensors and microsystems.

Nicole Jaffrezic-Renault, Gaelle Lissorgues
Special Issue Editors

sensors

MDPI

Article

Integration of P-CuO Thin Sputtered Layers onto Microsensor Platforms for Gas Sensing

Lionel Presmanes [1],*, Yohann Thimont [1], Imane el Younsi [1], Audrey Chapelle [2], Frédéric Blanc [2], Chabane Talhi [2], Corine Bonningue [1], Antoine Barnabé [1], Philippe Menini [2],* and Philippe Tailhades [1]

[1] CIRIMAT, Université de Toulouse, CNRS, INPT, UPS, 118 Route de Narbonne,
    F-31062 Toulouse CEDEX 9, France; thimont@chimie.ups-tlse.fr (Y.T.); ielyounsi@gmail.com (I.e.Y.);
    bonning@chimie.ups-tlse.fr (C.B.); barnabe@chimie.ups-tlse.fr (A.B.); tailhades@chimie.ups-tlse.fr (P.T.)
[2] LAAS-CNRS, Université de Toulouse, UPS, INSA, 7 avenue du colonel Roche, F-31031 Toulouse, France;
    chapelle@laas.fr (A.C.); frederic.blanc@laas.fr (F.B.); chabane.talhi@laas.fr (C.T.)
*   Correspondence: presmane@chimie.ups-tlse.fr (L.P.); Philippe.Menini@laas.fr (P.M.);
    Tel.: +33-5-6155-7751 (L.P.); +33-5-6133-6218 (P.M.)

Academic Editors: Nicole Jaffrezic-Renault and Gaelle Lissorgues
Received: 31 March 2017; Accepted: 14 June 2017; Published: 16 June 2017

Abstract: P-type semiconducting copper oxide (CuO) thin films deposited by radio-frequency (RF) sputtering were integrated onto microsensors using classical photolithography technologies. The integration of the 50-nm-thick layer could be successfully carried out using the lift-off process. The microsensors were tested with variable thermal sequences under carbon monoxide (CO), ammonia (NH_3), acetaldehyde (C_2H_4O), and nitrogen dioxide (NO_2) which are among the main pollutant gases measured by metal-oxide (MOS) gas sensors for air quality control systems in automotive cabins. Because the microheaters were designed on a membrane, it was then possible to generate very rapid temperature variations (from room temperature to 550 °C in only 50 ms) and a rapid temperature cycling mode could be applied. This measurement mode allowed a significant improvement of the sensor response under 2 and 5 ppm of acetaldehyde.

Keywords: gas sensor; RF sputtering; thin film; CuO; tenorite; photolithography; metal oxide microsensor; micro-hotplate; pulsed temperature

1. Introduction

Metal-oxide (MOS) gas sensors based on a micromachined silicon substrate [1] were a disruptive development which led to a mature and robust form of technology [2]. There are a few examples of devices on the market, which are notably based on SnO_2 and WO_3 metal oxides. To lower the resistivity of the sensitive film and improve the kinetics of the chemical reactions, commercial MOS gas sensors are operated in constant temperature mode (isothermal) knowing that the interactions between the sensitive material and the surrounding gases are temperature-dependent. Because the temperature dependence is not similar for all gases, the operation of a sensor at different temperatures can provide discrimination of several gases with one single sensor [3,4]. It has also been shown that with very short temperature pulses, transient sensor responses are strongly dependent on the ambient mixture of gases, which provides a good opportunity to enhance sensor selectivity [5–8]. These micro-hotplates can now be elaborated with different types of semiconducting sensitive layers, with the very interesting possibility of modulating the operational temperature and integrating the electronics with the sensor silicon chip. Current technologies allow temperature cycling up to several millions of cycles without failure. Despite their increased system-level complexity, microsensors have many advantages, such as, for example, high performance, small size, low cost, and low power consumption [9]. The latter requires on the order of a few or tens of mW for continuous operation, but sub-mW consumption can be reached by using a pulsed operating temperature [3]. Such microsensors are particularly suitable for air quality control systems in automotive cabins.

The literature therefore shows many examples of microsensors onto which sensitive layers have been deposited by using various methods, such as for example micropipetting [10–13], sputtering [14–21], precipitation–oxidation [22,23], stepwise-heating electrospinning [24], flame spray pyrolysis [25], spin coating [26], carbo-thermal route [27], evaporation [28], metal-assisted chemical etching [29], or organic binder printing [30]. Radio-frequency sputtering is a method compatible with the industrial fabrication of miniaturized sensors by microelectronics and MEMS (microelectromechanical systems) technologies. Radio-frequency (RF) sputtering has many other advantages, like the possibility of obtaining thin films with nanometric-scale grain sizes and very easy control of the inter-granular porosity by varying the deposition parameters [31,32]. Such films with controlled nanostructure are of great interest as sensitive layers [33–35] and can be integrated in gas sensing devices.

Cupric oxide (copper(II) oxide: CuO) is an intrinsically p-type semiconductor [36,37]. Among all other p-type semiconducting oxides it is the most studied for gas sensing applications [38] due to its low-cost, high stability and non-toxicity. Many researchers have focused on the development of novel CuO nanostructures for the detection of a large range of gases, such as for example organic gases [39], hydrogen sulfide (H_2S) [40–46], CO [47–50], NO_2 [50,51], ethanol gas [52–54], or NH_3 [55].

In this work, we show the interest of using fully compatible micromachining technologies to elaborate microheaters and deposit CuO-sensitive layers to obtain sensors at the micronic scale. Elaboration of micro-hotplates, as well as photolithographic steps for layer integration, were carried out using the micromachining facilities of the Laboratory for Analysis and Architecture of Systems (LAAS-CNRS). The sensor is based on a p-type CuO resistive layer that was deposited by radio-frequency (RF) sputtering in the Interuniversity Center of Materials Research and Engineering (CIRIMAT). The microsensors were tested with variable thermal sequences with CO, NH_3, C_2H_4O, and NO_2, which are among the main pollutant gases found in automotive cabins. Many works have focused on the sensing properties of pure or doped CuO as the main sensitive material or as an additive for other semiconducting oxides. There are fewer articles related to the sensing properties of CuO layers integrated on microhotplates. For example, Walden et al. [56] tested inkjet-printed CuO layers for NH_3 detection in rapid temperature cycled mode. Kneer et al. [57] used similar inkjet-printed CuO nanoparticles deposited on a microsensor and obtained good H_2S selectivity in the NO_2, NH_3 and SO_2 atmosphere but with a time pulsation of few minutes. However, there are no articles relating to acetaldehyde detection with a CuO-sensitive layer deposited on a microsensor and operated in pulsed temperature mode, although temperature cycling gives good results for acetaldehyde detection with other sensitive oxides [58].

2. Experimental

Thin sensitive films were deposited with an Alcatel SCM 400 apparatus using sintered ceramic targets of pure CuO with a relative density around 75% (10 cm in diameter). The RF power was lowered at 50 W to avoid target reduction [59] and the pressure inside the chamber was lower than 2×10^{-5} Pa before deposition. During the deposition of the films, the target-to-substrate distance was fixed at 7 cm (Table 1). The thicknesses of the deposited films were set to 50 nm on microsensors and 300 nm for the structural characterizations on fused silica substrates. In the case of copper oxides that can have multiple valences of copper, like in tenorite CuO (Cu^{II}), paramélaconite Cu_4O_3 (mixed Cu^{I}/Cu^{II}) or cuprite Cu_2O (Cu^{I}), high deposition pressure could lead to a reduction [60] of the CuO target and the deposition of a phase with lower valences states. Moreover, as the layer had to be integrated by a wet process, a low deposition pressure of 0.5 Pa was preferred to obtain dense [61] oxide layer. These deposition conditions are then adequate to avoid the filling of the intergranular porosity of the sensitive layer with dye or any residue obtained during photolithographic process but should lead to layers with not optimized sensitivities.

Table 1. Deposition parameters of thin sensitive films.

Target material	CuO
Magnetron	Yes
Substrates	Fused silica and micro-hotplate
Power	50 W
Argon pressure	0.5 Pa
Target to substrate distance	7 cm
Deposition rate	6.1 nm/min

Thickness calibrations were performed with a DEKTAT 3030ST profilometer. The structure properties were determined by grazing incidence X-ray diffraction (GI-XRD) using a Bruker-AXS D8-Advance X-ray diffractometer equipped with a copper source ($\lambda CuK_{\alpha 1}$ = 1.5405 Å and $\lambda CuK_{\alpha 2}$ = 1.5445 Å) at 1° incidence, a Göbel mirror and Bruker LynxEye detector used in 0 D mode. The GI-XRD data were analyzed with the Bruker-EVA software and the JC-PDF database, and refined with the Rietveld method implemented in the FullProf-Suite program. Raman spectra were collected under ambient conditions using a LabRAM HR 800 Jobin Yvon spectrometer with a fiber coupled 532-nm laser. Spectra acquisition was carried out for 150 s using a ×100 objective lens and 600 gr/mm grating. During the measurement, the resulting laser power at the surface of the sample was adjusted to 1.7 mW. Examination of multiple spots showed that the samples were homogeneous. Microscopic studies were realized with a Veeco Dimension 3000 Atomic Force Microscope (AFM) in tapping mode equipped with a super sharp TESP-SS Nanoworld tip (nominal resonance frequency 320 KHz, nominal radius curvature 2 nm). The scanning rate was fixed at 1 Hz (1000 nm/s).

For sensing measurements, the sensors were placed into a chamber flown by different gases. The composition and humidity of the gas mixture were controlled by mass flow controllers (MFC). The heating and the sensing resistors of each sensor were connected to a source measurement unit (SMU). The whole test bench was automatically controllable thanks to a suitable interface and dedicated software. After a period of stabilization of 2 h under synthetic air, the target gases were introduced alternatively. The global flow (200 sscm) and the relative humidity (30%) remained constant during both air and target gas sequences. Response of gas sensor toward the four gases (CO, NH$_3$, C$_2$H$_4$O and NO$_2$) has been calculated according to the formula (1).

$$S(\%) = \left(\frac{R_{gas} - R_{air}}{R_{air}} \right) \times 100 \tag{1}$$

where, R_{air} and R_{gas}, are resistances in air and test gas, respectively.

3. Preparation of Microheaters

The devices have been developed on an optimized microheater that can work at high temperature and low power consumption (500 °C and ~55 mW, respectively). In order to avoid edge effects, circular membrane geometry (Figure 1a) was chosen. Figure 1b shows the resulting thermal distribution simulated by Comsol Multiphysics software in such geometry. It can be observed that the temperature is homogeneous in the center of the heated area onto which the measurement electrodes are placed (Figure 1a).

Micro-hotplates have been elaborated by photolithographic process. The detailed microfabrication steps are presented in Figure 2. The platform consists of a silicon bulk on which a thermally resistive bilayer SiO$_2$/SiN$_x$ membrane was grown. Afterwards, Pt metallization was carried out by lift-off process to obtain the heating resistor. A passivation layer was then deposited (a 0.7-μm thin PECVD SiO$_2$ layer) and contacts were opened. Finally, a new lift-off step was used to elaborate the electrodes necessary for measuring the resistance of the sensing layer, and the rear side of the bulk was etched to release the membrane in order to increase the thermal resistance and then to limit thermal dissipation. Figure 1a shows the top view of the final membrane. Figure 1d shows a sensor mounted in its housing (TO5).

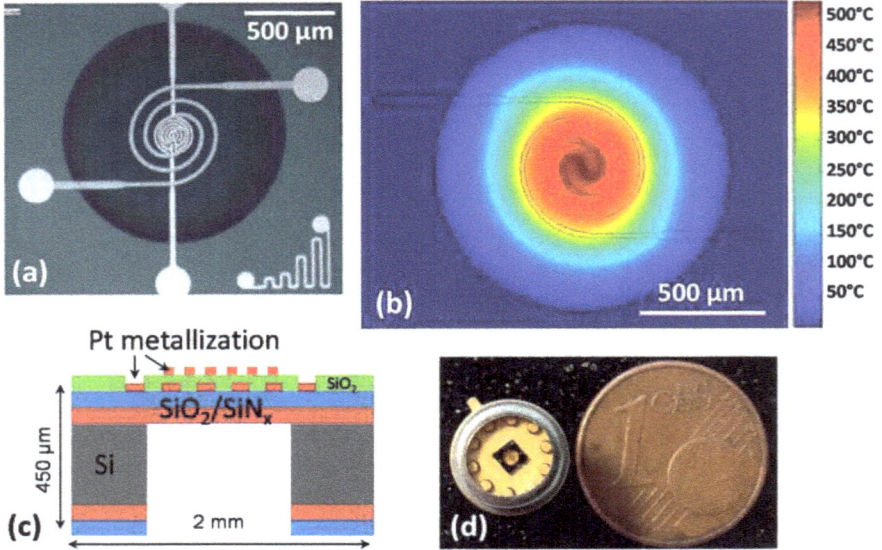

Figure 1. (a) Top view of the micro hotplate elaborated onto a membrane; (b) thermal simulation made with Comsol Multiphysics; (c) schematic view of a platform; (d) sensor packaged on TO5 housing.

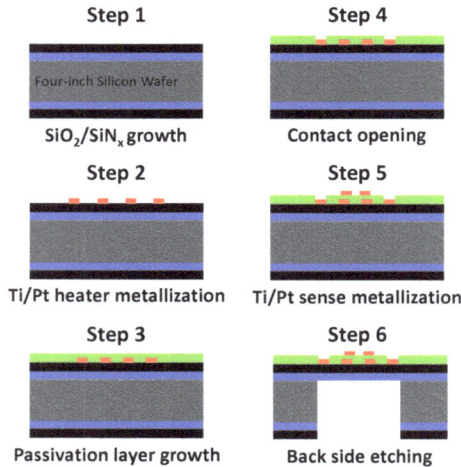

Figure 2. Main steps of platform elaboration process.

Thermal measurement of the surface of the platform performed with a Jade MWIR infra-red camera (CEDIP) allowed the calibration between the power applied and the resulting heating temperature onto the membrane. The results given in the Table 2 show a good linear relation between the power applied and the temperature measured. The heating platform makes it possible to heat from room temperature to 550 °C in 50 ms and the cooling time is of the same order of magnitude [62]. This type of platform can thus generate very rapid temperature variations, which is suitable for operating the sensor in pulsed mode. At the end of the step 6 and before dicing the chips, it is possible to locally deposit a metal-oxide layer onto the electrodes to form the sensing thin film resistor. This will be described below in the Section 4.

Table 2. Temperature reached in the center of the center of the microheater vs applied heating power.

Power (mW)	Temperature (°C)
55	500.7
45	402.7
35	304.8
30	255.8
25	206.9

4. Integration of P-Type CuO Layer by Photolithography Process

4.1. Structural Characterizations of CuO Layer

Figure 3 shows XRD pattern of copper oxide thin film deposited on a fused silica substrate and annealed at 500 °C. Measurement was carried out with a 50-nm-thick sample, similar to that deposited on the microsensor for gas sensing tests. The X-ray diffractogram clearly shows the presence of a pure tenorite phase (CuO: JCPDF 45-0937). The XRD patterns do not show any presence of extra phases with copper oxidation state lower than +II (like for example paramelaconite $Cu_4^{II/I}O_3$, cuprite Cu_2^IO, or metallic copper Cu).

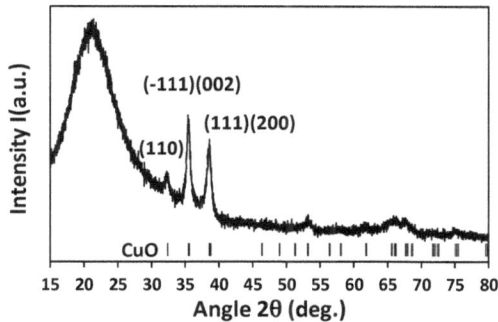

Figure 3. X-ray diffraction (XRD) pattern of CuO thin film annealed at 500 °C (thickness = 50 nm).

The Raman spectrum of a sample annealed at 500 °C is presented in Figure 4. The 50-nm-thick sample was too thin to be measured and a 300-nm-thick sample deposited with the same deposition conditions was characterized. Raman spectrum shows three vibration modes at 296, 346 and 636 cm^{-1}, which are characteristic of the CuO phase [54,63,64] and can be attributed to Ag, B(1)g, and B(2)g modes, respectively. Raman spectra of the reduced phases containing Cu(I) such as paramelaconite Cu_4O_3 or cuprite Cu_2O may be easily differentiated from those of tenorite phase CuO. In particular, paramelaconite provides a characteristic Raman peak at about 520–530 cm^{-1} and cuprite at 110 cm^{-1} and 220 cm^{-1}. In the Raman spectra of as-deposited and annealed films, none of these signals are visible, confirming the absence of such phases.

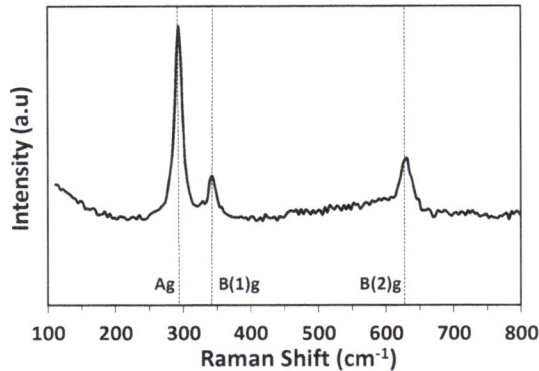

Figure 4. Raman spectra of CuO thin film annealed at 500 °C (thickness = 300 nm).

In conclusion, the data obtained by XRD and Raman spectroscopy show a pure CuO phase in the deposited films.

The image of the surface of CuO thin film, obtained by AFM, has been reported in Figure 5a. To ensure consistency with the layer used for sensing tests, a 50-nm thin film was observed. The surface consists of circular grains with surface domes (top of the grown column) which is a typical morphology in the case of the sputtered thin films. The distribution of the grains size shown in Figure 5b was estimated by an immersion threshold thanks to the Gwyddion software. The median grain size (d_{50}) was found to be equal to 27.6 nm, which is close to the half thickness of the sample.

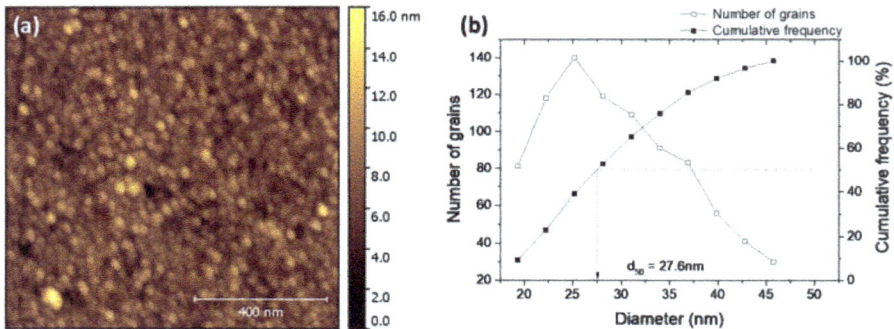

Figure 5. (a) Atomic force microscope (AFM) image of a 50-nm-thick CuO film annealed at 400 °C for 1 h under air atmosphere; (b) Grains size distribution deduced from the image analysis.

4.2. Description of the Integration Process

The integration of the copper oxide layer was performed using a classical photolithographic process (Figure 6). The lift-off resist was deposited in step 1. In the second and third steps the photoresist layer was exposed and then developed. The deposition of the 50-nm-thick CuO layer was undertaken using RF sputtering in step 4. Finally, all the unwanted parts were removed in step 6 by dissolution of the resist, thus leaving the sensitive layer in the desired areas.

The deposition of the sensitive layer is a critical step, as the bombardment occurring during the sputtering process is able to damage the photoresist used to mask the part that does not have to be covered by the oxide layer. Figure 7 shows the successful integration of copper oxide onto a microheater. The diameter of the area covered by the sensitive CuO layer is around 400 μm. The diameter of the electrodes used for the electrical measurement was approximately 200 µm, which was then totally covered by the sensitive layer.

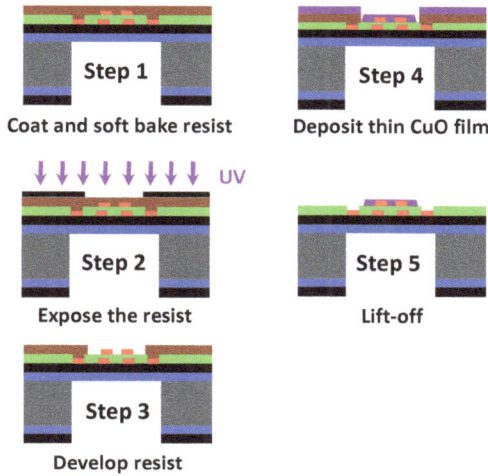

Figure 6. Main steps in the integration process of CuO-sensitive layers.

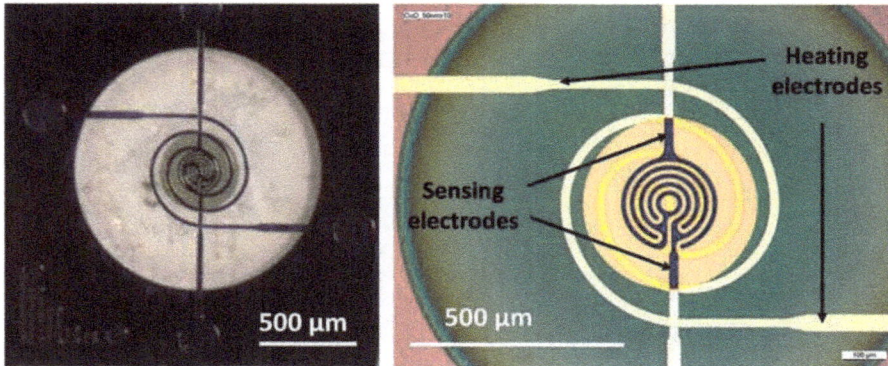

Figure 7. Images obtained by optical microscopy of a micro heater coated in the center with a p-type CuO semiconducting layer.

5. Sensing Tests

At first, the sensing device with integrated CuO layer (as-deposited) was annealed from 0 mW (room temperature) to 55 mW (~500 °C) in step at 5 mW/10 min and kept for 120 min at 55 mW to stabilize both the sensing layer and the microheater (Figure 8). XRD showed (Figure 3) that the CuO phase is stable up to 500 °C, and then only microstructural reorganization and a possible formation of slight over-stoichiometry in copper oxide ($CuO_{1+\delta}$) are expected.

Figure 8b shows the evolution of the layer resistance during initialization in the 50–55 mW heating power range. A decrease of the resistance was observed during annealing with 50 and 55 mW heating power. After 120 min at 55 mW the resistance of the sensitive layer was stabilized.

The microsensor based on the CuO semiconducting layer (thickness 50 nm) was tested under carbon monoxide CO, ammonia NH_3, acetaldehyde C_2H_4O, and nitrogen dioxide NO_2 according to the gas concentrations shown in Table 3. Each gas concentration was chosen close to the threshold concentration given by the various national (ANSES, French Agency for Food, Environmental and Occupational Health and Safety) [65] and international (WHO, World Health Organization) [66] health-based guidelines and guidance values for short time exposure in the case of indoor polluting gases. This is the reason why the concentration ranges are different for the four target gases.

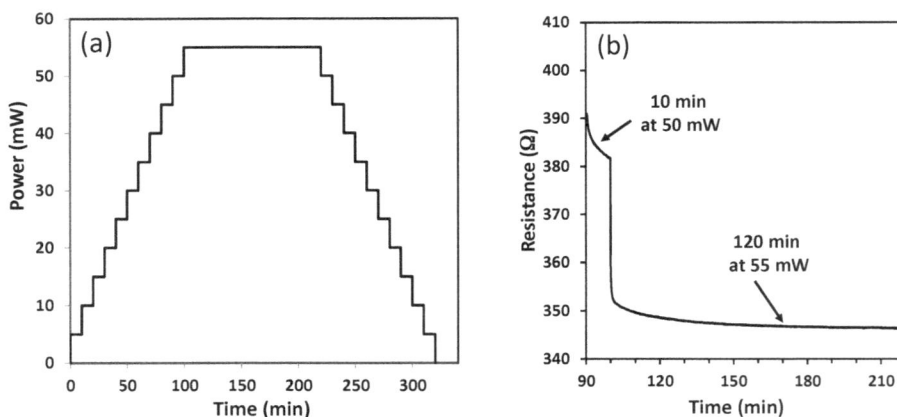

Figure 8. (a) Initialization program and (b) variation of the electrical resistance of the sensitive layer during the initialization (zoom in the heating power range of 50–55 mW, ~450–500 °C).

Table 3. Gases and concentrations used during the sensing tests with the CuO-sensitive layer.

Gas	Concentration (ppm)	
CO	100	200
NH_3	2	5
C_2H_4O	2	5
NO_2	0.2	0.5

In a first step, a "classical" constant temperature profile has been used. In this case the temperature is maintained at a constant temperature, while the gas composition and its concentration are alternated. The Figure 9 shows the response obtained at 400 °C (45-mW heating power) and at 200 °C (25-mW heating power) with the gases and the concentrations presented in the Table 3. Before starting the gas alternation, the resistance was stabilized under air for 2 h. For each gas, two concentrations were used for 15 min, and this sequence was repeated twice. Before changing the composition of the gas, the sensor was returned to air for 30 min.

The CuO layer showed decreases in resistance upon exposure to oxidizing gas (NO_2) and increases in resistance upon exposure to reducing gases (C_2H_4O, NH_3, and CO). This is consistent with the p-type semiconducting behavior of copper oxide. The results show that the response of the sensor is very low for carbon monoxide and almost zero for ammonia. On another hand, significant response values were obtained for acetaldehyde at 400 °C and for nitrogen dioxide at 200 °C. It can be noted in contrast that the response of the CuO layer under acetaldehyde at 200 °C and under nitrogen dioxide at 400 °C is roughly equal to zero.

These results are in accordance with the bibliography which shows that the sensitivity of CuO toward C_2H_4O and NO_2 in constant temperature mode is dependent on the temperature measurement. For NO_2 gas sensing, various studies [50,57,67] have shown that higher sensitivity is obtained at low or moderate temperature (150–250 °C). For acetaldehyde it is more difficult to refer to sensing studies as there is no article related to the detection of this gas by CuO. However, Cordi et al. [68] have carried out temperature-programmed oxidation (TPO) measurements on CuO for catalytic applications, and they showed that C_2H_4O is oxidized at higher temperatures (340 °C). By alternating high and low measurement temperature, the discrimination of C_2H_4O and NO_2 can then be improved. Even if the response values remain quite low, it should be borne in mind that the gas concentrations used for the test are also very low.

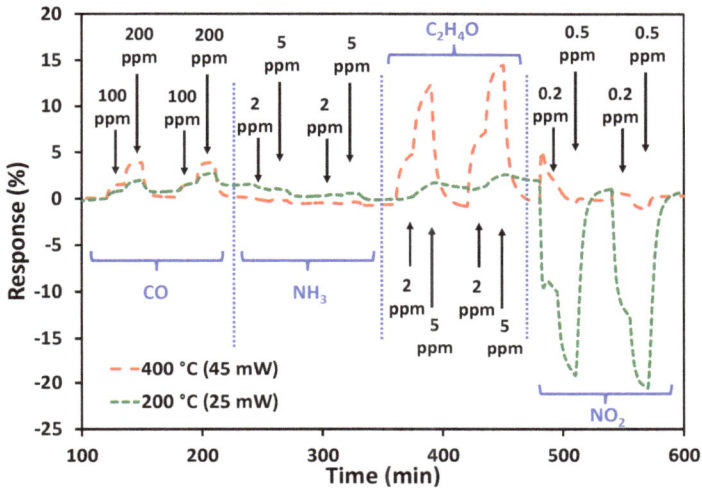

Figure 9. Response of the CuO-sensitive layer at 400 °C and 200 °C with constant temperature mode.

In a second step, a dynamic test profile was carried out with the same gas and concentrations. Many works have already shown the interest of operating the sensor with temperature cycling by using different profile shape and plateau duration in order to rapidly change its sensitivity and then the selectivity after suitable data treatment. Among the thermal cycle approaches, one consists of making the temperature vary with stair shape from ambient to high temperature or vice-versa. Another consists of using short heating or cooling pulses from a reference temperature which can be set at ambient, high or intermediate temperature [69]. The pattern we chose in this study is the latter, with two-second steps at each target temperature and a baseline fixed at 500 °C. The measurement pattern had already been optimized in the past and the temperature profile that allowed a good reproducibility, the fastest stabilization, and the best discrimination was selected [70]. The short plateau duration is well-adapted to only observe transient phenomena, while the high baseline temperature allows regular cleaning of the surface of the sensitive layer to obtain good reversibility and reproducibility. Moreover, this profile is easy to implement in embedded electronics. The temperature profiles are presented in the Figure 10a,b: each step lasts 2 s; the temperature baseline is the highest operating temperature (500 °C), while the other steps are at lower temperatures (400, 300, and 200 °C for high temperature measurements and 50, 30, and 20 °C for low-temperature measurements), and a complete cycle lasts 12 s. This profile is repeated throughout the test under various gaseous atmospheres.

The Figure 11 shows an example of resistance measurement under pulsed temperature mode over one hour with a relative humidity (RH) of 30%. The sensor has been set successively under four different "ambiances": (1) synthetic air; (2) 2 ppm of acetaldehyde; (3) 5 ppm of acetaldehyde; and finally (4) synthetic air. During these "sequences", the sensor is periodically powered (on the heater) with the four different two-second steps (0, 5, 10 and 55 mW) as seen before in Figure 10b, considering that 55 mW is the reference or the base-line. In this view the resistance values at all temperature steps are multiplexed, and it is hardly possible to guess all the step transitions and the behavior of the resistance at each power step. For this reason, it is necessary to extract or separate results from the different power-steps in order to calculate their associated normalized response in the same way as it was done for the static operating mode.

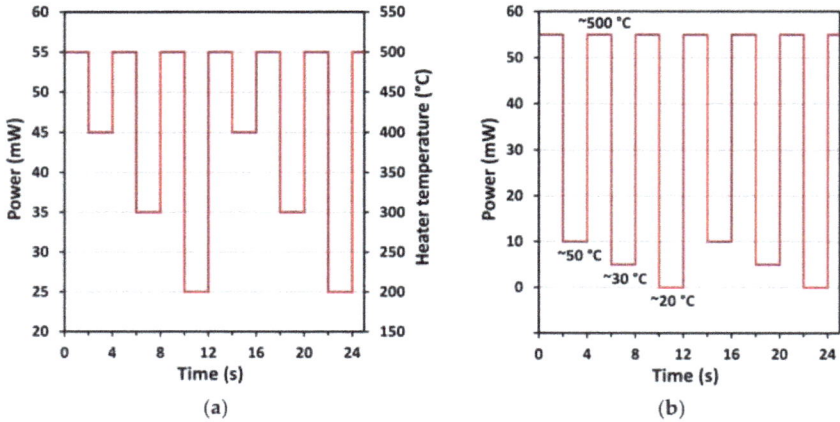

Figure 10. Power profiles in (**a**) high temperature range and (**b**) low temperature range.

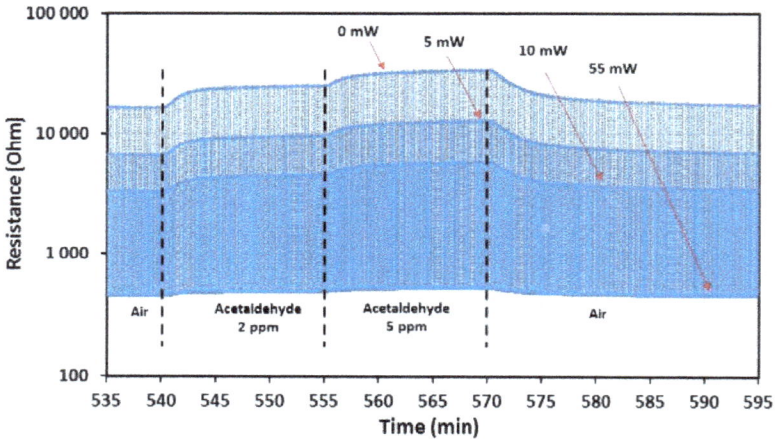

Figure 11. CuO sensor operated with dynamic temperature cycling mode (temperature cycle of the Figure 10b has been used). Resistance measurement under 30% RH for 1 h with a first injection of 2 ppm of acetaldehyde (15 min), then a second injection of 5 ppm (15 min). The resistance values corresponding to the four temperatures (or heating power) are multiplexed due to two-second step temperature cycling.

The detailed views of the resistance variation under the same pulsed temperature cycles are shown in Figure 12a for the last three cycles just before the injection of acetaldehyde and in Figure 12b for the last three cycles at the end of the gas sequence under 5 ppm of acetaldehyde. The small red lines correspond to the penultimate point of each step in the last cycle before a gas transition. For the calculation of the response $\Delta R/R$ given in the Formula (1), the reference values (R_{air}) have been taken in the last cycle under air for each heating power, while the R_{gas} values have been taken in the last cycle under target gas (acetaldehyde in the example of the Figure 12).

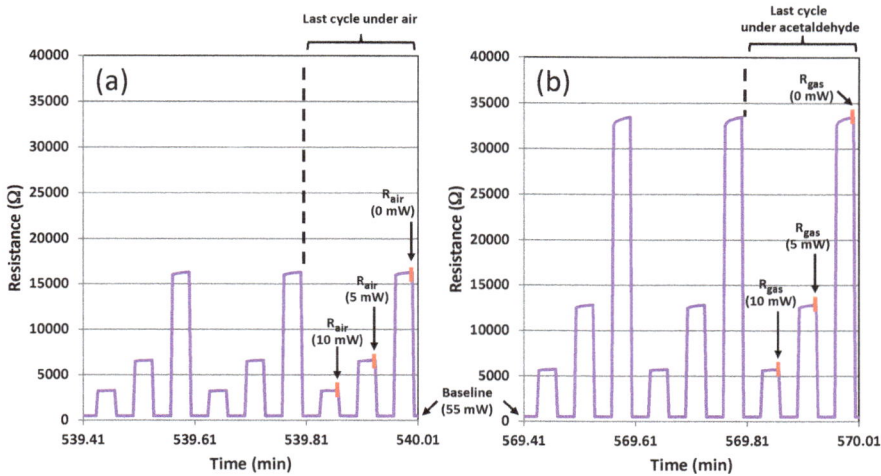

Figure 12. Detailed view of resistance variation during (**a**) the last three cycles under air and (**b**) the last three cycles under 5 ppm of acetaldehyde. The sensor was operated with dynamic temperature cycling mode (the sequence presented in the Figure 10b has been applied), under 30% relative humidity, and a 55 mW baseline. R_{air}: resistance in air; R_{gas}: resistance in test gas.

Pulsed temperature cycling mode has been carried out under the same gases and concentrations which were used in static mode previously (Table 3), during 15 min for each target gas and with 30% RH. The response has been calculated according to the procedure explained just before. The sequence from the Figure 10a was used first. Every 12 s (i.e., one complete cycle) the power was switched between 100 µA and 1 mA to study, in addition to the gas pulse, the influence of the current applied during the electrical resistance measurement. Because all this information was multiplexed in the same experimental file, the normalized responses for each gas, concentration and current were extracted and presented as a synthetic result in the form of bar-graph in the Figure 13. It can be seen that the values of the bias current have almost no influence on the response. On the contrary, the measurement temperature has a strong effect and it can be seen that the best results have been obtained for the lower power step (25 mW~200 °C). Even if the response under CO and NH$_3$ has been slightly increased in dynamic mode, the maximal change in resistance remains less than about 10% for 200 ppm of CO and 5% for 5 ppm of NH$_3$. The strongest improvement has been obtained for the measurements under C$_2$H$_4$O, which show an improvement of the response that has been multiplied by 4 in comparison with the tests carried out with constant temperature profiles. Moreover, the trend is reversed between the two measurement modes. In constant temperature mode the response decreases when the measurement temperature (i.e., the heating power) is lowered, whereas the response increases strongly in dynamic mode.

The use of a pulsed-temperature operating mode promotes the transient chemical reactions. The difference observed between constant and modulated temperature mode can be due to the fact that the adsorption/desorption and reaction phenomena occur in out of equilibrium conditions in this last mode [71]. The effective adsorption, which is the result of the competition between OH$^-$, O$_2^-$, O^{2-}, O$^-$ and the target gas, is thus modified when the sensitive layer is brought to high temperature and then cooled very rapidly [72]. Another consequence of this rapid thermal cycling mode is the total disappearance of the resistance variation under NO$_2$, which is not observed even at low heating power. It has already been shown in the literature that NO$_2$ has a complex interaction with oxide surface [73] which can lead in some cases to a transition from reducing to oxidizing behavior with the operating temperature [74]. The lack of response of NO$_2$ on the surface of CuO in quick pulsed temperature mode can be due to kinetic reactions longer than pulse duration and/or opposite reactions that counteract each other in the high and low temperature alternate steps.

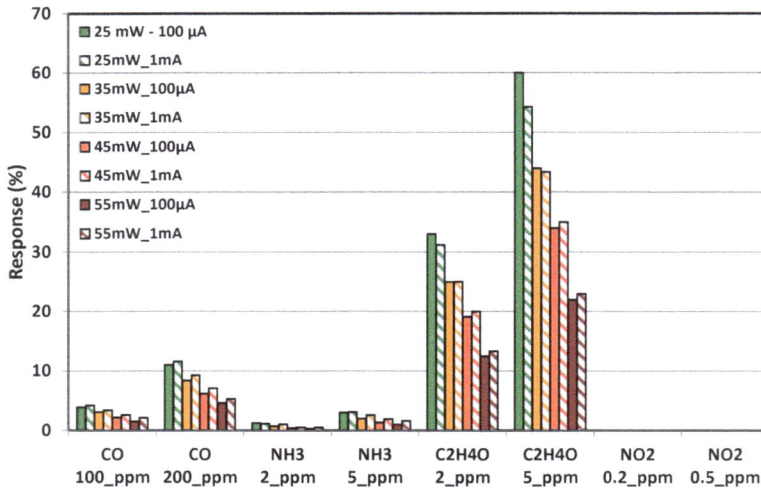

Figure 13. Comparison of the response obtained under dynamic tests at high temperature range (200–500 °C) by using two different bias currents (100 µA and 1 mA).

To further explore the effect of the decrease in the measurement temperature, three additional powers (10, 5 and 0 mW) have been used according to the cycle shown in Figure 10b. The comparison of all measurements carried out between 20 °C (0 mW) and 500 °C (55 mW) with 100-µA bias current are shown in the Figure 14. The results show that the increase of the response observed when the measuring heating power was decreased from 55 mW to 25 mW continues to occur when it is lowered down to 0 mW. Finally, the response under C_2H_4O could be multiplied by 7 in comparison with constant temperature mode when a thermal cycling mode was applied. This last measurement also confirms that cycled mode allows total selectivity with respect to NO_2, which is not detected whatever the temperature applied.

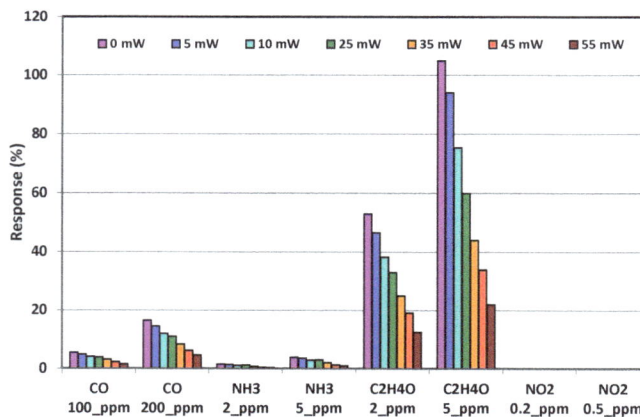

Figure 14. Comparison of the response obtained under dynamic tests in the full temperature range (20–500 °C). Bias current has been fixed at 100 µA.

6. Conclusions

Micro-hotplates were first prepared by using silicon microtechnologies. P-type semiconducting CuO layers deposited by RF sputtering were integrated onto these microsensors by using classical photolithography technologies that were used for the preparation of micro-hotplates. Even if this route has the disadvantage of exposing the sensitive layer to chemical products which are able to attack it (acidic or basic solutions), the integration of the copper oxide layer could be successfully carried out. Because the microheater was designed on a membrane, it was then possible to generate very rapid temperature variations and a rapid temperature cycled mode could be applied. This measurement mode showed a strong improvement, by of a factor of 7, in the sensor response under 2 and 5 ppm of acetaldehyde and by a factor 2 in the case of carbon monoxide.

Acknowledgments: This work was partly supported by LAAS-CNRS micro and nano technolgies platform member of the French RENATECH network and by University of Toulouse in the frame of NeoCampus program.

Author Contributions: L.P. has written the article; with P.M. they have both coordinated the research work, processed and analyzed the gas sensing results. I.e.Y. has deposited CuO thin films and Y.T. has characterized their microstructure. C.B. has prepared the target devoted for the sputtering of the active material. A.B. and P.T. have characterized the structure of the CuO thin film. A.C. worked on the fabrication of microthotplate platforms, photolithography integration, and on the sensors' characterizations under controlled atmospheres. F.B. and C.T. have developed the gas sensing setup.

Conflicts of Interest: The authors declare no conflict of interest.

References

1. Demarne, V.; Grisel, A. An integrated low-power thin-film CO gas sensor on silicon. *Sens. Actuators* **1988**, *13*, 301–313. [CrossRef]

2. Courbat, J.; Canonica, M.; Teyssieux, D.; Briand, D.; de Rooijet, N.F. Design and fabrication of micro-hotplates made on a polyimide foil: Electrothermal simulation and characterization to achieve power consumption in the low mW range. *J. Micromech. Microeng.* **2010**, *21*, 015014. [CrossRef]

3. Briand, D.; Courbat, J. Chapter 6: Micromachined semiconductor gas sensors. In *Semiconductor Gas Sensors*; Jaaniso, R., Tan, O.K., Eds.; Woodhead Publishing: Cambridge, UK, 2013; pp. 220–260. ISBN 9780857092366.

4. Sears, W.M.; Colbow, K.; Consadori, F. General characteristics of thermally cycled tin oxide gas Sensors. *Semicond. Sci. Technol.* **1989**, *4*, 351–359.

5. Ratton, L.; Kunt, T.; McAvoy, T.; Fuja, T.; Cavicchi, R.; Semancik, S. A comparative study of signal processing techniques for clustering microsensor data (a first step towards an artificial nose). *Sens. Actuators B Chem.* **1997**, *41*, 105–120. [CrossRef]

6. Rogers, P.H.; Semancik, S. Development of optimization procedures for application-specific chemical sensing. *Sens. Actuators B Chem.* **2012**, *163*, 8–19. [CrossRef]

7. Llobet, E.; Brezmes, J.; Ionescu, R.; Vilanova, X.; Al-Khalifa, S.; Gardner, J.W.; Bârsan, N.; Correig, X. Wavelet transform and fuzzy ARTMAP-based pattern recognition for fast gas identification using a micro-hotplate gas sensor. *Sens. Actuators B Chem.* **2002**, *83*, 238–244. [CrossRef]

8. Parret, F.; Ménini, Ph.; Martinez, A.; Soulantica, K.; Maisonnat, A.; Chaudret, B. Improvement of Micromachined SnO_2 Gas Sensors Selectivity by Optimised Dynamic Temperature Operating Mode. *Sens. Actuators B Chem.* **2006**, *118*, 276–282. [CrossRef]

9. Faglia, G.; Comini, E.; Cristalli, A.; Sberveglieri, G.; Dori, L. Very low power consumption micromachined CO sensors. *Sens. Actuators B Chem.* **1999**, *55*, 140–146. [CrossRef]

10. Fong, C.-F.; Dai, C.-L.; Wu, C.-C. Fabrication and Characterization of a Micro Methanol Sensor Using the CMOS-MEMS Technique. *Sensors* **2015**, *15*, 27047–27059. [CrossRef] [PubMed]

11. Martinez, C.J.; Hockey, B.; Montgomery, C.B.; Semancik, S. Porous tin oxide nanostructured microspheres for sensor applications. *Langmuir* **2005**, *21*, 7937–7944. [CrossRef] [PubMed]

12. Liao, W.-Z.; Dai, C.-L.; Yang, M.-Z. Micro Ethanol Sensors with a Heater Fabricated Using the Commercial 0.18 μm CMOS Process. *Sensors* **2013**, *13*, 12760–12770. [CrossRef] [PubMed]

13. Yang, M.-Z.; Dai, C.-L. Ethanol Microsensors with a Readout Circuit Manufactured Using the CMOS-MEMS Technique. *Sensors* **2015**, *15*, 1623–1634. [CrossRef] [PubMed]

14. Behera, B.; Chandra, S. An innovative gas sensor incorporating ZnO–CuO nanoflakes in planar MEMS technology. *Sens. Actuators B Chem.* **2016**, *229*, 414–424. [CrossRef]

15. Stankova, M.; Ivanov, P.; Llobet, E.; Brezmes, J.; Vilanova, X.; Gràcia, I.; Cané, C.; Hubalek, J.; Malysz, K.; Correig, X. Sputtered and screen-printed metal oxide-based integrated micro-sensor arrays for the quantitative analysis of gas mixtures. *Sens. Actuators B Chem.* **2004**, *103*, 23–30. [CrossRef]

16. Lee, C.-Y.; Chiang, C.-M.; Wang, Y.-H.; Ma, R.-H. A self-heating gas sensor with integrated NiO thin-film for formaldehyde detection. *Sens. Actuators B Chem.* **2007**, *122*, 503–510. [CrossRef]

17. Stankova, M.; Vilanova, X.; Calderer, J.; Llobet, E.; Ivanov, P.; Gràcia, I.; Cané, C.; Correig, X. Detection of SO_2 and H_2S in CO_2 stream by means of WO_3-based micro-hotplate sensors. *Sens. Actuators B Chem.* **2004**, *102*, 219–225. [CrossRef]

18. Tang, Z.; Fung, S.K.H.; Wong, D.T.W.; Chan, P.C.H.; Sin, J.K.O.; Cheung, P.W. An integrated gas sensor based on tin oxide thin-film and improved micro-hotplate. *Sens. Actuators B Chem.* **1998**, *46*, 174–179. [CrossRef]

19. Sheng, L.Y.; Tang, Z.; Wu, J.; Chan, P.C.H.; Sin, J.K.O. A low-power CMOS compatible integrated gas sensor using maskless tin oxide sputtering. *Sens. Actuators B Chem.* **1998**, *49*, 81–87. [CrossRef]

20. Takács, M.; Dücső, C.; Pap, A.E. Fine-tuning of gas sensitivity by modification of nano-crystalline WO_3 layer morphology. *Sens. Actuators B Chem.* **2015**, *221*, 281–289. [CrossRef]

21. Zappa, D.; Briand, D.; Comini, E.; Courbat, J.; de Rooij, N.F.; Sberveglieri, G. Zinc Oxide Nanowires Deposited on Polymeric Hotplates for Low-power Gas Sensors. *Procedia Eng.* **2012**, *47*, 1137–1140. [CrossRef]

22. Yang, M.Z.; Dai, C.L.; Shih, P.J.; Chen, Y.C. Cobalt oxide nanosheet humidity sensor integrated with circuit on chip. *Microelectron. Eng.* **2011**, *88*, 1742–1744. [CrossRef]

23. Dai, C.L.; Chen, Y.C.; Wu, C.C.; Kuo, C.F. Cobalt oxide nanosheet and CNT micro carbon monoxide sensor integrated with readout circuit on chip. *Sensors* **2010**, *10*, 1753–1764. [CrossRef] [PubMed]

24. Tang, W.; Wang, J. Methanol sensing micro-gas sensors of SnO_2-ZnO nanofibers on Si/SiO_2/Ti/Pt substrate via stepwise-heating electrospinning. *J. Material. Sci.* **2015**, *50*, 4209–4220. [CrossRef]

25. Kuhne, S.; Graf, M.; Tricoli, A.; Mayer, F.; Pratsinis, S.E.; Hierlemann, A. Wafer-level flame-spray-pyrolysis deposition of gas-sensitive layers on microsensors. *J. Micromech. Microeng.* **2008**, *18*, 035040. [CrossRef]

26. Wan, Q.; Li, Q.H.; Chen, Y.J.; Wang, T.H.; He, X.L.; Li, J.P.; Lin, C.L. Fabrication and ethanol sensing characteristics of ZnO nanowire gas sensors. *Appl. Phys. Lett.* **2004**, *84*, 3654–3656. [CrossRef]

27. Nguyen, H.; Quy, C.T.; Hoa, N.D.; Lam, N.T.; Duy, N.V.; Quang, V.V.; Hieu, N.V. Controllable growth of ZnO nanowire grown on discrete islands of Au catalyst for realization of planar type micro gas sensors. *Sens. Actuators B Chem.* **2014**, *193*, 888–894. [CrossRef]

28. Pandya, H.J.; Chandra, S.; Vyas, A.L. Integration of ZnO nanostructures with MEMS for ethanol sensor. *Sens. Actuators B Chem.* **2011**, *161*, 923–928. [CrossRef]

29. Peng, K.Q.; Wang, X.; Lee, S.T. Gas sensing properties of single crystalline porous silicon nanowires. *Appl. Phys. Lett.* **2009**, *95*, 243112. [CrossRef]

30. Dong, K.Y.; Choi, J.K.; Hwang, I.S.; Lee, J.W.; Kang, B.H.; Ham, D.J.; Lee, J.H.; Ju, B.K. Enhanced H_2S sensing characteristics of Pt doped SnO_2 nanofibers sensors with micro heater. *Sens. Actuators B Chem.* **2011**, *157*, 154–161. [CrossRef]

31. Oudrhiri-Hassani, F.; Presmanes, L.; Barnabé, A.; Tailhades, P.H. Microstructure, porosity and roughness of RF sputtered oxide thin films: Characterization and modelization. *Appl. Surf. Sci.* **2008**, *254*, 5796–5802. [CrossRef]

32. Sandu, I.; Presmanes, L.; Alphonse, P.; Tailhades, P. Nanostructured cobalt manganese ferrite thin films for gas sensor application. *Thin Sol. Films* **2006**, *495*, 130–133. [CrossRef]

33. Chapelle, A.; El Younsi, I.; Vitale, S.; Thimont, Y.; Nelis, T.; Presmanes, L.; Barnabé, A.; Tailhades, P.H. Improved semiconducting $CuO/CuFe_2O_4$ nanostructured thin films for CO_2 gas sensing. *Sens. Actuators B Chem.* **2014**, *204*, 407–413. [CrossRef]

34. Chapelle, A.; Yaacob, M.; Pasquet, I.; Presmanes, L.; Barnabe, A.; Tailhades, P.H.; Du Plessis, J.; Kalantar, K. Structural and gas-sensing properties of $CuO-Cu_xFe_{3-x}O_4$ nanostructured thin films. *Sens. Actuators B Chem.* **2011**, *153*, 117–124. [CrossRef]

35. Presmanes, L.; Chapelle, A.; Oudrhiri-Hassani, F.; Barnabe, A.; Tailhades, P.H. Synthesis and CO Gas-Sensing Properties of CuO and Spinel Ferrite Nanocomposite Thin Films. *Sens. Lett.* **2013**, *9*, 587–590. [CrossRef]

36. Jeong, Y.K.; Choi, G.M. Nonstoichiometry and electrical conduction of CuO. *J. Phys. Chem. Solids* **1996**, *57*, 81–84. [CrossRef]

37. Koffyberg, F.P.; Benko, F.A. A photoelectrochemical determination of the position of the conduction and valence band edges of p-type CuO. *J. Appl. Phys.* **1982**, *53*, 1173–1177. [CrossRef]
38. Kim, H.J.; Lee, J.H. Highly sensitive and selective gas sensors using p-type oxide semiconductors: Overview. *Sens. Actuators B Chem.* **2014**, *192*, 607–627. [CrossRef]
39. Zhu, G.; Xu, H.; Xiao, Y.; Liu, Y.; Yuan, A.; Shen, X. Facile fabrication and enhanced sensing properties of hierarchically porous CuO architectures. *ACS Appl. Mater. Interfaces* **2012**, *4*, 744–751. [CrossRef] [PubMed]
40. Chen, J.; Wang, K.; Hartman, L.; Zhou, W. H$_2$S detection by vertically aligned CuO nanowire array sensors. *J. Phys. Chem. C* **2008**, *112*, 16017–16021. [CrossRef]
41. Steinhauer, S.; Brunet, E.; Maier, T.; Mutinati, G.C.; Köck, A. Suspended CuO nanowires for ppb level H$_2$S sensing in dry and humid atmosphere. *Sens. Actuators B Chem.* **2013**, *186*, 550–556. [CrossRef]
42. Ramgir, N.S.; Kailasa Ganapathi, S.; Kaur, M.; Datta, N.; Muthe, K.P.; Aswal, D.K.; Gupta, S.K.; Yakhmi, J.V. Sub-ppm H$_2$S sensing at room temperature using CuO thin films. *Sens. Actuators B Chem.* **2010**, *151*, 90–96. [CrossRef]
43. Qin, Y.; Zhang, F.; Chen, Y.; Zhou, Y.; Li, J.; Zhu, A.; Luo, Y.; Tian, Y.; Yang, J. Hierarchically porous CuO hollow spheres fabricated via a one-pot template-free method for high-performance gas sensors. *J. Phys. Chem. C* **2012**, *116*, 11994–12000. [CrossRef]
44. Li, X.; Wang, Y.; Lei, Y.; Gu, Z. Highly sensitive H$_2$S sensor based on template-synthesized CuO nanowires. *RSC Adv.* **2012**, *2*, 2302–2307. [CrossRef]
45. Zhang, F.; Zhu, A.; Luo, Y.; Tian, Y.; Yang, J.; Qin, Y. CuO nanosheets for sensitive and selective determination of H$_2$S with high recovery ability. *J. Phys. Chem. C* **2010**, *114*, 19214–19219. [CrossRef]
46. Yang, C.; Su, X.; Xiao, F.; Jian, J.; Wang, J. Gas sensing properties of CuO nanorods synthesized by a microwave-assisted hydrothermal method. *Sens. Actuators B Chem.* **2011**, *158*, 299–303. [CrossRef]
47. Steinhauer, S.; Brunet, E.; Maier, T.; Mutinati, G.C.; Köck, A.; Freudenberg, O.; Gspan, C.; Grogger, W.; Neuhold, A.; Resel, R. Gas sensing properties of novel CuO nanowire devices. *Sens. Actuators B Chem.* **2013**, *187*, 50–57. [CrossRef]
48. Aslani, A.; Oroojpour, V. CO gas sensing of CuO nanostructures synthesized by an assisted solvothermal wet chemical route. *Physica B Condens Matter* **2011**, *406*, 144–149. [CrossRef]
49. Liao, L.; Zhang, Z.; Yan, B.; Zheng, Z.; Bao, Q.L.; Wu, T.; Li, C.M.; Shen, Z.X.; Zhang, J.X.; Gong, H.; et al. Multifunctional CuO nanowire devices: P-type field effect transistors and CO gas sensors. *Nanotechnology* **2009**, *20*, 085203. [CrossRef] [PubMed]
50. Kim, Y.S.; Hwang, I.S.; Kim, S.J.; Lee, C.Y.; Lee, J.H. CuO nanowire gas sensors for air quality control in automotive cabin. *Sens. Actuators B* **2008**, *135*, 298–303. [CrossRef]
51. Kim, K.M.; Jeong, H.M.; Kim, H.R.; Choi, K.I.; Kim, H.J.; Lee, J.H. Selective detection of NO$_2$ using Cr-doped CuO nanorods. *Sensors* **2012**, *12*, 8013–8025. [CrossRef] [PubMed]
52. Raksa, P.; Gardchareon, A.; Chairuangsri, T.; Mangkorntong, P.; Mangkorntong, N.; Choopun, S. Ethanol sensing properties of CuO nanowires prepared by an oxidation reaction. *Ceram. Int.* **2009**, *35*, 649–652. [CrossRef]
53. Hsueh, H.T.; Chang, S.J.; Hung, F.Y.; Weng, W.Y.; Hsu, C.L.; Hsueh, T.J.; Lin, S.S.; Dai, B.T. Ethanol gas sensor of crabwise CuO nanowires prepared on glass substrate. *J. Electrochem. Soc.* **2011**, *158*, J106–J109. [CrossRef]
54. Zoolfakar, A.S.; Ahmad, M.Z.; Rani, R.A.; Ou, J.Z.; Balendhran, S.; Zhuiykov, S.; Latham, K.; Wlodarski, W.; Kalantar-zadeh, K. Nanostructured copper oxides as ethanol vapour sensors. *Sens. Actuators B Chem.* **2013**, *185*, 620–627. [CrossRef]
55. Mashock, M.; Yu, K.; Cui, S.; Mao, S.; Lu, G.; Chen, J. Modulating gas sensing properties of CuO nanowires through creation of discrete nanosized p–n junctions on their surfaces. *ACS Appl. Mater. Interfaces* **2012**, *4*, 4192–4199. [CrossRef] [PubMed]
56. Walden, P.; Kneer, J.; Knobelspies, S.; Kronast, W.; Mescheder, U.; Palzer, S. Micromachined Hotplate Platform for the Investigation of Ink-Jet Printed, Functionalized Metal Oxide Nanoparticles. *J. Microelectromech. Syst.* **2015**, *24*, 1384–1390. [CrossRef]
57. Kneer, J.; Knobelspies, S.; Bierer, B.; Wöllenstein, J.; Palzer, S. New method to selectively determine hydrogen sulfide concentrations using CuO layers. *Sens. Actuators B Chem.* **2016**, *222*, 625–631. [CrossRef]
58. Presmanes, L.; Thimont, Y.; Chapelle, A.; Blanc, F.; Talhi, C.; Bonningue, C.; Barnabé, A.; Menini, P.H.; Tailhades, P.H. Highly Sensitive Sputtered ZnO:Ga Thin Films Integrated by a Simple Stencil Mask Process on Microsensor Platforms for Sub-ppm Acetaldehyde Detection. *Sensors* **2017**, *17*, 1055. [CrossRef] [PubMed]

59. Bui, M.A.; Le Trong, H.; Presmanes, L.; Barnabé, A.; Bonningue, C.; Tailhades, P.H. Thin films of $Co_{1.7}Fe_{1.3}O_4$ prepared by radio-frequency sputtering—First step towards their spinodal decomposition. *CrystEngComm* **2014**, *16*, 3359–3365. [CrossRef]

60. Le Trong, H.; Bui, T.M.A.; Presmanes, L.; Barnabé, A.; Pasquet, I.; Bonningue, C.; Tailhades, P.H. Preparation of iron cobaltite thin films by RF magnetron sputtering. *Thin Solid Films* **2015**, *589*, 292–297. [CrossRef]

61. Shang, C.; Thimont, Y.; Barnabe, A.; Presmanes, L.; Pasquet, I.; Tailhades, P.H. Detailed microstructure analysis of as-deposited and etched porous ZnO films. *Appl. Surf. Sci.* **2015**, *344*, 242–248. [CrossRef]

62. Menini, P. *Habilitation à Diriger les Recherches*; Université de Toulouse: Toulouse, French, 2011; (Figure 51 page 97, and Figure 54 page 101); Available online: https://tel.archives-ouvertes.fr/tel-00697471/document (accessed on 12 May 2017).

63. Debbichi, L.; Marco de Lucas, M.C.; Pierson, J.F.; Krüger, P. Vibrational properties of CuO and Cu_4O_3 from first-principles calculations, and Raman and infrared spectroscopy. *J. Phys. Chem. C* **2012**, *116*, 10232–10237. [CrossRef]

64. Meyer, B.K.; Polity, A.; Reppin, D.; Becker, M.; Hering, P.; Klar, P.J.; Sander, T.H. Binary copper oxide semiconductors: From materials towards devices. *Phys. Status Solidi* **2012**, *249*, 1487–1509. [CrossRef]

65. ANSES Website. Available online: https://www.anses.fr/en (accessed on 16 May 2017).

66. World Health Organization (WHO) website. *WHO Guidelines for Indoor Air Quality: Selected Pollutants.* 2010. Available online: http://www.who.int/en/ (accessed on 16 May 2017).

67. Kim, K.-M.; Jeong, H.-M.; Kim, H.-R.; Choi, K.-I.; Kim, H.-J.; Lee, J.-H. Selective Detection of NO_2 Using Cr-Doped CuO Nanorods. *Sensors* **2012**, *12*, 8013–8025. [CrossRef] [PubMed]

68. Cordi, E.M.; O'Neill, P.J.; Falconer, J.L. Transient oxidation of volatile organic compounds on a CuO/Al_2O_3 catalyst. *Appl. Catal. B Environ.* **1997**, *14*, 23–36. [CrossRef]

69. Kunt, T.A.; McAvoy, T.J.; Cavicchi, R.E.; Semancik, S. Optimization of temperature programmed sensing for gas identification using micro-hotplate sensors. *Sens. Actuators B Chem.* **1998**, *53*, 24–43. [CrossRef]

70. Tropis, C.; Menini, P.; Martinez, A.; Yoboue, N.; Franc, B.; Blanc, F.; Fadel, P.; Lagrange, D.; Fau, P.; Maisonnat, A. Characterization of dynamic measurement with nanoparticular SnO_2 gas sensors. In Proceedings of the 20th MicroMechanics Europe Workshop (MME 2009), Toulouse, France, 20–22 September 2009.

71. Ionescu, R.; Llobet, E.; Al-Khalifa, S.; Gardner, J.W.; Vilanova, X.; Brezmes, J.; Correig, X. Response model for thermally modulated tin oxide-based microhotplate gas sensors. *Sens. Actuators B Chem.* **2003**, *95*, 203–211. [CrossRef]

72. Ducéré, J.-M.; Hemeryck, A.; Estève, A.; Rouhani, M.D.; Landa, G.; Ménini, P.; Tropis, C.; Maisonnat, A.; Fau, P.; Chaudret, B. A computational chemist approach to gas sensors: Modeling the response of SnO_2 to CO, O_2, and H_2O Gases. *J. Comput. Chem.* **2012**, *33*, 247–258. [CrossRef] [PubMed]

73. Ruhland, B.; Becker, T.H.; Müller, G. Gas-kinetic interactions of nitrous oxides with SnO_2 surfaces. *Sens. Actuators B Chem.* **1998**, *50*, 85–94. [CrossRef]

74. Kneer, J.; Wöllenstein, J.; Palzer, S. Manipulating the gas–surface interaction between copper(II) oxide and mono-nitrogen oxides using temperature. *Sens. Actuators B Chem.* **2016**, *229*, 57–62. [CrossRef]

sensors

MDPI

Article

A Multi-Parametric Device with Innovative Solid Electrodes for Long-Term Monitoring of pH, Redox-Potential and Conductivity in a Nuclear Waste Repository

Jordan Daoudi [1],*, Stephanie Betelu [1],*, Theodore Tzedakis [2], Johan Bertrand [3] and Ioannis Ignatiadis [1]

[1] Water, Environment and Eco-technologies, BRGM French Geological Survey, 45060 Orléans, France; i.ignatiadis@brgm.fr

[2] Laboratory of Chemical Engineering, Université de Toulouse III Paul Sabatier, 31062 Toulouse, France; tzedakis@chimie.ups-tlse.fr

[3] Monitoring and Data Processing Department (DRD/MTD), ANDRA French National Radioactive Waste Management Agency, 92290 Châtenay Malabry, France; johan.bertrand@andra.fr

* Correspondence: j.daoudi@brgm.fr (J.D.); s.betelu@brgm.fr (S.B.); Tel.: +33-238-64-3268 (S.B.)

Academic Editors: Nicole Jaffrezic-Renault and Gaelle Lissorgues
Received: 7 April 2017; Accepted: 2 June 2017; Published: 13 June 2017

Abstract: We present an innovative electrochemical probe for the monitoring of pH, redox potential and conductivity in near-field rocks of deep geological radioactive waste repositories. The probe is composed of a monocrystalline antimony electrode for pH sensing, four AgCl/Ag-based reference or Cl^- selective electrodes, one Ag_2S/Ag-based reference or S^{2-} selective electrode, as well as four platinum electrodes, a gold electrode and a glassy-carbon electrode for redox potential measurements. Galvanostatic electrochemistry impedance spectroscopy using AgCl/Ag-based and platinum electrodes measure conductivity. The use of such a multi-parameter probe provides redundant information, based as it is on the simultaneous behaviour under identical conditions of different electrodes of the same material, as well as on that of electrodes made of different materials. This identifies the changes in physical and chemical parameters in a solution, as well as the redox reactions controlling the measured potential, both in the solution and/or at the electrode/solution interface. Understanding the electrochemical behaviour of selected materials thus is a key point of our research, as provides the basis for constructing the abacuses needed for developing robust and reliable field sensors.

Keywords: multi-parametric probe; all-solid-state electrodes; reference electrodes; pH sensor; redox potential; conductivity; galvanostatic electrochemistry impedance spectroscopy (GEIS); nuclear waste disposal monitoring

1. Introduction

Near-neutral pH and low redox potential (E_h) are considered to be favourable conditions for nuclear waste disposal in clay formations, because most radionuclides, including actinides, have a low solubility under such conditions [1]. Radioactive waste-management programmes today mainly focus on deep geological storage as this is currently the most appropriate strategy for ensuring the long-term safety of people and environment. "Cigeo" is the name of a future deep geological disposal facility for radioactive waste, to be built between 2020 and 2025 in France, at 500 m depth within the clayey Callovian-Oxfordian (COx) formation.

The COx formation is an assembly of mineral complexes dating back to 160 million years ago and lying at a depth of 400 to 600 m. It is a water-saturated environment with extremely low permeability, porosity and hydraulic conductivity. The temperature, pH and CO_2 partial pressure of the COx pore-water solution are constant at 25 °C, 7.3 (±0.1) and 8×10^{-3} atm, respectively [2]. Anoxic conditions prevail in the COx formation. Within the mineralogical assemblage [3,4], geochemical models predict E_H values ranging from -180 to -200 mV, corresponding to an equilibrium between pyrite and pore-water sulphate $[S^{(+VI)}]$ concentrations, and iron-bearing phases such as Fe-bearing carbonates or nanogoethite [4–7].

The French National Radioactive Waste Management Agency (Andra) is in charge of the long-term radioactive waste management in France. Its technical specifications for the development of monitoring techniques are based on: (i) Requirements due to the specific nature of parameters that need to be measured on key thermal-hydraulic-mechanical-chemical and radiological (THMCR.) processes; and (ii) Requirements due to the minimum accuracy and long-term stability of the monitoring methods—considering that there will be little or no access for re-calibrating the sensors—for the accurate monitoring of the evolution of the near-field around the radioactive waste.

Some constraints specific to on-site conditions must be considered for developing the sensors:

1. The progressive alkalization of the COx pore water due to degradation of concrete casings (pH up to 11.7).
2. The wide range of redox potentials over the Pourbaix diagram due to: (i) gas emissions such as O_2 due to excavation, H_2 due to release from radioactive waste and metal corrosion, CO_2 due to organic-matter degradation, H_2S due to the activity of sulphate-reducing bacteria (SRB), or CH_4 due to the activity of methanogenic bacteria; (ii) sulphide (HS^-/S^{2-}) production due to the activity of SRB; and (iii) nitrate (NO_3^-) production from the chemical and bacterial denitrification of the bitumen coating the concrete casings.
3. The temperature increase due to radioactive disintegration (25 °C $\leq$ T $\leq$ 90 °C).

The three key parameters for monitoring the above parameters are thus pH, conductivity and redox potential [8]. The objective was to design, create and optimize a robust multi-parameter probe for on-site monitoring of pH (±1 pH unit), redox potential (±100 mV) and electrical conductivity (±50 mS·cm^{-1}), in order to ensure the long-term safety of the operation.

We present an innovative electrochemical multi-parameter probe device carrying up to 20 electrodes for such long-term monitoring. To achieve our objective, various types of electrode made of different sensitive materials were studied: a monocrystalline antimony electrode investigated as pH sensor; four silver-chloride-coated silver (AgCl/Ag) based electrodes and a silver-sulphide-coated silver (Ag_2S/Ag) one investigated as reference electrodes and Cl^-/S^{2-}-selective electrodes; four platinum electrodes, one gold electrode and one glassy-carbon electrode, all investigated as redox-potential electrodes. AgCl/Ag and platinum electrodes were also used for conductivity measurements by galvanostatic electrochemistry impedance spectroscopy (GEIS).

We calibrated the developed sensors under conditions similar to those that will be met on-site. Overall performance, reliability and robustness were examined by electrochemical measurements at 25 °C, at atmospheric pressure and/or in a glove box (PCO$_2$ = 8×10^{-3} atm; PO$_2 \approx 10^{-6}$ atm).

2. Methodology

Using a multi-parameter probe offers an important advantage, because it provides redundant information, based on the simultaneous behaviour under identical conditions of different electrodes of the same material as well as that of electrodes made of different materials. This information identifies the redox reactions controlling the measured potential, both in solution and/or at the electrode/solution interface. The objective is the development of robust and "frustrated" electrodes: departing from a calibrated state, all sensors show potentiometric drift due to aging and alteration of the sensitive materials; such drift must be known and mastered for a correct interpretation of the recorded signals at any time. Knowledge of the electrochemical behaviour of the selected materials described below thus is a key point of such research, as it will help in establishing reliable abacuses.

2.1. Antimony-Based Electrodes for pH Sensing

Based upon reversible interfacial redox processes involving H^+, metal/metal-oxide electrodes are a promising technology for the monitoring of pH in underground nuclear-waste disposal sites, due to their physical and chemical stability [9] concerning temperature, pressure and aggressive environments [10,11]. Metal/metal-oxide electrodes present the additional advantage of being easily miniaturized [12–14].

Among the metal/metal-oxide group, the antimony/antimony-oxide system, whose properties were improved by using monocrystalline antimony [15–17], has been the first [18,19] and most investigated (and disputed) one for pH sensing. The fact remains that it is the most commonly used system for practical pH measurements [8,20,21]. The antimony electrode potential is governed by the Sb_2O_3/Sb couple, since the electrode surface is spontaneously oxidized to form a thin antimony oxide film in the presence of oxygen [8,15,22].

The merits and applications of this electrode have been described for bio-medical applications [12,13,23–26], gas-sensing [27], industrial applications such as water-treatment systems, soda ash neutralization, sulphite solutions, etc. [28], or for environmental measurements [14]. It is thus of great interest to use this type of electrode for on-site pH monitoring of the near-field of a nuclear waste disposal site. However, numerous conflicting data concern the disturbance of its potential by various physical and chemical parameters [20,28–30]. Several authors published the influence of oxygen or certain anions, such as carbonates and phosphates, on the open circuit potential (OCP) of the antimony electrode [12], indicating that the electrode must be calibrated under conditions similar to on-site ones.

Three types of antimony-based electrodes are reported in the literature: cast electrode, plated electrode, and antimony powder electrode. The last, made of compacted of $Sb^{(0)}$ and $Sb^{(+III)}$ powders, has been little investigated [26], probably because of the need of a supporting electrolyte saturated by Sb_2O_3 to stabilize the electrode potential, which is unsuitable for continuous sensing.

Cast antimony-based electrodes in the form of a metal stick are the most common ones. The inner diameter of the glass capillary used for fabricating governs the size (macro/micro) of the electrodes. Moreover, cast antimony-based electrodes have the advantage of being relatively rugged and present a low electrical resistance, provided the electrode surface area is not too small. The electrical resistance of an electrode is an important characteristic since it is directly linked to the electrode response time [12]. Although some authors prefer to add some $Sb^{(+III)}$ oxide to the antimony melt before casting the electrode as a cylindrical rod [20], the use of pure antimony is considered to improve the electrode characteristics with respect to reproducibility and stability of the electrode potential over long periods. The way of casting the electrode is important as well: antimony needs to be melted (T ≥ 631 °C, Sb melting point), sucked into a glass capillary and cooled down to solidify [10,12,31]. Slowly cooled electrodes are said to have a faster response than rapidly cooled ones. Furthermore, oxygen should be excluded during casting to avoid oxides in the bulk material. In addition, most methods include polishing the metal surface [12], followed by a treatment to obtain superficial oxidation, like the cumbersome oxidation method via KNO_3 powder heated to 500 °C in a furnace for two hours under air atmosphere [10,31]. However, it must be noted that antimony spontaneously oxidizes by the oxygen contained in air [22,32]. In case of malfunction, the electrode surface can be renewed by polishing and re-oxidizing the surface. A resting period of a few days is recommended afterwards in order to relieve strain in the metal surface [33].

The literature also describes the making of antimony-coated electrodes via cathodic polarization of the sensor immersed in a solution containing $Sb^{(+III)}$ ions [8]. Several supports such as copper, platinum and mercury-coated platinum electrodes were investigated [33], but the literature also mentions a pronounced non-reproducibility as well as fragility of the coating, limiting the lifetime of the system to about five measurements.

The first aim of our work was to investigate the reliability of monocrystalline antimony-based electrodes for their possible incorporation into a multi-parameter device for the observation and monitoring of the near field of a nuclear waste disposal site. The choice of such sensors was motivated by the high binding energy of monocrystalline antimony [23], leading to a low corrosion rate. Therefore the surface is only slowly changed and an occluding oxide is almost completely avoided [23]. The uniform binding energy promotes a general surface corrosion. Thus, only the next crystallographic plane with the same orientation and characteristics as those of the original one is exposed to corrosion. At the same time, experiments were also conducted on antimony-based screen-printed electrodes (SPEs). The reliability of SPEs has already been demonstrated for various applications in earlier papers: semi-continuous monitoring of trace metals [34], and the development of screen-printed pH electrodes based on ruthenium dioxide [32,35], cobalt oxide [36], phenanthraquinone [37] and, more recently, cerium-oxide [38]. Moreover, screen-printing is a simple and fast method for the large-scale production of reproducible low-cost sensors, which allows quick generation of reliable data [39,40]. In this study, antimony-based screen-printed electrodes were used as tools for demonstrating the reliability of monocrystalline antimony electrodes. To ensure accurate measurements, SPEs were renewed every 48 h.

2.2. Inert Electrodes for Redox Measurements

Platinum is considered as an inert indicator electrode for redox-potential measurements of Ox/Red systems in a solution [41–43]; note that platinum is easily oxidized into PtO, a well-conductive oxide that completely covers its surface. The results from platinum electrodes in well-defined concentrations of dissolved species in a solution, are stable and precise; however, in the absence of a well-defined Ox/Red system, platinum results strongly depend on its surface properties, typically the nature of the oxide covering it [41–43].

As part of subsurface redox monitoring in and around a nuclear waste disposal site, gold (Au) and glassy carbon (GC) as inert indicator electrodes were investigated in addition to platinum. Platinum differs from gold in that it has a higher exchange current density ($10 \, \text{mA/cm}^2$ for Pt, $0.3 \, \text{mA/cm}^2$ for Au regarding the O_2/H_2O redox couple) [44]. But gold is known as a more inert material since Pt can catalyse reactions, form oxides and adsorb H_2, which gold cannot [9]. Gold also has a greater potential range (-0.8 to $1.8 \, \text{V/NHE}$) towards positive potentials ($\approx 400 \, \text{mV}$) than Pt [45]. In comparison, glassy carbon has a larger potential range (-1.0 to $1.0 \, \text{V/NHE}$) toward negative potentials ($\approx -200 \, \text{mV}$) than Au and Pt.

2.3. Ag-Based Electrodes Acting as Reference or Selective Electrodes

Reference electrodes are as important as indicator electrodes [46–48]. Nevertheless, compared to all-solid-state indicator electrodes, the research effort into all-solid-state reference electrodes is smaller [46,48]. In their publication, Blaz et al. [48] listed the different types of existing reference electrodes. However, the interest in conducting polymer-based all-solid-state reference electrodes—equitransferent salts dispersed in polymer or compensated cationic and anionic response in polymer—grows each day [48–50].

Existing electrodes essentially use AgCl/Ag-based systems because of the invariability of their potential to pH changes, or to the presence of redox species, unless the temperature and/or the chloride ions vary. In the absence of AgCl/Ag-based systems, it is reported that the OCP of the developed electrode is influenced by redox couples, such as O_2/H_2O [48,50]. Whatever the investigated methodology, Ag-based electrodes predominate and seem essential. Comparatively speaking, macro-metric all-solid-state Ag-based electrodes appear stronger in the design field, as they are based on an important reserve of raw material. Moreover, Ag electrodes can either be coated by AgCl for making Ag/AgCl-based electrodes or by Ag_2S for making Ag/Ag_2S-based electrodes.

Given the quasi-invariability of the on-site Cl^- concentration (0.04 M) [5] of the COx pore-water and the very low permeability of the COx formation, the $AgCl/Ag/Cl^-$ 0.04 M electrode could prove to be of primary interest for monitoring the physical- and the chemical parameter variations within and around a radioactive waste disposal site. A priori, such electrodes present a certain robustness because of: (i) the absence of feeling electrolyte; (ii) the very weak solubility product constants of AgCl ($K_s = 10^{-9.75}$); and (iii) the invariability of their potential to pH changes or to the presence of redox species, unless the temperature and/or the chloride ions vary.

In comparison with $AgCl/Ag$-based electrodes, an Ag_2S/Ag-based electrode (Ag_2S ($K_s = 10^{-49.2}$)) should be less soluble because of its higher pK_s value. In the absence of $S^{(-II)}$, its potential will be a constant OCP. The comparison of potential values measured with both of these electrodes could help in demonstrating any changes over time of chloride and sulphide concentrations in the medium. Understanding of the electrochemical behaviour of selected materials is the second key point of our research, as it will help in constructing abacuses ($[Cl^-] = 0.04$ M; $S^{(-II)} = 0$, in this study) that are needed for monitoring of the physical and chemical parameters of the clay barrier.

2.4. GEIS for Conductivity Measurements

Geophysical methods are standard tools for obtaining information on the volumetric distribution of subsurface physical properties of rocks and fluids. One of several electrical methods that measure electrical properties of the ground is electrical resistivity (ER). In a typical ER measurement [51], four electrodes are used for measuring potential differences between pairs of electrodes, where the potentials result from a current applied between two other electrodes. By measuring at different locations, an electrical resistivity section is reconstructed as a 2D slice of the porous material [51]. Geophysical electrical methods are similar to galvanostatic electrochemistry impedance spectroscopy (GEIS) techniques. In the context of monitoring the surroundings of the radioactive waste disposal site, GEIS was selected as an alternative robust technique for conductivity measurements. A potentiostat/galvanostat was used as this investigates a larger frequency domain, in the range from mHz to MHz.

3. Materials and Methods

3.1. Materials

3.1.1. Description of the All-Solid-State Electrode Surface Materials

Silver Chloride/Silver-, Silver Sulphide/Silver-Based Electrodes for the Development of Reference or Specific Electrodes

A three-electrode cell (100 cm^3) was used for creating AgCl coatings on bare Ag electrodes by oxidation. Chronopotentiometry applied a fixed anodic current density of around 0.5 mA/cm^2 (below the chloride diffusion limited current of the oxidation of silver) between the working Ag electrodes and a platinum counter electrode, using a PAR 273A potentiostat/galvanostat (Princeton Applied Research, Oak Ridge, TN, USA). The electrodes were immersed in a 0.1 M HCl solution. The potential was monitored over the time versus a saturated calomel electrode (SCE) [52,53]. Similar experiments used silver immersed in a 0.1 M NaOH solution containing 0.1 M of Na_2S for making Ag/Ag_2S-based electrodes; the working electrode potential was measured with respect to a mercury/mercurous sulphate electrode (MSE).

Antimony-Based All-Solid-State pH Electrode

We used a monocrystalline antimony (99.999%, m = 500 mg and d = 6.7) electrode without any pre-treatment or treatment over 16 months, in order to investigate its long-term robustness for pH monitoring. Carbon-based screen-printed electrodes (SPEs) [34,38,40,54], with a working surface of 9.6 mm^2, were conditioned in a stirred solution containing $Sb^{(III)}$ (10^{-2} M $SbCl_3$) and HNO_3 (pH = 0) by applying four cycles of cyclic voltammetry (potential range -0.1 V to $+0.8$ V, scan rate 100 mVs^{-1}). The Sb crystals (Sb^{3+} + 3e^{-} → Sb) were then deposited at -0.5 V/SCE during 1600 s. These two steps were performed without removing oxygen from the solution. Figure 1 shows surface analyses carried out by scanning electron microscopy coupled to energy dispersive X-ray (SEM/EDX) on a freshly antimony-coated SPE. Experiments were performed using a TESCAN MIRA XMU scanning electron microscope (TESCAN ORSAY FRANCE, Fuveau, France). Figure 1A exhibits a representative picture of the working surface. Approximately 80% of its surface is coated with antimony. As shown on Figure 1B, shape and size (≈ 13 μm) of grains are homogeneous. The EDX spectrum presented on Figure 1C demonstrates the attainment of a composite material consisting in Sb-coated SPE.

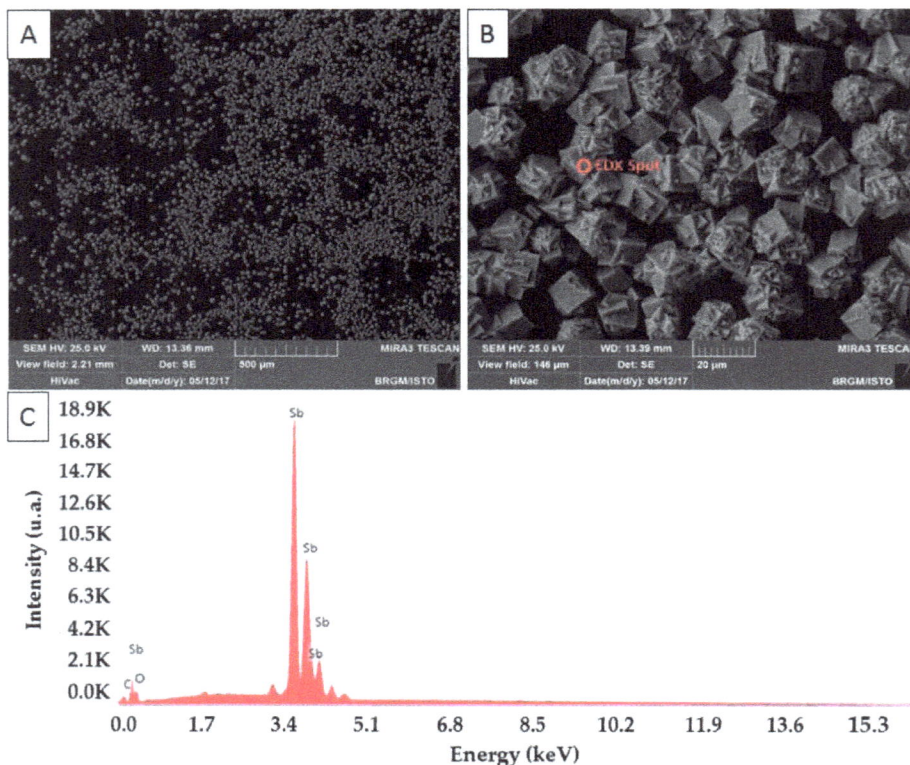

Figure 1. SEM-EDX analyses carried out on a freshly antimony-coated SPE. (**A**) Global view of the Sb-coated SPE working surface; (**B**) Close view of the Sb-coated SPE working surface; (**C**) EDX spectrum resulting of the analysis of the spot shown on Figure 1B.

Whatever the Sb-based sensors used, no preliminary anodic process took place on the electrode surface. Sb metal is slowly oxidized in air to form antimony oxide (Sb$_2$O$_3$) [55].

Platinum, Gold and Glassy Carbon as Inert Electrodes for Redox Potential Measurements

The experiments used a 10 mm disk-shaped Pt electrode (78.54 mm^2), a 10 mm disk-shaped Au electrode (3.14 mm^2), and a 5 mm disk-shaped glassy carbon electrode (3.14 mm^2).

3.1.2. Description of the Experimental Device and Its Components

A schematic representation of the electrode mounting is presented on Figure 2A. Two seals are first inserted on the sensitive element. The sensitive material is welded to the cable. Then, the assembly is slipped until the stop of the electrode body coated with glue. Finally, the body is filled with a liquid resin (molding step) which stiffens the interior and infiltrates the two molding grooves. The sensitive element is thus anchored to the body. A real view of the electrode, built according to the mounting process described above, is shown on Figure 2B. Electrode bodies are made of polyether ether ketone (PEEK), a semi-crystalline thermoplastic with excellent mechanical and chemical resistance properties, resistant to aging over several decades and stable at temperatures up to 100 °C. A maximum of 20 electrodes can be introduced within the probe-holder that is entirely made of polyvinylidene fluoride (PVDF, also called Kynar, made by Arkema, Colombes, France), which is a highly non-reactive and durable thermoplastic fluoropolymer. The internal volume of the column is about 210 cm^3. To ensure water- and air-proofing between intra-column fluid and the exterior, each electrode has a screw-thread with stuffing box and O-ring (O-ring VITON®, Dieppe, France).

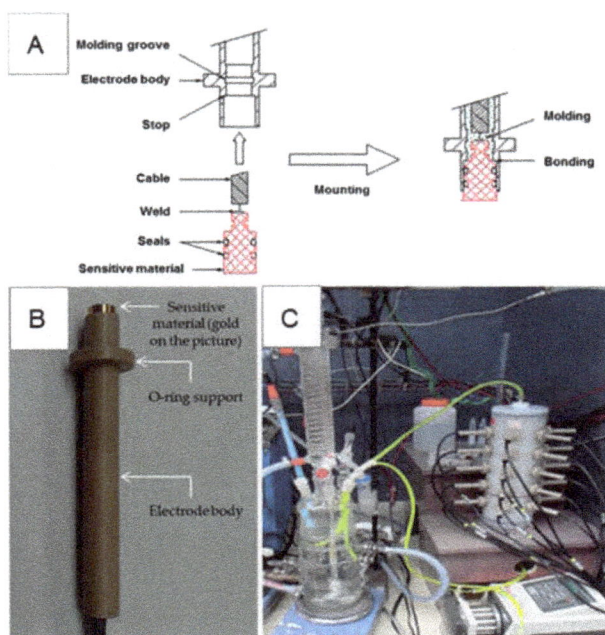

Figure 2. (**A**) Schematic representation of electrode construction; (**B**) Actual view of a gold-based electrode; (**C**) View of the entire experimental set-up.

The whole experimental device, composed of probe holder, electrodes and the other operational elements is shown on Figure 2C. The experiments took place under dynamic conditions. The probe-holder was fed from an electrochemical cell using a peristaltic pump (Figure 2C); the fluid flowed from bottom to top of the probe-holder to avoid bubble formation and thus two-phase flow. The flow rate of the pump was set at an average of 20 mL/min to reduce long-term electrode erosion.

3.1.3. Supporting Electrolytes: Buffers and Solutions

The experiments were performed at a constant temperature (25.0 ± 0.1 °C), either at atmospheric pressure or in a thermo-regulated glove box under N_2/CO_2 (99/1%) atmosphere ($PCO_2 = 8 \times 10^{-3}$ atm, $PO_2 \approx 10^{-6}$ atm).

pH buffers were prepared either using milliQ water (18 MΩ) or 0.1 M NaCl solution. The different conjugate acid-base pairs with their effective pH range are listed in Table 1. For all experiments investigating the influence of pH on electrode OCP, measurements were successively made by increasing then decreasing the pH values in order to highlight hysteresis effects.

Table 1. The different buffer species with their effective pH range.

Buffer Species	Effective pH Range
NH_4^+/NH_3	7.2–11.0
HCO_3^-/CO_3^{2-}	9.1–11.1
$H_2PO_4^-/HPO_4^{2-}$	5.5–7.8

Measurements were also carried out in the presence of a synthetic solution, called reconstituted COx pore water, whose major-element composition and pH at 25 °C were representative of the COx pore water (Table 2) [4,5,56,57].

Table 2. Composition of the reconstituted COx solution at 25 °C.

Species in Solution	Concentration (M)	Species in Solution	Concentration (M)
Ca^{2+}	0.0074	SO_4^{2-}	0.0156
Mg^{2+}	0.0067	Cl^-	0.0400
Sr^{2+}	0.0002	Total Carbon	0.0032
Na^+	0.0450	pH	7.0–7.4
K^+	0.0010	Ionic strength	0.1

3.2. Methods

3.2.1. Potentiometric and/or pH Measurements

OCP values were recorded continuously with a data acquisition device (Keithley Instruments, model 2700, Cleveland, OH, USA). Three different reference electrodes were used: (i) An internal reference electrode of a combined pH electrode (Fischer brand, AgCl/Ag/KCl sat./AgCl sat. E = 197.0 mV/SHE); (ii) SCE inserted in a lugging capillary containing KCl 3 M (SCE/KCl sat. E = 244.4 mV/SHE); and (iii) An MSE electrode inserted in a lugging capillary containing saturated K_2SO_4 (MSE/K_2SO_4 sat. E = 640 mV/SHE). This lugging capillary introduces a junction potential of 1 mV at 25 °C. All potential values were converted with respect to the standard hydrogen electrode (SHE). During experiments, pH was also monitored with a commercial glass electrode that was calibrated daily using commercial standard buffer solutions (4, 7 and 10).

3.2.2. Conductivity Measurements

GEIS measurements used the previously defined PAR 273A. Measuring of the conductivity with the multi-parameter probe was based on the same principle as a 4-pole conductivity meter, which applies a known alternating current (AC) value of 10 µA between two electrodes in the frequency range from 0.1 to 10^5 Hz and measures the induced potential between two other electrodes. From these two parameters, the resistance of the solution (R) is obtained via the ohm law. The solution resistivity (ρ), which is the reciprocal of the conductivity, $\sigma = 1/\rho$, was determined from the relation $R_{(\Omega)} = \rho_{(\Omega \times m)} \times l_{(m)}/S_{(m^2)}$, where $l_{(m)}/S_{(m^2)}$ represents the geometric factor k (m). Since this factor k depends on the inter-electrode distance, it has to be determined for each electrode couple studied.

A schematic representation of the multi-parameter probe with the electrodes used for measuring the conductivity is shown on Figure 3. This also shows the four electrode couples that were tested measuring the conductivity. The first electrode couple is I-F4F1_E-F3F2 (AgCl/Ag-based electrodes); the second is I-F5F2_E-F4F3 (AgCl/Ag-based electrodes); the third is I-F5F1_E-F4F2 (AgCl/Ag-based electrodes); and the fourth is I-F4F1_E-F3F2 (Pt electrodes). The letter "I" is associated to the electrodes used for current injection and the letter "E" is associated to those used for measuring the induced potential.

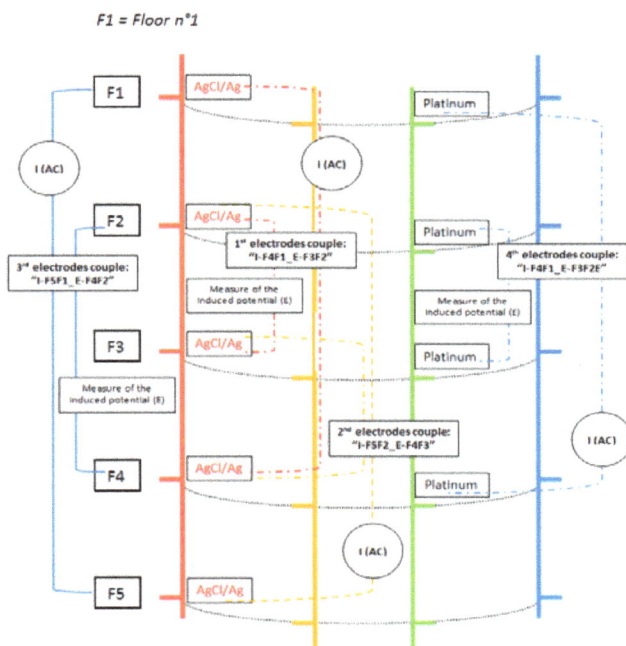

Figure 3. Schematic representation of the multi-parameter probe as well as the different electrode couples used for conductivity measurements.

3.2.3. Geochemical Modelling

The PHREEQC® (USGS, Denver, CO, USA) geochemical code (see also the PHREEQC web site: https://wwwbrr.cr.usgs.gov/projects/GWC_coupled/phreeqc/) was used for thermodynamic investigation of the E_h of the measured sample with the appropriate associated THERMODEM® [58] thermodynamic database generated by BRGM (Orléans, France). Redox potential values were calculated by using speciation data provided by UV-spectrometry (ISO 6332:1988) [59].

4. Results

4.1. All-Solid-State Monocrystalline Antimony pH Electrode

The performances of the monocrystalline antimony electrode (Sb(s)) were investigated by monitoring OCPs on pH values ranging from 5 to 12, in accordance with those anticipated in the COx formation once radioactive waste is buried. The reliability and robustness were investigated in the presence of several anions such as chloride, phosphate, nitrate and hydrogenocarbonate/carbonate, since these are or are susceptible to be present on site and are likely to cause OCP drift. Throughout the 18 months of experiments, monocrystalline antimony electrodes were used without any treatment.

For a better insight, data were compared to those acquired with antimony-based screen-printed electrodes (Sb(SPEs)) that were renewed every 48 h. First, several experiments were conducted in ammonia pH buffer solutions from pH 7.3 to 11, which includes the major pH values that are anticipated in the COx formation during its evolution as radioactive waste repository. The influence of either chloride (NaCl, 0.1 M) or nitrate (NaNO$_3$, from 1.0×10^{-3} to 6.0×10^{-3} M) was investigated in the same way.

Afterwards, phosphate pH buffers were used to calibrate the electrodes from pH 5.5 to 7.7. To investigate their influence, measurements made in a phosphate pH buffer solution were compared to those made in an ammonia pH buffer solution. Because the pH of the clay-rock's pore-water is controlled by carbonate-system equilibria, electrodes were also tested in hydrogenocarbonate/carbonate buffers of various ionic strengths (IS from 0.05 to 0.2 M), from pH 9.2 to 10.2.

The influence of oxygen was investigated by comparing experiments run at atmospheric pressure to experiments run in the glove box, comparing the effect of oxygen on antimony and platinum electrodes. Finally, the monocrystalline antimony electrode was immersed in the reconstituted COx pore-water solution during one month in the glove box. The average potential recorded during this month was compared to our calibration curves, to investigate the reliability and the accuracy of the electrode.

4.2. Influence of Anions at Atmospheric Pressure

4.2.1. Feasibility Study and Influence of Chloride on the OCP of Antimony-Based Electrodes

The influence of chloride on the antimony electrode OCP was investigated in ammonia pH buffer solutions. Results obtained from both the monocrystalline antimony electrode and antimony-based screen-printed electrodes are shown on Figure 4.

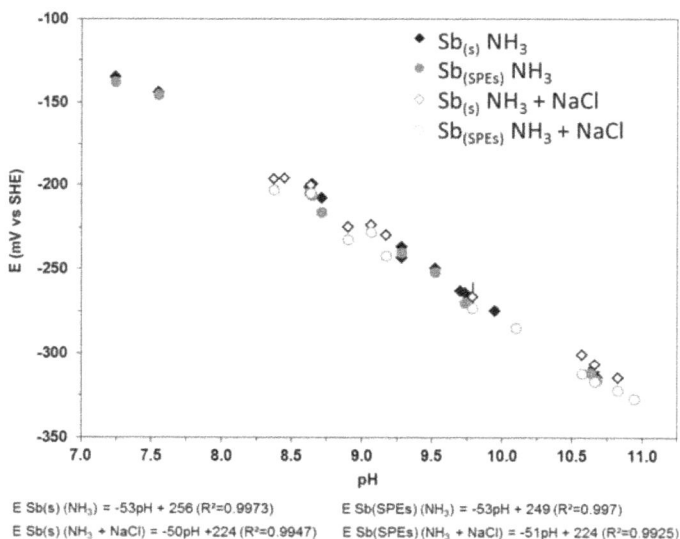

E Sb(s) (NH$_3$) = -53pH + 256 (R²=0.9973)   E Sb(SPEs) (NH$_3$) = -53pH + 249 (R²=0.997)
E Sb(s) (NH$_3$ + NaCl) = -50pH +224 (R²=0.9947)   E Sb(SPEs) (NH$_3$ + NaCl) = -51pH + 224 (R²=0.9925)

Figure 4. Eh-pH diagram of both the monocrystalline antimony electrode (Sb(s)) and antimony-based screen-printed electrodes (Sb(SPEs)) in ammonia pH buffer solutions, in the absence or presence of 0.1 M NaCl (25 °C, at atmospheric pressure).

The general convergence of the stabilization potential of the two types of electrode showed that an equilibrium state was reached under the experimental conditions. The potential-pH dependence of all electrodes over the investigated range is similar and linear. The fact that there are no significant differences between the monocrystalline-antimony and screen-printed electrodes demonstrates the reliability and robustness of the monocrystalline-antimony electrode. Overall, results agree with previous studies on the potential-pH relationship of antimony electrodes [8,20,33]: Ives [33] obtained $E° = 0.245$ V and Glab et al. [8] obtained a slope of 52 mV/pH. The Eh-pH slopes obtained from Sb(s) and Sb(SPEs) are both close to the theory (-59.1 mV/pH, according to Nernst equation, T = 25 °C [9]). As Sb spontaneously oxidizes in air or in water in the presence of oxygen, the electrode potential should be governed by the Sb_2O_3/Sb couple ($E°(Sb_2O_3/Sb) = 0.152$ mV/SHE). Our results confirm the metal-cell action theory of metal corrosion [8,33,55] that mentions that a small portion of the antimony electrode surface is the siege of oxygen reduction, causing positive drift of the antimony electrode potential ($E°(O_2/H_2O) = 1.23$ V/SHE). Those results not only show that monocrystalline Sb is an electrode material of interest for measuring pH, but also that it is not affected by the presence of chloride in solution, which agrees with previous results obtained by Uhl and Kestranek in 1923 [18].

4.2.2. Influence of Nitrate on the OCP of Antimony-Based Electrodes

The presence of oxidizing reagents in solution is highly likely to induce drift in electrode potential. Consequently, experiments tested both Sb(s) and Sb(SPEs) electrodes in the presence of different amounts of nitrate (NO_3^-, $N^{(+V)}$) in solution. Nitrate was chosen as oxidizing reagent since it is present in bituminized sludge as nitrate salts, such as $NaNO_3$. Bituminized sludges are going to be stored as long-lived intermediate-level radioactive waste within the COx formation and could diffuse in the long-term. The results are shown on Figure 5A,B.

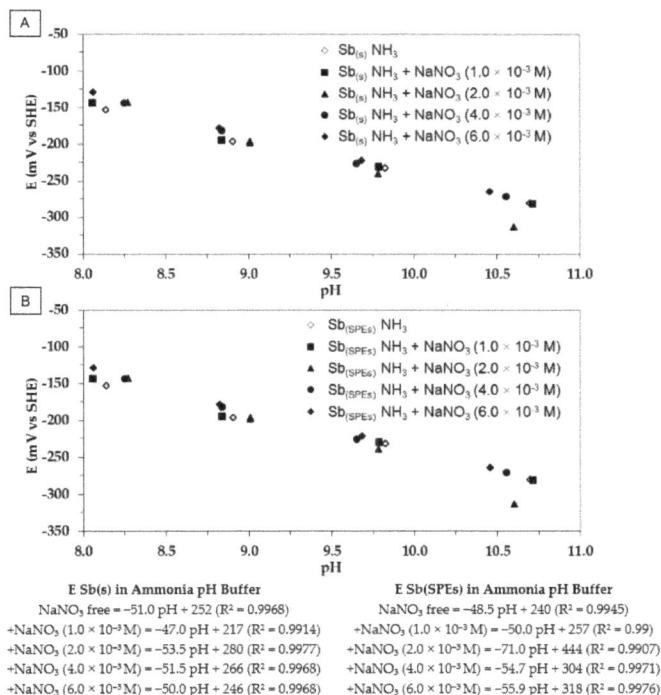

E Sb(s) in Ammonia pH Buffer	E Sb(SPEs) in Ammonia pH Buffer
$NaNO_3$ free = -51.0 pH + 252 (R^2 = 0.9968)	$NaNO_3$ free = -48.5 pH + 240 (R^2 = 0.9945)
+$NaNO_3$ (1.0 × 10^{-3} M) = -47.0 pH + 217 (R^2 = 0.9914)	+$NaNO_3$ (1.0 × 10^{-3} M) = -50.0 pH + 257 (R^2 = 0.99)
+$NaNO_3$ (2.0 × 10^{-3} M) = -53.5 pH + 280 (R^2 = 0.9977)	+$NaNO_3$ (2.0 × 10^{-3} M) = -71.0 pH + 444 (R^2 = 0.9907)
+$NaNO_3$ (4.0 × 10^{-3} M) = -51.5 pH + 266 (R^2 = 0.9968)	+$NaNO_3$ (4.0 × 10^{-3} M) = -54.7 pH + 304 (R^2 = 0.9971)
+$NaNO_3$ (6.0 × 10^{-3} M) = -50.0 pH + 246 (R^2 = 0.9968)	+$NaNO_3$ (6.0 × 10^{-3} M) = -55.9 pH + 318 (R^2 = 0.9976)

Figure 5. Eh-pH diagram of both Sb(s) (**A**) and Sb(SPEs) (**B**). Experiments were done at 25 °C, at atmospheric pressure, in ammonia pH buffer solutions (IS = 0.1) containing $NaNO_3$ (0, 1.0 × 10^{-3}, 2.0 × 10^{-3}, 4.0 × 10^{-3} and 6.0 × 10^{-3} M).

Antimony electrodes (Sb(s) and Sb(SPEs)) were not significantly influenced by the presence of nitrate. The monocrystalline antimony electrode behaves in the same way as in ammonia (with or without chloride) pH buffer solutions, i.e., its Eh-pH slopes as well as its standard potential remain in the same range of values. Conversely, Eh-pH slopes and standard potential values measured from antimony-based screen-printed electrodes show a tendency to increase in the presence of nitrate. The higher OCP value of the Sb(SPEs) over the whole pH range can be explained by interfacial Sb oxidation via NO_3^- such as:

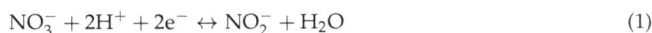

$$NO_3^- + 2H^+ + 2e^- \leftrightarrow NO_2^- + H_2O \tag{1}$$

$$E'^{\circ}_{NO_3^-/NO_2^-} = 0.4 + \frac{RT}{F} \times \ln\left(10^{-pH}\right) \tag{2}$$

$$Sb_2O_{3(s)} + 6H^+ + 6e^- \leftrightarrow 2Sb_{(s)} + 3H_2O \tag{3}$$

$$3\,NO_3^- + 2Sb_{(s)} \leftrightarrow Sb_2O_{3(s)} + 3NO_2^- \tag{4}$$

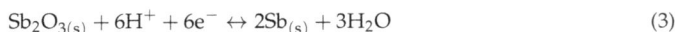

A small Sb(SPEs) part of the antimony electrode surface is the seat of nitrate reduction, which leads to a slight positive drift of the OCP. The measured values can be interpreted as a mixed potential between $Sb_2O_{3(s)}/Sb_{(s)}$ and NO_3^-/NO_2^- redox couples. These results demonstrate the robustness and the usefulness of the massive monocrystalline antimony electrode for monitoring pH within repositories dedicated to nuclear waste disposal.

4.2.3. Influence of Phosphate on the OCP of Antimony-Based Electrodes

The influence of the presence of phosphate $P^{(V)}$ on the electrode OCP was studied by means of phosphate buffer species (NaH_2PO_4/Na_2HPO_4 − IS = 0.1) from pH 5.6 to 7.7, comparing the data to those obtained from ammonia chloride free pH buffer solutions. The results are shown on Figure 6.

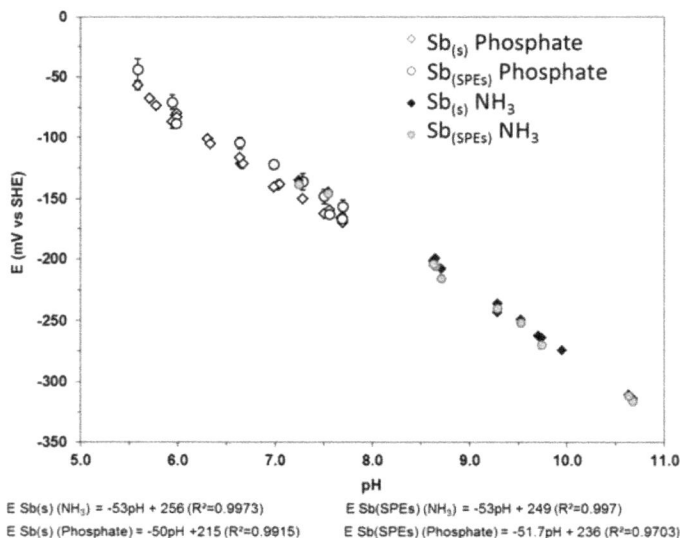

E Sb(s) (NH₃) = -53pH + 256 (R²=0.9973)   E Sb(SPEs) (NH₃) = -53pH + 249 (R²=0.997)

E Sb(s) (Phosphate) = -50pH +215 (R²=0.9915)   E Sb(SPEs) (Phosphate) = -51.7pH + 236 (R²=0.9703)

Figure 6. Eh-pH diagram of both the Sb(s) and Sb(SPEs) acquired at 25 °C, at atmospheric pressure, in phosphate and ammonia pH buffer solutions (IS = 0.1).

The electrodes present a near-Nernstian behaviour over the investigated range. The standard potential is slightly lowered in the presence of phosphate. One of the early investigations of phosphate influence was made by Gysinck [60], who observed a kink in the calibration curve between pH 7.1 and 8.2. Green and Giebish [29] and Perley [28] also investigated the influence of phosphate on the Sb electrode, and observed electrode potential drift. Nevertheless, our results show that the presence of phosphate species does not significantly influence the analytical response of the Sb electrodes. Work carried out by Glab et al. (1981) [16] is in better agreement with our results, showing very little dependence of the Sb electrode to phosphate concentration.

4.2.4. Influence of Carbonate and Ionic Strength on the OCP of Antimony-Based Electrodes

In natural water samples, ionic strength is a variable chemical parameter [61]. Its influence on Sb-based-developed electrodes was investigated in $NaHCO_3/Na_2CO_3$ at different ionic strengths ranging from 0.05 to 0.2 M. The results are compared to those obtained in ammonia pH buffer solutions and shown on Figure 7. Measurements in a carbonate pH buffer solution at different ionic strengths are close to those measured in ammonia buffers, demonstrating the insignificant influence of carbonate species on electrode OCP. Green and Giebisch [29] studied the influence of ionic strength on a micro antimony electrode by means of a phosphate buffer and showed that, independently of hydrogen activity, electrode potential variations occurred when varying the ionic strength. Our results are in better agreement with those published by Caflisch et al. [12], demonstrating that antimony electrodes are not quantitatively influenced by the presence of hydrogenocarbonate/carbonate anions in solution (from 0.05 to 0.2 M).

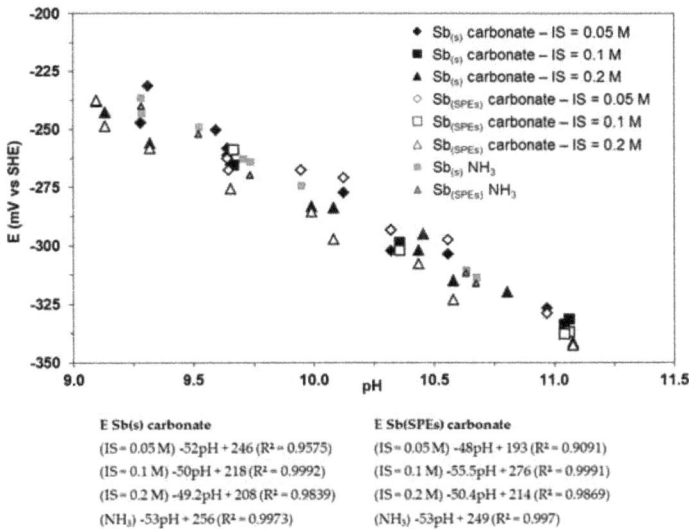

E Sb(s) carbonate
(IS = 0.05 M) -52pH + 246 (R^2 = 0.9575)
(IS = 0.1 M) -50pH + 218 (R^2 = 0.9992)
(IS = 0.2 M) -49.2pH + 208 (R^2 = 0.9839)
(NH_3) -53pH + 256 (R^2 = 0.9973)

E Sb(SPEs) carbonate
(IS = 0.05 M) -48pH + 193 (R^2 = 0.9091)
(IS = 0.1 M) -55.5pH + 276 (R^2 = 0.9991)
(IS = 0.2 M) -50.4pH + 214 (R^2 = 0.9869)
(NH_3) -53pH + 249 (R^2 = 0.997)

Figure 7. Eh-pH diagram of both Sb(s) and Sb(SPEs). Experiments were run in carbonate pH buffer solutions with various ionic strength (IS) and ammonia pH buffer solution under atmospheric pressure at 25 °C.

All our results show that the potential of a massive monocrystalline antimony electrode is not influenced by chlorides (from 0 to 0.1 M), or phosphates (from 0 to 0.1 M), or nitrates (from 0 to 6 mM), or carbonates (from 0 to 0.2 M), demonstrating its robustness and thus its interest for pH monitoring within repositories dedicated to nuclear radioactive waste disposal, such as the one at Bure in France. Moreover, no hysteresis effects affected the antimony electrodes.

Overall, our results demonstrate the interest of screen-printed electrodes for generic studies of electrode materials. It also should be noted that the antimony-based screen-printed electrodes made for this study present a better behaviour than other Sb-based screen-printed electrodes described in the literature [8,32]. Koncki and Mascini [32] mentioned a certain irreproducibility of antimony-based sensors made by screen-printing as well as hysteresis effects. Our antimony-based SPEs were renewed every 48 h and, as shown by the different calibration curves, there are no doubts regarding the reproducibility of the measurements or the absence of hysteresis effects. Moreover, our results in various buffers showed no effects of complexing ligands as was the case for some authors [8,16,32]. Compared to the −40/−45 mV/pH sensitivity obtained by Koncki and Mascini [32] under specific conditions (i.e., the only suitable results were obtained in TRIS buffer and in the absence of chloride), the sensitivity of our antimony-based SPEs is around −50/−55 mV/pH and does not depend on the nature of buffer species. Our antimony-based SPEs show a near-Nernstian response (an average of −53.6 mV/pH) and thus appear to be suitable electrodes for accurate pH measurements. As pH sensors, the performance level of Sb-based screen-printed electrodes can be compared to that obtained with CeO_2-based [38] or RuO_2-based [32] SPEs.

4.3. Calibration Curve of the Monocrystalline Sb Electrode at Amospheric Pressure and in the Glove Box. Comparison with a Pt Electrode. Investigation of Measurement Accuracy in the Reconstituted COx Pore Water

In the same way as the experiments conducted at atmospheric pressure, experiments were also performed in the glove box. In parallel and for comparison, the same measurements were performed with four Pt electrodes. The results are presented on Figure 8, where "Sb" represents all data obtained from Sb(s) and Sb-based SPEs.

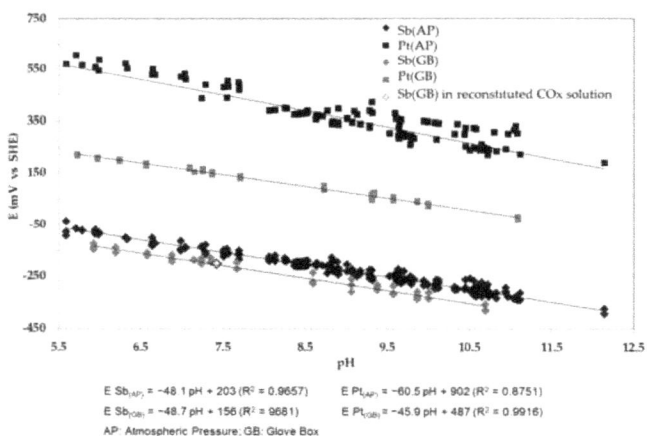

E $Sb_{(AP)}$ = −48.1 pH + 203 (R^2 = 0.9657)   E $Pt_{(AP)}$ = −60.5 pH + 902 (R^2 = 0.8751)
E $Sb_{(GB)}$ = −48.7 pH + 156 (R^2 = 9681)   E $Pt_{(GB)}$ = −45.9 pH + 487 (R^2 = 0.9916)
AP: Atmospheric Pressure; GB: Glove Box

Figure 8. Comparison of calibration curves obtained from antimony (Sb(s) and Sb(SPEs)) and platinum electrodes obtained at 25 °C, under atmospheric pressure and in the glove box. An antimony potential value (vs. SHE) representing the average potential of the electrode in the reconstituted COx solution over one month was added.

As well as Sb electrodes, Pt electrodes present a Nernstian behaviour as a function of pH at atmospheric pressure. In absence of any other redox couple except O_2/H_2O, the potential of the Pt electrode is governed by the PtO/Pt couple ($E°$ (PtO/Pt) = 900 mV/SHE).

Measurements acquired with Pt electrodes are in good agreement with published data [43,61]. At a given pH at atmospheric pressure and in the absence of any species able to fix the redox potential except O_2/H_2O ($PO_2 \approx 0.2$ atm), a difference of 700 mV (at pH 0) exists between $E°_{PtO/Pt}$ and $E°_{Sb2O3/Sb}$. This difference is reduced to about 330 mV in the glove box. The Sb-based sensitivity to pH change is not affected (−48.1 pH vs. −48.7 pH) with regard to the investigated conditions. The intercept at pH = 0 differs by approximately 50 mV, demonstrating the low dependence of the Sb electrode versus oxygen and, consequently, the interest of electrodes based on Sb for pH measurement.

To verify the reliability of a Sb(s)-based electrode in our context of pH monitoring within a radioactive waste-disposal site, the monocrystalline antimony electrode was tested for about one month in the reconstituted COx pore-water solution in the glove box. Measurements using the four Pt electrodes were done every 5 min in the same reconstituted COx solution, for a total of 8640 measures over the month. Average pH values determined from the Sb(s) electrode and Pt electrodes are summarized in Table 3.

Table 3. Average pH and OCP of Sb electrode measured over 1 month in the reconstituted COx solution in the glove box.

Electrode	Average Value (pH Units)	Standard Deviation
Commercial pH electrode	7.42	±0.03
Monocrystalline Sb-electrode	7.37	±0.06
Pt electrodes (mV/SHE)	7.32	±0.07

The average potential of the Sb-electrode over one month was added on Figure 8 (-203.1 mV vs. SHE). The calculated standard deviation is low (2.9 mV), again demonstrating the robustness of the electrode. From the figure it is clear that the measured potential over one month in the reconstituted COx solution in the glove box agrees very well with our calibration curve and the same is true for the Pt electrodes whose average potential is also very stable (151.2 ± 2.88 mV/SHE).

Concerning the relative standard deviation of the commercial pH electrode, the monocrystalline Sb electrode and the platinum electrodes, the measured pH values are in line with our expectations, which confirms the reliability and so the interest of the monocrystalline Sb electrode for monitoring pH within a future radioactive waste disposal site.

4.4. Investigation of Inert Pt, Au and GC Electrodes for Redox Potential Measurements

This experiment consisted in investigating the behaviour of gold, glassy carbon and platinum electrodes in samples where iron dominated redox reactions. The investigation was conducted in a 0.1 M NaCl solution containing 100 mg of Fe° powder at pH 8 (not balanced system, Figure 9) in the glove box. Prior to measurements, the solution was stirred during 24 h.

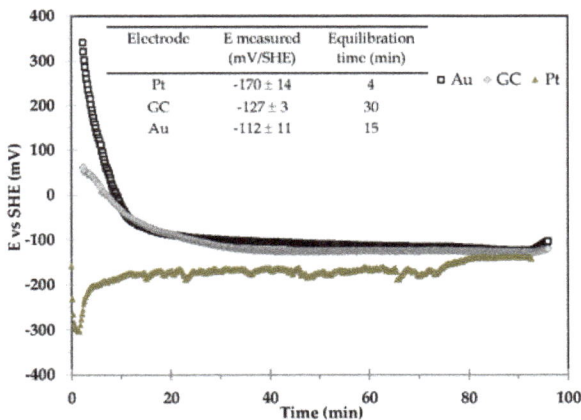

Electrode	E measured (mV/SHE)	Equilibration time (min)
Pt	-170 ± 14	4
GC	-127 ± 3	30
Au	-112 ± 11	15

Figure 9. Evolution of Au, GC and Pt OCP electrodes as a function of time. The experiment was conducted in 0.1 M NaCl solution containing 100 mg of Fe° powder at pH 8 and 25 °C, in the glove box.

4.4.1. Electrode Performances and Robustness—Inert Electrodes

Potentials recorded by all immersed inert electrodes converge to a value of the same order of magnitude. This fact demonstrates that, under the same experimental conditions, the same equilibrium state has been reached on all inert electrodes.

4.4.2. Comparison between Measured Potential, Speciation Measurements and Geochemical Modelling

At pH 8, the redox couple fixing the potential should be $H^+/H_{2(g)}$. There is no thermodynamical equilibrium between $Fe°$ and H_2O, as $Fe°$ corrodes under anaerobic conditions. In the presence of O_2, $Fe°$ also corrodes to form Fe^{2+}, which can then be oxidized in $Fe^{(3+)}$ that can precipitate. Thus, $Fe^{(III)}/Fe^{(2+)}$ progressively becomes the predominant redox couple leading to a progressive increase of the redox potential. The presence of dissolved $Fe^{(3+)}$ (provided by speciation measurements obtained by UV-visible spectroscopy ($Fe^{(3+)}$ 3×10^{-7} M, $Fe^{(2+)}$ 2.95×10^{-6} M) corroborates the presence of oxygen. The convergence to a stabilized potential close to -140 mV/SHE at pH 8 and 25 °C (Figure 9) demonstrates the limitation of the O_2 intrusion (absence of total oxidation), and the qualitative agreement between the acquired OCPs of all inert electrodes argues for an influence of O_2 trace concentrations originating from the gas phase in the glove-box. Considering the disturbance by O_2 in the sample, the geochemical modelling predicts:

1. $E_h = -165$ mV/SHE in the presence of lepidocrocite $+ 0.10 \times 10^{-6}$ M Fe^{2+} in solution.
2. $E_h = -157$ mV/SHE in the presence of goethite $+ 0.07 \times 10^{-6}$ M Fe^{2+} in solution.

Nevertheless, problems of solubility uncertainty are assumed in the database. The redox potential was probably fixed by the $Fe(OH)_3/Fe^{2+}$ redox couple, which depends upon $Fe(OH)_3$ solubility as a function of pH. While limitations to the interpretation of E_h remain, the interest in continuous monitoring of voltage measurements using multiple redox electrodes is clearly shown in order to ensure reliable qualitative measurements.

This example highlights that for not well-balanced oxygen-sensitive systems, speciation should absolutely be preserved to provide qualitative information on redox conditions. Nevertheless, preserving the initial conditions is extremely difficult. A complementary approach by successively coupling amperometric and potentiometric measurements, in which an electro-active redox mediator improved the rate of electron transfer from the redox active solid phases to the electrode, has been developed for redox-potential measurements of minerals [62].

4.5. All-Solid-State AgCl/Ag- and Ag$_2$S/Ag-Based Electrodes as Reference or Selective Electrodes

The long-term behaviour of AgCl/Ag- and Ag$_2$S/Ag-based electrodes was investigated in the reconstituted COx pore water in the glove box, at constant pH (7.4) and by varying the pH by means of addition of NaOH and K_2CO_3 solutions in the range 7.4 to 11. Measurements were made with respect to MSE.

4.5.1. Experiments Performed at Constant pH (7.4)

Figure 10 shows the potentiometric response of three AgCl/Ag/[Cl$^-$] = 0.04 M electrodes over almost two months in the reconstituted COx solution at a constant pH of 7.4. It also shows the theoretical OCP calculated from the Nernst equation by considering a chloride concentration of 0.04 M as well as the ionic strength of the solution (I = 0.1 M). The activity coefficient of chloride (γ_{Cl^-}) was calculated from the Debye-Hückel model and is equal to 0.75. An increase of chloride-ion activity leads to a decrease of the AgCl/Ag-based electrode potential, according to the Nernst equation.

The electrochemical behaviour at zero current of the three AgCl/Ag/[Cl$^-$] = 0.04 M electrodes is very similar. A very slight decrease of the potential is observed over time: 4 mV loss over 1200 h for all electrodes, corresponding to an increase in chloride activity of 14.4%. This small deviation is consistent with the 13.7% water loss measured in the cell due to evaporation, leading to an increase of the chloride activity over time.

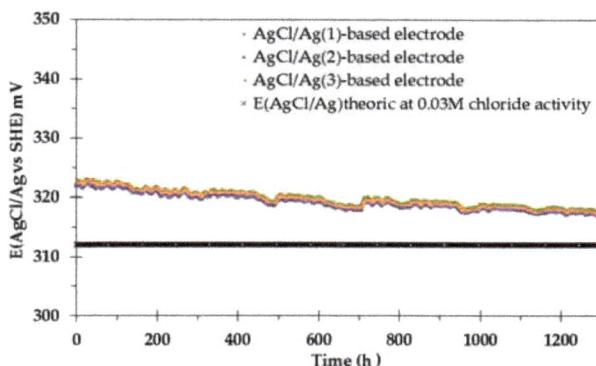

Figure 10. Evolution of the potentiometric response of the all-solid-state AgCl/Ag-based electrodes at constant pH (7.4) as a function of time over 54 days. Measurements were made in the reconstituted COx solution, at 25 °C in the glove box.

In comparison to the behaviour observed at constant pH when using a solid-state AgCl/Ag/KCl reference electrode based on carbon nanotubes and polyacrylate membranes [63], our results obtained on massive AgCl/Ag/[Cl$^-$] = 0.04 M electrodes thus showed encouraging results: 4 mV loss over 1200 h due to evaporation compared to an average 1 mV loss per hour obtained by [63]. The potential response of AgCl/Ag/[Cl$^-$] = 0.04 M electrodes when subjected to pH variations was then investigated.

4.5.2. Experiments Performed under pH Variations

To investigate the influence of pH on the potentiometric response of AgCl/Ag/[Cl$^-$] = 0.04 M electrodes, NaOH was added continuously to the reconstituted COx solution. At the same time, another experiment was conducted on an Ag$_2$S/Ag-based electrode (Figure 11).

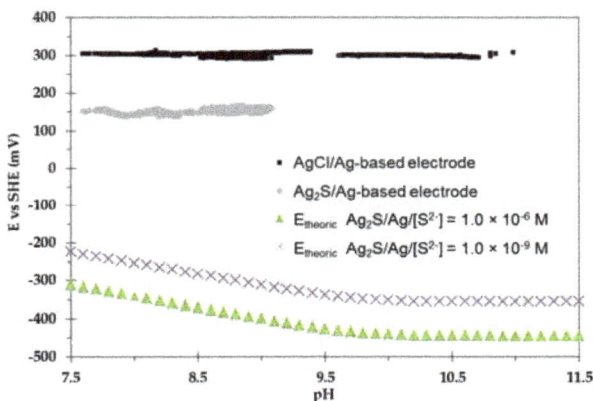

Figure 11. Eh-pH diagram of both AgCl/Ag- and Ag$_2$S/Ag-based electrodes. Experiments were done in the reconstituted COx solution at 25 °C, in the glove box. NaOH was used to increase pH. Theoretical Eh-pH variation curves of the Ag$_2$S/Ag-based electrode in the presence of [S^{2-}] at total concentrations of 1.0×10^{-6} M and 1.0×10^{-9} M were added for comparison.

As well as at constant pH, the potential of the $AgCl/Ag/[Cl^-]$ = 0.04 M electrodes was not significantly influenced by pH variations. This is in good agreement with the theoretical electrode behaviour (19) and encourages their utilization as reference electrodes for further experiments such as to calibrate the electrodes for monitoring the near field of a nuclear waste disposal site.

In comparison to $AgCl/Ag/[Cl^-]$ = 0.04 M, the Ag_2S/Ag-based electrode is characterized by an OCP that is constant (E = 151 ± 4.6 mV) in the investigated pH range.

Similar to the $AgCl/Ag/[Cl^-]$ = 0.04 M electrodes, the results obtained from an Ag_2S/Ag-based electrode are encouraging for further experiments in the presence of $S^{(-II)}$.

4.6. Measurements of the Conductivity in Solution from the Multi-Parameter Probe

Multiplying electrode couples for accurate conductivity measurements appears to be crucial. For this purpose, the accuracy of the conductivity measurements was investigated on the four electrode couples I-F4F1_E-F3F2(AgCl), I-F5F2_E-F4F3(AgCl), I-F5F1_E-F4F2(AgCl) and I-F4F1_E-F3F2(Pt), as shown on Figure 3.

Prior to any measurement on a real sample, it is necessary to determine the geometric factor (k) of each investigated electrode couple. The calibration step of the multi-parameter probe was done by measuring the impedance ($|Z|$) of four sodium chloride solutions whose concentrations ranged from 10^{-4} to 10^{-1} M. The resistivity values of these solutions were measured by means of a commercial conductivity probe, and are reported in Table 4.

Table 4. Conductivity and resistivity values of the four sodium-chloride solutions used to determine the geometric factor value of each electrode couple.

NaCl (M)	σ (S·m^{-1})	ρ (Ω·m)
10^{-1}	1.2155	0.8227
10^{-2}	0.12	8.3333
10^{-3}	0.0126	79.3651
10^{-4}	0.001274	784.9294

The influence of the frequency of the AC current applied between the two injection electrodes was investigated by calibrating the four electrode couples at various frequencies ranging from 0.1 to 10^5 Hz. The objective of this investigation was to determine the optimal frequency range for measuring the conductivity in the range from 10^{-4} to 10^{-1} M (i.e., from 1.28×10^{-3} S·m^{-1} to 1.22 S·m^{-1}).

4.6.1. Influence of the Alternating Current Frequency

In order to investigate the frequency influence on our experiments, impedance measurements were carried out at different frequencies ranging from 0.1 to 10^5 Hz for each of our electrode couples and in each sodium-chloride solution (10^{-4}, 10^{-3}, 10^{-2}, 10^{-1} M). The results and conclusions being the same for all electrode couples, only the one obtained from the second electrode couple (I-F5F2_E-F4F3_AgCl) will be discussed (Figure 12).

As shown on Figure 12, impedance values measured in low-conductive solutions (Figure 12A,B) are fairly stable at low frequencies, but vary significantly at high frequencies. The inverse is observed in more conductive solutions (Figure 12C,D), where the impedance values are fairly stable at high frequencies, but not at low frequencies. This shows that the less the solution is conductive, the lower the frequency should be. As it was decided to calibrate the device in a concentration range of 10^{-4} M to 10^{-1} M, i.e., to determine a geometric factor "*k*" that would allow the conversion of impedance ($|Z|$) into resistivity (ρ) over this concentration range, the alternating current (AC) frequency had to be chosen carefully, considering the impedance variation as a function of AC frequency. For instance, regardless of the concentration of the solution, the impedance remains constant as a function of the frequency from 10 Hz to about 1500 Hz (Figure 12). Therefore, the geometric factor was determined with a 1373 Hz AC frequency (10 μA). The determination and comparison of the geometric factors of each electrode couple is presented in the next section.

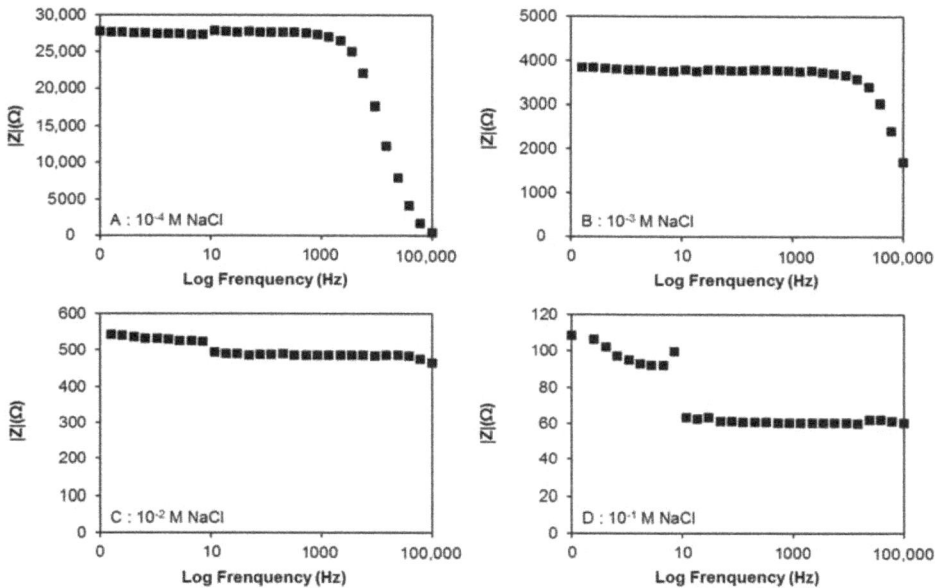

Figure 12. Variation of the impedance values as a function of alternating current frequency resulting from measurements carried out on the second electrode couple (I-F5F2_E-F4F3_AgCl) in sodium chloride solutions of different concentrations: (**A**) 10^{-4} M, (**B**) 10^{-3} M, (**C**) 10^{-2} M, (**D**) 10^{-1} M.

4.6.2. Determination of the Geometric Factors

By applying the procedure described earlier in the Section 4.6 (Measurements of the conductivity in solution from the multi-parameter probe), the geometric factors of each electrode couple were calculated. Table 5 summarizes the measurements at different concentrations ranging from 10^{-4} to 10^{-1} M. The geometric factor "k" was obtained by plotting the resistivity of the solutions (Table 5) as a function of impedance, since it corresponds to the slope.

Table 5. Summary of the impedance measures realized on the different electrodes couples for the determination of the geometric factors. An AC of 10 μA at a frequency of 1373 Hz has been applied during these experiments.

N° Couple	Injection Electrodes (I)	Induced Potential Electrodes (E)	$|Z|$ (Ω) at Different NaCl Content (M) and at Frequency = 1373 Hz				k (m^{-1})
			10^{-1} M	10^{-2} M	10^{-3} M	10^{-4} M	
1	F4F1_AgCl	F3F2_AgCl	58.98	475.55	3684.48	26,849.22	0.0296
2	F5F2_AgCl	F4F3_AgCl	60.37	484.18	3736.20	26,975.55	0.0294
3	F5F1_AgCl	F4F2_AgCl	119.39	957.7	7377.56	53,710.79	0.0148
4	F4F1_Pt	F3F2_Pt	58.45	472.20	3661.99	26,417.52	0.0301

Comparing the geometric factors obtained from couples 1 and 2 (AgCl/Ag-based electrodes) shows that measurement reproducibility is good.

Comparing the results obtained with silver chloride/silver electrodes and platinum electrodes, i.e., couples 1, 2 and 4, shows that the geometric factor does not depend on the type of electrode.

Finally, comparing the results of electrode couple 3 with those of the other electrode couples shows that the geometric factor value depends on the distance between electrodes. However, this dependence on the distance between electrodes seems only to concern the distance between electrodes for measuring the induced potential. As shown on Table 5, when the distance between these electrodes is doubled, the geometric factor increases by a factor of two as well.

Geometric factors were also calculated for a lower AC frequency (79 Hz) following the same approach as above. The results (not presented) are very similar to those just described, showing a relative gap of the geometric factors ranging from 0.75 to 3%, depending on the electrode couple considered. The calibration of the device allowed measuring the resistivity of the reconstituted COx pore water from impedance measurements at 1373 Hz. The good agreement between the conductivity value measured with the multi-parametric probe (0.78 ± 0.02) $S \cdot m^{-1}$, the mean value of the data acquired by the four electrode couples) and that acquired with a conventional four-cell conductivity electrode (0.85 $S \cdot m^{-1}$) shows the accuracy of our new probe (relative gap of 8.2%), which can be used for punctual conductivity measurements. It can also be used for establishing a vertical conductivity profile of the solution flowing through the multi-parametric probe, by injecting the current between floors 1 and 5, and by successively measuring the induced potential between floors 5–4, 4–3, 3–2 and 2–1. If the conductivity of the solution changes, then the flow of the solution through the device can be estimated.

5. Conclusions

We present an innovative electrochemical device for pH, redox potential and conductivity monitoring in near-field rock of deep geological radioactive waste repositories. A monocrystalline antimony electrode was tested over 16 months. Both reliability and robustness of this electrode were clearly demonstrated for pH monitoring in the range of 5.5–12 within a radioactive nuclear-waste disposal site.

OCP measurements provided by Pt, Au and glassy carbon (GC) electrodes for determining the E_h value confirmed the robustness of platinum as an indicator electrode for this purpose. Voltage measurements by gold and GC electrodes tended towards those provided by platinum ones (same order of magnitude), demonstrating the analytical feasibility of redox measurement via other inert electrodes.

All-solid-state AgCl/Ag/[Cl$^-$] = 0.04 M electrodes showed a constant OCP value (E = 319.2 $\pm$ 1.4 mV) over one month of analyses of reconstituted COx pore water in the 7.4–11 pH range. Under the investigated conditions, the electrodes did not show potential drift when subjected to pH variations which opens the possibility of using them as reference electrodes. Under the same conditions and in the absence of S$^{(-II)}$, the OCP of an Ag$_2$S/Ag-based electrode acquires the constant OCP value of E = 151 $\pm$ 4.6 mV in the pH range of 7.4 to 9.

The multi-parametric device can also be used for conductivity measurements by GEIS. The good agreement between the conductivity value measured with the multi-parametric probe (0.78 ± 0.02 $S \cdot m^{-1}$), which is a mean of the data acquired by the four electrode couples, and the one acquired with a conventional four-cell conductivity meter (0.85 $S \cdot m^{-1}$) testifies to the accuracy of the method.

Overall, the bundle of electrodes as designed by us appears suitable for monitoring the COx formation during its envisaged use for hosting a nuclear waste repository. Further is ongoing to develop abacuses for an accurate calibration of the new probe. Experiments for estimating corrosion rates of the different electrode materials in the reconstituted COx solution are planned as well.

Acknowledgments: This work was funded by a BRGM-ANDRA partnership (CAPTANDRA project 2009–2018). The authors thank Société HEITO from Paris, France (Jean Pierre Heitzmann and Frédéric Bota) for fabricating the electrodes and SCODIP from Orléans, France (Yannik Legueunic) for manufacturing the multi-parametric device. Marinus Kluijver from Olivet France, corrected the English of the manuscript.

Author Contributions: All authors have given approval to the final version of the manuscript. S.B. and I.I. have selected, conceived and designed the electrodes and the experiments. S.B. and J.D. performed the experiments and wrote the paper, and all authors contributed to its improvement. I.I. and J.B. are responsible for the CAPTANDRA project respectively from BRGM and ANDRA; I.I. and T.T. are co-directors of the Doctoral Thesis of J.D.

Conflicts of Interest: The authors declare no conflict of interest.

References

1. Altmann, S. "Geo"chemical research: A key building block for nuclear waste disposal safety cases. *J. Contam. Hydrol.* **2008**, *102*, 174–179. [CrossRef] [PubMed]
2. Lassin, A.; Marty, N.C.M.; Gailhanou, H.; Henry, B.; Trémosa, J.; Lerouge, C.; Madé, B.; Altmann, S.; Gaucher, E.C. Equilibrium partial pressure of CO_2 in Callovian-Oxfordian argillite as a function of relative humidity: Experiments and modelling. *Geochim. Cosmochim. Acta* **2016**, *186*, 91–104. [CrossRef]
3. Gaucher, E.C.; Robelin, C.; Matray, J.M.; Négrel, G.; Gros, Y.; Heitz, J.F.; Vinsot, A.; Rebours, H.; Cassagnabère, A.; Bouchet, A. ANDRA underground research laboratory: Interpretation of the mineralogical and geochemical data acquired in the Callovian–Oxfordian formation by investigative drilling. *Phys. Chem. Earth* **2004**, *29*, 55–77. [CrossRef]
4. Pearson, F.J.; Tournassat, C.; Gaucher, E.C. Biogeochemical processes in a clay formation in situ experiment: Part E—Equilibrium controls on chemistry of pore water from the Opalinus Clay, Mont Terri Underground Research Laboratory, Switzerland. *Appl. Geochem.* **2011**, *26*, 990–1008. [CrossRef]
5. Gaucher, E.C.; Tournassat, C.; Pearson, F.J.; Blanc, P.; Crouzet, C.; Lerouge, C.; Altmann, S. A robust model for pore-water chemistry of clayrock. *Geochim. Cosmochim. Acta* **2009**, *73*, 6470–6487. [CrossRef]
6. Kars, M.; Lerouge, C.; Grangeon, S.; Aubourg, C.; Tournassat, C.; Madé, B.; Claret, F. Identification of nanocrystalline goethite in reduced clay formations: Application to the Callovian-Oxfordian formation of Bure (France). *Am. Mineral.* **2015**, *100*, 1544–1553. [CrossRef]
7. Tournassat, C.; Vinsot, A.; Gaucher, E.C.; Altmann, S. Chemical conditions in clay-rocks. *Dev. Clay Sci.* **2015**, *6*, 71–100.
8. Glab, S.; Hulanicki, A.; Edwall, G.; Ingman, F. Metal-metal oxide and metal oxide electrodes as pH sensors. *Crit. Rev. Anal. Chem.* **1989**, *21*, 29–47. [CrossRef] [PubMed]
9. Pourbaix, M. *Atlas D'équilibres Electrochimiques*; Gauthier-Villars & Cie.: Paris, France, 1963.
10. Wang, M.; Ha, Y. An electrochemical approach to monitor pH change in agar media during plant tissue culture. *Biosens. Bioelectron.* **2007**, *22*, 2718–2723. [CrossRef] [PubMed]
11. Capelato, M.D.; dos Santos, A.M.; Fatibello-Filho, O.; Gama, R. Flow injection potentiometric determination of coke acidity and acetic acid content in vinegar using an antimony electrode. *Anal. Lett.* **1996**, *29*, 711–724. [CrossRef]
12. Caflisch, C.R.; Pucacco, L.R.; Carter, N.W. Manufacture and utilization of antimony pH electrodes. *Kidney Int.* **1978**, *14*, 126–141. [CrossRef] [PubMed]
13. Huang, G.F.; Guo, M.K. Resting dental plaque pH values after repeated measurements at different sites in the oral cavity. *Proc. Natl. Sci. Counc. Repub. China B* **2000**, *24*, 187–192. [PubMed]
14. Baghdady, N.H.; Sommer, K. Improved construction of antimony micro-electrodes for measuring pH-changes at the soil-root interface (rhizosphere). *J. Plant Nutr.* **1987**, *10*, 1231–1238. [CrossRef]
15. Kinoshita, E.; Ingman, F.; Edwall, G.; Thulin, S.; Głab, S. Polycrystalline and monocrystalline antimony, iridium and palladium as electrode material for pH-sensing electrodes. *Talanta* **1986**, *33*, 125–134. [CrossRef]
16. Glab, S.; Edwall, G.; Jöngren, P.A.; Ingman, F. Effects of some complex-forming ligands on the potential of antimony pH-sensors. *Talanta* **1981**, *28*, 301–311. [PubMed]
17. Edwall, G. Improved antimony-antimony(III)oxide pH electrodes. *Med. Biol. Eng. Comput.* **1978**, *16*, 661–669. [CrossRef] [PubMed]
18. Uhl, S.; Kestranek, W. Die Elektrometrische Titration Von Säuren Und Basen Mit Der Antimon-Indikatorelektrode. *Monatsh. Chem. Teile Wiss.* **1923**, *44*, 29–34. [CrossRef]
19. Buytendijk, F.J.J. The Use of Antimony Electrode in the Determination of pH In Vivo. *Arch. Neerl Physiol.* **1927**, *12*, 319–321.
20. Stock, J.T.; Purdy, W.C.; Garcia, L.M. The Antimony-Antimony Oxide Electrode. *Chem. Rev.* **1958**, *58*, 611–626. [CrossRef]

21. Fog, A.; Buck, R.P. Electronic semiconducting oxides as pH sensors. *Sens. Actuators* **1984**, *5*, 137–146. [CrossRef]
22. Galster, H. *pH Measurement: Fundamentals, Methods, Applications, Instrumentation*; Wiley-VCH: New York, NY, USA, 1991.
23. Ask, P.; Edwall, G.; Johansson, K.E.; Tibbling, L. On the use of monocrystalline antimony pH electrodes in gastro-oesophageal functional disorders. *Med. Biol. Eng. Comput.* **1982**, *20*, 383–389. [CrossRef] [PubMed]
24. Nilsson, E.; Edwall, G. Arterial pH monitoring with monocrystalline antimony sensors. A study of sensitivity for PO_2 variations. *Scand. J. Clin. Lab. Investig.* **1982**, *42*, 323–329.
25. Sjöberg, F.; Edwall, G.; Lund, N. The oxygen sensitivity of a multipoint antimony electrode for tissue pH measurements. A study of the sensitivity for in vivo PO_2 variations below 6 kPa. *Scand. J. Clin. Lab. Investig.* **1987**, *47*, 11–15.
26. Fenwick, F.; Gilman, E. The use of the antimony-antimony trioxide electrode for determining the dissociation constants of certain local anesthetics and related compounds. *J. Biol. Chem.* **1929**, *84*, 605–628.
27. Mascini, M.; Cremisini, C. A new pH electrode for gas-sensing probes. *Anal. Chim. Acta* **1977**, *92*, 277–283. [CrossRef]
28. Perley, G.A.; Company, N. Characteristics of the antimony electrode. *Ind. Eng. Chem. Anal. Ed.* **1939**, *11*, 319–322. [CrossRef]
29. Green, R.; Giebisch, G. Some problems with the antimony microelectrode. In *Ion Selective Microelectrodes*; Berman, H.J., Herbert, N.S., Eds.; Plenum Press: New York, NY, USA, 1974; pp. 43–53.
30. Quehenberger, P. The influence of carbon dioxide, bicarbonate and other buffers on the potential of antimony microelectrodes. *Pflügers Arch.* **1977**, *368*, 141–147. [CrossRef] [PubMed]
31. Ha, Y.; Wang, M. Capillary melt method for micro antimony oxide pH electrode. *Electroanalysis* **2006**, *18*, 1121–1125. [CrossRef]
32. Koncki, R.; Mascini, M. Screen-printed ruthenium dioxide electrodes for pH measurements. *Anal. Chim. Acta* **1997**, *351*, 143–149. [CrossRef]
33. Ives, D.J.G. Oxide, oxygen, and sulfide electrodes. In *Reference Electrodes: Theory and Practice*; Ives, D.J.G., Janz, G.J., Eds.; Academic Press: London, UK; New York, NY, USA, 1961; pp. 322–391.
34. Betelu, S.; Parat, C.; Petrucciani, N.; Castetbon, A.; Authier, L.; Potin-Gautier, M. Semicontinuous monitoring of cadmium and lead with a screen-printed sensor modified by a membrane. *Electroanalysis* **2007**, *19*, 399–402. [CrossRef]
35. McMurray, H.N.; Douglas, P.; Abbot, D. Novel thick-film pH sensors based on ruthenium dioxide-glass composites. *Sens. Actuators* **1995**, *28*, 9–15. [CrossRef]
36. Qingwen, L.; Guoan, L.; Youqin, S. Response of nanosized cobalt oxide electrodes as pH sensors. *Anal. Chim. Acta* **2000**, *409*, 137–142. [CrossRef]
37. Kampouris, D.K.; Kadara, R.O.; Jenkinson, N.; Banks, C.E. Screen printed electrochemical platforms for pH sensing. *Anal. Methods* **2009**, *1*, 25–28. [CrossRef]
38. Betelu, S.; Polychronopoulou, K.; Rebholz, C.; Ignatiadis, I. Novel CeO_2-based screen-printed potentiometric electrodes for pH monitoring. *Talanta* **2011**, *87*, 126–135. [CrossRef] [PubMed]
39. Parat, C.; Betelu, S.; Authier, L.; Potin-Gautier, M. Determination of labile trace metals with screen-printed electrode modified by a crown-ether based membrane. *Anal. Chim. Acta* **2006**, *573*, 14–19. [CrossRef] [PubMed]
40. Betelu, S.; Vautrin-Ul, C.; Ly, J.; Chaussé, A. Screen-printed electrografted electrode for trace uranium analysis. *Talanta* **2009**, *80*, 372–376. [CrossRef] [PubMed]
41. Stumm, W.; Morgan, J.J. *Aquatic Chemistry: Chemical Equilibria and Rates in Natural Waters*; Wiley & Son: New York, NY, USA, 1996.
42. Stumm, W.; Morgan, J.J. *Aquatic Chemistry: An Introduction Emphasizing Chemical Equilibria in Natural Waters*, 2nd ed.; Wiley & Son: New York, NY, USA, 1981.
43. Schüring, J.; Schulz, H.D.; Fischer, W.R.; Böttcher, J.; Duijnisveld, W.H.M. *Redox: Fundamentals, Processes and Applications*; Springer: Berlin, Heidelberg, Germany, 2000.
44. Brookins, D.G. *Eh-pH Diagrams for Geochemistry*; Springer: New York, NY, USA, 1988.
45. Saban, S.B.; Darling, R.B. Multi-element heavy metal ion sensors for aqueous solutions. *Sens. Actuators* **1999**, *61*, 128–137. [CrossRef]

46. Michalska, A. All-solid-state ion selective and all-solid-state reference electrodes. *Electroanalysis* **2012**, *24*, 1253–1265. [CrossRef]
47. Hu, J.; Stein, A.; Bühlmann, P. Rational design of all-solid-state ion-selective electrodes and reference electrodes. *Trends Anal. Chem.* **2016**, 102–114. [CrossRef]
48. Blaz, T.; Migdalski, J.; Lewenstam, A.; Lewenstam, A.; Ivaska, A.; Strong, T.D.; Brown, R.B.; VanKessel, A.L.; Zijlstra, W.G. Junction-less reference electrode for potentiometric measurements obtained by buffering pH in a conducting polymer matrix. *Analyst* **2005**, *130*, 637–643. [CrossRef] [PubMed]
49. Kwon, N.H.; Lee, K.S.; Won, M.S.; Shim, Y.B. An all-solid-state reference electrode based on the layer-by-layer polymer coating. *Analyst* **2007**, *132*, 906–912. [CrossRef] [PubMed]
50. Kisiel, A.; Marcisz, H.; Michalska, A.; Maksymiuk, K. All-solid-state reference electrodes based on conducting polymers. *Analyst* **2005**, *130*, 1655–1662. [CrossRef] [PubMed]
51. Hayashi, M. Temperature-electrical conductivity relation of water for environmental monitoring and geophysical data inversion. *Environ. Monit. Assess.* **2004**, *96*, 119–128. [CrossRef] [PubMed]
52. Brewer, P.J.; Leese, R.J.; Brown, R.J.C. An improved approach for fabricating Ag/AgCl reference electrodes. *Electrochim. Acta* **2012**, *71*, 252–257. [CrossRef]
53. Stoica, D.; Brewer, P.J.; Brown, R.J.C.; Fisicaro, P. Influence of fabrication procedure on the electrochemical performance of Ag/AgCl reference electrodes. *Electrochim. Acta* **2011**, *56*, 10009–10015. [CrossRef]
54. Betelu, S.; Vautrin-Ul, C.; Chaussé, A. Novel 4-carboxyphenyl-grafted screen-printed electrode for trace Cu(II) determination. *Electrochem. Commun.* **2009**, *11*, 383–386. [CrossRef]
55. Edwall, G. Influence of crystallographic properties on antimony electrode potential-II. Monocrystalline material. *Electrochim. Acta* **1979**, *24*, 605–612. [CrossRef]
56. Gaucher, E.C.; Blanc, P.; Bardot, F.; Braibant, G.; Buschaert, S.; Crouzet, C.; Gautier, A.; Girard, J.-P.; Jacquot, E.; Lassin, A.; et al. Modelling the porewater chemistry of the Callovian–Oxfordian formation at a regional scale. *C. R. Geosci.* **2006**, *338*, 917–930. [CrossRef]
57. Tournassat, C.; Vinsot, A.; Gaucher, E.C.; Altmann, S. Chapter 3—Chemical conditions in clay-rocks. In *Natural and Engineered Clay Barriers*; Tournassat, C., Steefel, C.I., Bourg, I.C., Bergaya, F., Eds.; Developments in Clay Science; Elsevier: Amsterdam, The Netherlands, 2015; Volume 6, pp. 71–100.
58. Blanc, P.; Lassin, A.; Piantone, P.; Azaroual, M.; Jacquemet, N.; Fabbri, A.; Gaucher, E.C. Thermoddem: A geochemical database focused on low temperature water/rock interactions and waste materials. *Appl. Geochem.* **2012**, *27*, 2107–2116. [CrossRef]
59. AFNOR. *ISO 6332:1988—Qualite De L'eau. Dosage Du Fer. Methode Spectrometrique a La Phenanthroline-1,10*; AFNOR: La Plaine Saint-Denis, France, 1988.
60. Gysinck, A. The use of antimony electrode for determining the degree of acidity: Suikerind. *Chem. Abstr.* **1933**, *27*, 2325–2327.
61. Edwall, G. Influence of crystallographic properties on antimony electrode potential-I. Polycrystalline material. *Electrochim. Acta* **1979**, *24*, 595–603. [CrossRef]
62. Betelu, S.; Ignatiadis, I.; Tournassat, C. Redox potential measurements in a claystone. *Environ. Sci. Technol. Lett.* **2017**, under review.
63. Rius-Ruiz, F.X.; Kisiel, A.; Michalska, A.; Maksymiuk, K.; Riu, J.; Rius, F.X. Solid-state reference electrodes based on carbon nanotubes and polyacrylate membranes. *Anal. Bioanal. Chem.* **2011**, *399*, 3613–3622. [CrossRef] [PubMed]

MDPI

Article

High Dynamic Range Spectral Imaging Pipeline For Multispectral Filter Array Cameras

Pierre-Jean Lapray [1], Jean-Baptiste Thomas [2,3,*] and Pierre Gouton [3]

[1]   MIPS Laboratory, Université de Haute Alsace, 68093 Mulhouse, France; pierre-jean.lapray@uha.fr
[2]   The Norwegian Colour and Visual Computing Laboratory, NTNU-Norwegian University of Science and Technology, 2815 Gjøvik, Norway
[3]   Le2i Laboratory, FRE CNRS 2005, Université de Bourgogne Franche-Comté, 21000 Dijon, France; pgouton@u-bourgogne.fr
*   Correspondence: jean.b.thomas@ntnu.no; Tel.: +47-6113-5246

Academic Editors: Nicole Jaffrezic-Renault and Gaelle Lissorgues
Received: 12 April 2017; Accepted: 31 May 2017; Published: 3 June 2017

Abstract: Spectral filter arrays imaging exhibits a strong similarity with color filter arrays. This permits us to embed this technology in practical vision systems with little adaptation of the existing solutions. In this communication, we define an imaging pipeline that permits high dynamic range (HDR)-spectral imaging, which is extended from color filter arrays. We propose an implementation of this pipeline on a prototype sensor and evaluate the quality of our implementation results on real data with objective metrics and visual examples. We demonstrate that we reduce noise, and, in particular we solve the problem of noise generated by the lack of energy balance. Data are provided to the community in an image database for further research.

Keywords: spectral imaging; spectral filter arrays; high dynamic range; image database

1. Introduction

Spectral filter arrays (SFA) technology [1] provides a compact and affordable means to acquire multispectral images (MSI). Such images have been proven to be useful in countless applications, but there extended use to general computer vision was limited due to complexity of imaging set-up, calibration and specific imaging pipelines and processing. In addition, spectral video are not easily handled either. SFA, however, developed around a very similar imaging pipeline to color filter arrays (CFA), which is rather well understood and already implemented in many solutions. Indeed, SFA, similarly to CFA, is a spatio-spectral sampling of the scene captured by a single shot of a solid-state, single, image sensor. In this sense, SFA may provide a conceptual solution that improves vision systems by trading spatial resolution for more spectral information.

Until recently, only simulations of SFA cameras were available, which made its experimental evaluation and validation difficult. Recent works on optical filters [2–4] in parallel to the development of SFA camera prototypes in the visible electromagnetic range [5], in the near infrared (NIR) [6] and in combined visible and NIR [7,8] permitted the commercialization of solutions, e.g., Imec [9], Silios [10], Pixelteq [11]. In addition, several color cameras include custom filter arrays that are in-between CFA and SFA (e.g., [12,13]). Recent applications in medical imaging [14], agriculture and environment [15] have been published. This indicates that we could consider the application of this technology to large scale use soon after the development of standard imaging pipelines and drivers.

We define and demonstrate the imaging pipeline in this communication. One strong remaining limitation of SFA is to preserve the energy balance between channels [16,17] while capturing a scene. Indeed, due to the large number of filters and their spectral characteristics, i.e., narrow band sensitivities and inadequacy with the scene and illumination, or large inhomogeneity between filter shapes, it is frequent to observe one or several channels under- or over-exposed for a given integration time and illumination, which is common to all filters. This may be solved in theory, by optimizing the filters before to create the sensor [17]. But filter realization is not yet very flexible. Another way to solve this issue would be to develop sensors with per-pixel integration control. This is in development within some 3D silicon sensor concepts [18,19], but this technology is at its very beginning, despite recent advances.

On the other hand, in gray-level and color imaging, the problem of under and over-exposure of parts of the scene is addressed by means of high dynamic range (HDR) imaging [20,21]. HDR imaging permits to potentially recover the radiance of the scene independently of the range of intensities present in the scene. Since the dynamic range of a given sensor is limited, the quantization of the radiance values is a source of problems. The signal detection of very low intensity is limited by the dark noise. On the other hand, high intensities of the input signal cannot be completely recovered and are sometimes voluntarily ignored (saturated pixels). To overcome these problems, a low exposure image could be used to quantify the highest intensities, whereas a high exposure allows us to quantify relatively low light signals well. Such an approach may also be used to bring less and more sensitive channels to a common representation space with a reduced noise amount.

In an ideal configuration, an HDR image is created by bringing standard dynamic range (SDR, typically 8 bits per channel) images in the same domain by dividing each image by its particular exposure time, and then by summing the corresponding pixel values. However, due to the effect of electronic circuits, most of the cameras have a non-linear processing regarding to the digitization of intensities, leading to a finite pixel brightness range and definition. This non-linear transformation is materialized by the camera response function, denoted $g(i)$, where i indexes the pixel value. It is assumed that this curve is monotonic and smooth. Some algorithms have been developed to recover this characteristic [21–23]. The most common method is the non-parametric technique from Debevec and Malik [21]. For a given exposure time and camera pixel value, the relative radiance value is estimated by using the integration times, $g(i)$ and a weighting function $\omega(i)$. Debevec and Malik use a "hat" function as weighting function, based on the assumption that mid-range pixels (values close to 128 for 8-bits sensors) are the most reliable and the best exposed pixel for a given scene and integration time. In addition, recent advances have been done on the capture and processing of HDR video with low latency, using hardware-based platform [24–26]. For HDR video, merging images captured at different sequential moments could lead to ghost artifacts when there are moving objects. This has been largely studied in recent years [27,28]. So, we argue that such methodology could be embedded in the SFA imaging pipeline without breaking the advantages of SFA technology for computer vision.

HDR multispectral acquisition is already treated by, e.g., Brauers et al. [29] and Simon [30]. However, they consider the problem of HDR using individual bands acquired sequentially, so each band is treated independently for potentially different integration times. In the case of SFA, we may consider specific joined processes. We communicated preliminary qualitative results at the Scandinavian Conference on Image Analysis (SCIA 2017) [31], and this paper extends, generalizes, and evaluates widely our preliminary results. We compare the HDR resulting images with an SDR database of the same objects [32]. In addition, we make our HDR database available for further research as supplementary material.

In this paper, we first generalize the imaging CFA pipeline to SFA in Section 2. This new SFA imaging pipeline embeds the HDR concept; It is based on multiple exposure spectral raw images. Then, the experimental implementation is developed in Section 3, the implementation is based on real data acquired with a prototype state-of-the-art camera [8] that captures visible and NIR information. Description of the database of images are provided in Section 4. Results and analysis are based on objective metric scores and visual examples in Section 5.

2. Imaging Pipelines

2.1. CFA Imaging Pipeline

Several CFA imaging pipelines exist. We can classify them in two large groups: one concerns the hardware and real-time processing community [33–35], the other concerns the imaging community [36–38]. A very general distinction is that the former one often considers the problem from the sensor and signal point of view and demosaics the raw image in very early steps, rather the latter considers the problem from a visualization point of view, and demosaics after or jointly with other processing such as white balance. In this work, we design the pipeline after the generic version defined by Ramanath et al. [37], which is shown in Figure 1.

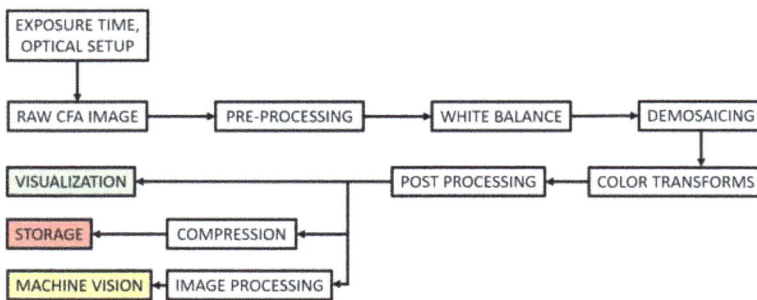

Figure 1. Color filter arrays (CFA) imaging pipeline similarly defined as in [37]. The pipeline contains pre-processing on raw data, which include for instance a dark noise correction and other denoising. Raw data would be corrected for illumination before to be demosaiced. Images are then projected into an adequate color space representation and followed by some post-processing, e.g., image enhancement, before coming out of the pipeline on a visualization media. Alternatively, this information could be compressed before archiving or be used for machine vision through adequate image processing.

2.2. HDR-CFA Imaging Pipeline

HDR imaging has been developed mostly within monochromatic sensors for the acquisition of HDR data. Indeed, the HDR capture is mostly an intensity process performed by channel [30]. However, there is a huge amount of work that developed the tone-mapping of HDR color images for visualization, (e.g., [39,40]).

We encapsulate a general HDR imaging process in the previous pipeline such as shown in Figure 2. This pipeline is based on sequence of images of the same scene having different integration times. The HDR pipeline may have distinguished outputs: one leads to HDR radiance images, which can be stored or used for automatic application. Another leads to a display-friendly visualization of color images. Note that these two outputs may overlap in specific applications.

2.3. SFA Imaging Pipeline

SFA sensors are currently investigated and developed, however beside demosaicing and application dedicated processing, the rest of the pipeline is not very well defined, thus understood, to our knowledge. We argue that a similar pipeline than CFA may be considered, which is then defined in Figure 3. In this pipeline, we still consider pre-processing as denoising and an equivalent to white balance as gain adjustment channel balance (referred to as multispectral constancy by some research). In the visualization pipeline, the color transform shall project the multispectral data into a color space representation.

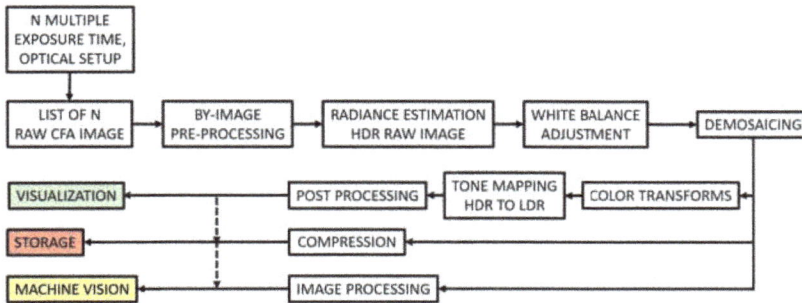

Figure 2. High dynamic range (HDR)-color filter arrays (CFA) imaging pipeline. In this case, the pre-processing is typically performed per image, similarly to the standard dynamic range (SDR)-CFA case. Then, radiance estimation is performed based on the multiple images, providing radiance raw images. White balancing and demosaicing are performed on this data. Then, the HDR image may be used as is for machine vision, or it continues into a visualization pipeline, where color transform, tone-mapping and image-enhancement may be applied before visualization. Bridge between the different output may occur if, for instance, the machine vision is designed to SDR content.

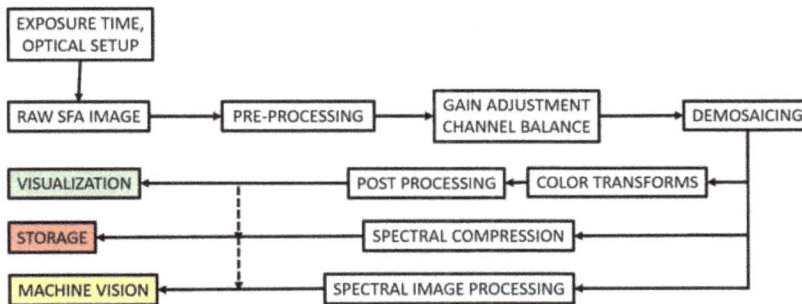

Figure 3. SFA imaging pipeline. At the instar of CFA, this pipeline defines some illumination discarding process and demosaicing. The spectral image would be typically used for application after demosaicing. However, these data may not be observable as they are, so the pipeline is prolonged for visualization. The color transform is ought to be slightly different than CFAs, for several channels are present and out of the visible range information, NIR, may be present in the spectral image. Compression of spectral data and spectral image processing, for, e.g., material identification or texture classification, are yet active research fields.

2.4. HDR-SFA Imaging Pipeline

According to the introductory discussion, we propose to extend the SFA pipeline to an HDR version. Beside the advantage of increasing the dynamic range of our images, we are also particularly interested in a better balance between channel sensitivities and to the reduction and an homogeneous distribution of noise by channel thanks to the increase of information. We propose to consider the raw SFA image as a gray-level image for relative radiance estimation, since this process is essentially a per-pixel operation. Thus, we perform all radiance reconstruction prior to any separation between bands. The pipeline is defined in Figure 4. We insist on the fact that this is only one possibility to consider the HDR pipeline for SFA, which has the advantage to imply only little modifications of the gray-level HDR pipeline and to permit the embedding of any individual algorithms in any of the boxes.

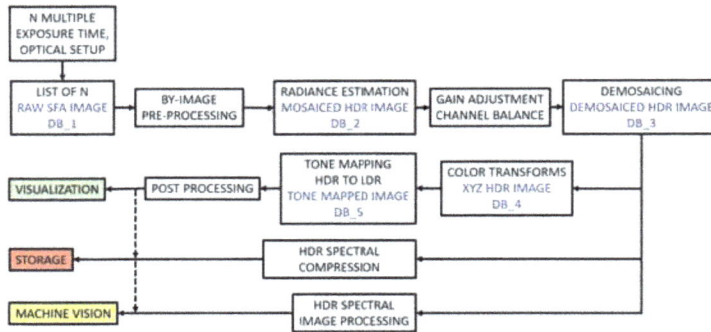

Figure 4. Our HDR-SFA imaging pipeline. The radiance estimation is performed on the list of raw image taken as a whole (DB_1 in the database), which permit to create the HDR raw images (DB_2 in the database). The raw HDR image may be corrected for illumination [41,42] and demosaiced by state-of-the-art methods (DB_3 in the database). After this, a visualization process projects the data into a HDR color representation CIEXYZ (DB_4 in the database), which is tone-mapped (DB_5 in the database) and processed for visualization on SDR media. Other outputs of the pipeline may be considered similarly to the previous pipelines.

3. Implementation of the HDR-SFA Imaging Pipeline

This section explicitly defines what processing is embedded in each of the pipeline box in our experiment. We selected well-established and understood methods from the state-of-the-art in order to provide benchmarking proposal and analysis and not go towards the evaluation of each of those methods individually. Those methods are combined into the pipeline. Our proposal is not exclusive in the sense that any method may be used and different order of processing or joint processing may also be considered in the future.

The prototype SFA camera from Thomas et al. [8] is used in this study. Detailed information on sensitivities, spatial arrangement and other aspects may be found in their article. The raw images are pre-processed and denoised according to what is performed in their article, which is essentially a dark noise removal. Then, following the pipeline, HDR data are computed. Section 3.1 covers the HDR radiance estimation. HDR images are balanced according to each channel and illumination and demosaiced, according to Miao et al. [43] algorithm, which form the full resolution HDR multispectral images. The output of the pipeline and the visualization procedure are developed in Section 3.2. Discussion on the role of illumination is provided in Section 3.2.1.

3.1. HDR Generation

Debevec and Malik radiance reconstruction [21] is probably the most understood HDR imaging pipeline. The model is based on the assumption that pixel values can be related to the physical quantity of radiance, by using a computed camera response function, which is recovered through a self calibration method. Due to the digitization process that converts radiance into pixel value in the image, this mapping is generally nonlinear and a calibration should be done before any estimation. To reconstruct HDR images, the camera response function (CRF) must be estimated. We captured 8 SDR bracketed images at different exposure times, from 0.125 ms to 16 ms with a one-stop increment, see Figure 5. This leads to a good response curve estimation in term of robustness to noise.

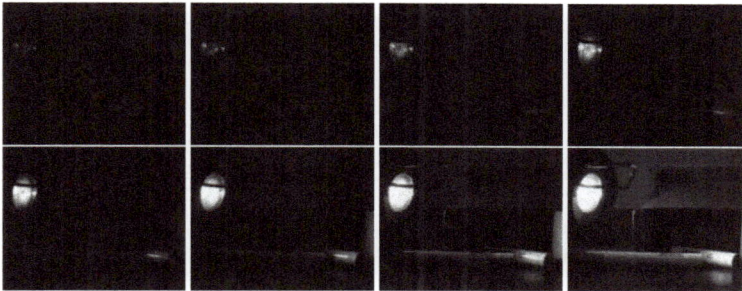

Figure 5. The set of SDR raw mosaiced images acquired with different exposure times: $\{0.125, 0.22, 0.5, 1, 2, 4, 8, 16\}$ ms (all spaced by one stop). These exposures are used to compute the global response curves of the prototype camera, shown in Figure 6a.

The algorithm to recover the CRF is based on the resolution of a set of linear equations by the singular value decomposition method. The algorithm is generally applied on RGB cameras, and it recovers 3 different response curves, one by channel. In our case, as we get eight spectral channels, we recover eight curves (see Figure 6a). We notice that the dispersion is relatively low between each of the channels, so in the following, we use the median of these curves for all channels, allowing us to work directly on the raw data at once to generate HDR values.

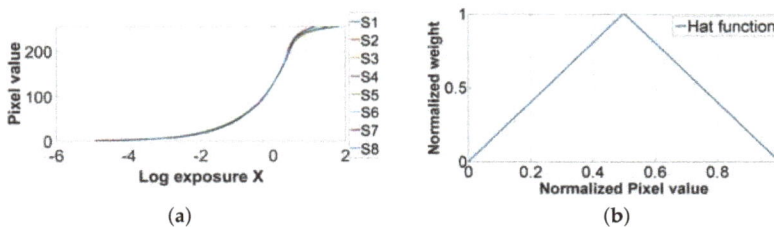

(a)

(b)

Figure 6. (a) Camera response functions to correct for the non-linearity between relative real radiance values and pixel intensities in the images. It is recovered from a complete image set shown in Figure 5. In the pipeline, the median of these curves is used to treat all of the pixels, independently of their spectral sensitivities; (b) The well-exposedness hat function used in this implementation.

As described in the pipeline, we recover the relative radiance values directly from the preprocessed raw data (mosaiced data). A number of 3 exposure times is selected. We chose only three exposures because it is a number commonly used in the literature [24,28], as it gives relatively high dynamic range and not too much ghost effects in case of video capture. The radiance values are recovered using the CRF, and by combining the pixel value with its corresponding exposure time [21]. A weighted sum of the radiance values computed from all exposure times is done using the hat weighting function, that gives more contribution to mid-range intensity pixels during the HDR reconstruction (see Figure 6b).

After the radiance is estimated by pixel, we apply a balance that compensates for the different spectral sensitivities of each band $i \in \mathbb{N}$ *and* $i \in [1,8]$ and for the illumination spectral power distribution. We implement a linear correction similarly to a white balance in color camera based on Equation (1).

$$\rho_i = \int_{400}^{1100} I_{Ill}(\lambda).S_i(\lambda)d\lambda, \tag{1}$$

where $I_{Ill}(\lambda)$ is the spectral emission of the illuminant used, $S_i(\lambda)$ is the measured camera spectral sensitivity of each channel (see Figure 7c). We then can compute the multiplication gain factors F to correct the data and balance their energy as in Equation (2).

$$F_i = \frac{\max\{\rho_1, ..., \rho_8\}}{\rho_i} \qquad (2)$$

We obtain 8 factors $F_{S_1 - S_8} = \{3.15, 2.86, 3.29, 2.90, 4.34, 5.39, 6.41, 1.00\}$ that are applied to each of the channels. This can be performed independently of the illumination by removing its contribution in Equation (1). In this case, illuminant compensation would not be taken into account and should be handled in another process.

The raw HDR image is then demosaiced to recover the spatial resolution of each of the HDR spectral channels. We then obtain the HDR multispectral image.

3.2. Visualization and Other Output

Visualization is the traditional use of HDR data. To this end, we project the radiance data into an HDR coded CIEXYZ color space according to a linear color transform computed on the 24 Macbeth ColorChecker reflectance patches, and the scene acquisition illumination measured in situ. This colorimetric image may be tone mapped by state-of-the-art algorithms. We used four tone-mapping techniques later for a representative illustration of visualization experience. We used the code furbished in the Matlab HDR Toolbox [44].

Although the visualization process has the very well defined goal of producing a pleasant and informative visual experience, machine vision output may target several purposes and specific HDR spectral image processing must be considered depending on the task. One particular challenge lies in the best way to handle 32-bit wide pixel information per spectral channel in real time applications. This is also a challenge to compress and store this information, but these aspects are not addressed in this communication.

3.2.1. Illumination Constraints on the Pipeline

The role of illumination is major in any imaging system. The first constraint on illumination is that its spectral distribution must be compatible with the camera sensitivities so that it maximizes the signal to noise ratio of the measurement. If there are a priori assumptions on the material surface, it may also be used to tune the illumination. The impact of illumination on the pipeline itself depends highly on the method that is implemented in each of the blocks. First, we assume here that there is no illumination change between the multiple frames that permits multi-exposure. Second, radiance estimation is sensitive to illumination as mentioned in Section 2.6 of Debevec and Malik [21] for the color case. That means that, if the illumination does not change, there is no issues with the reconstruction of radiance of multiple bands and means also that if there are changes in spectral distribution of the illumination, scaling terms should be adjusted for each of the bands. This is handled in the "Gain adjustment/Channel balance" block of our pipeline. In the article, we measured the light source. We could consider an unknown light, but then the channel balance would benefit from some additional tuning that may be scene dependent. This could be done in several ways: by having a white patch or calibration tile within the scene; by estimating and correct for the illumination based on a priori assumption and statistics of the image [41,42,45]. Demosaicing is not necessarily dependent on illumination, but usually methods based on learning are trained on white balanced images, which makes them sensitive to this aspect. Subsequent color transform and mapping are also dependent on the illumination to guarantee the neutrality of the color image appearance.

4. HDR SFA Database

We created a database of images of real scenes, while using this SFA HDR pipeline on the prototype camera defined on Figure 7.

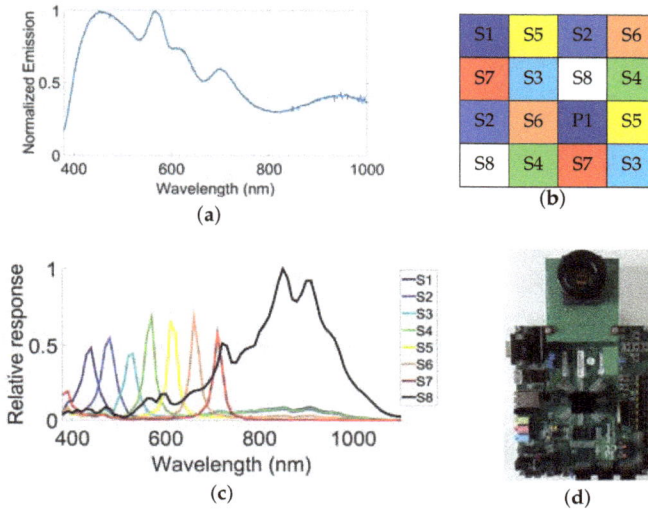

Figure 7. The hardware and acquisition procedure, from the illuminant source to the digitized output of the camera. (**a**) D65 simulator emission spectra used during the experiment; (**b**) Spatial distribution of filters over the sensor; (**c**) Joint spectral characteristics of optical filters and CMOS sensor [8]; (**d**) Camera and electronic architecture, composed of a FPGA (Field-Programmable Gate Array) board and an attached daughter card holding the SFA sensor.

A total of 18 scenes were captured using hardware and conditions shown in Table 1. Care was taken to select three adequate exposure times. A high exposure was selected for which only a few pixels are saturated in low intensities, and a low exposure was selected for which only a few pixels are saturated in high intensities. In the experiment, relatively good highlight conditions were found, and these exposure times were selected to 4, 8 and 16 ms (by doubling the amount of photons hitting the sensors between exposures).

Table 1. Summary of the global parameters and the SFA camera characteristics used during the acquisition.

Camera Sensor	E2V EV76C661 + MSFA-Global Shutter Mode
Camera resolution	1280 × 1024 (sensor native)–319 × 255 (image pre-processed)
Number of bands	8 (7 visible and 1 NIR)
Wavelength (calibrated)	380–1100 nm
Exposure time	3 exposure times: 4–8–16 ms
Illuminant	D65 simulator (see Figure 7a)
Optics/Aperture	Edmund optics 12 mm 58001–F/1.8
Focus	Fixed (20 cm)
Image format	Tiff 8 bits

All raw images have been pre-processed to remove the dark noise and the neighboring effects due to NIR blooming, by applying the procedure according to paper [8]. A file set from raw to tone-mapped processed data is available according to the following organization:

- DB_0: A raw HDR scene with image data stored as single page Tiff files. It contains a set of 8 images (see Figure 5), which are used to reconstruct the camera response (see Figure 6).
- DB_1: The raw image data for the 18 scenes at the three integration times, stored as single page Tiff files.
- DB_2: The mosaiced HDR data in .hdr files, which contain raw radiance recovered from the three exposures, following the method by Debevec and Malik [21].
- DB_3: The demosaiced HDR multispectral images in .mat files, which contain the demosaiced radiances by channel, recovered with the demosaicing algorithm by Miao et al. [43] after that channel sensitivities and illumination are discarded.
- DB_4: The HDR color CIEXYZ images in .hdr files are computed from the multispectral HDR image by a linear colorimetric calibration computed on the Macbeth ColorChecker.
- DB_5: The RGB tone mapped .png files for visualization. Four tone-mapping techniques were implemented for comparison, spanning different processing complexities.

Table 2 gives details about the content of the HDR SFA database, including the scene description, the file names and the file extensions. The demosaiced and color transformed HDR images can be, for example, visualized by any software implementation of the PFSTools framework by Mantiuk et al. [46] (e.g., Luminance HDR software). By using these tools, the user could visualize on his screen the 8-bits data per RGB channels, enables a HDR rendering and select a tone-mapping algorithm among several methods to visualize the file. Along with the complete database of images, a Matlab script, named *Script.m* is provided to load data into the workspace. The user can select which scene data to load amongst the 18 scenes available.

Table 2. The files can be downloaded as supplementary material at http://chic.u-bourgogne.fr [47], where the link *SFA_HDR* points out to a zip file that contains five directories, one directory for each stage of the pipeline called "DB_#" in Figure 4. The raw SDR, HDR mosaiced, HDR demosaiced, HDR CIEXYZ and RGB color tone mapped data are available to the community for further research.

Database		DB_1	DB_2	DB_3	DB_4	DB_5
Scene Name	Dynamic Range	File Name RAW	File Name HDR Mosaiced	File Name HDR Demosaiced	HDR XYZ	HDR Tone Mapped
CD	159.2	raw_preprocessed_cd_"exposure".tiff	hdr_mosaiced_cd.hdr	hdr_demosaiced_cd.mat	hdr_xyz_cd.hdr	hdr_tonemapped_"method"_cd.png
Knife	226.5	raw_preprocessed_knife_"exposure".tiff	hdr_mosaiced_knife.hdr	hdr_demosaiced_knife.mat	hdr_xyz_knife.hdr	hdr_tonemapped_"method"_knife.png
Water	147.3	raw_preprocessed_water_"exposure".tiff	hdr_mosaiced_water.hdr	hdr_demosaiced_water.mat	hdr_xyz_water.hdr	hdr_tonemapped_"method"_water.png
Train front	503.9	raw_preprocessed_train_front_"exposure".tiff	hdr_mosaiced_train_front.hdr	hdr_demosaiced_train_front.mat	hdr_xyz_train_front.hdr	hdr_tonemapped_"method"_train_front.png
Pens	145.6	raw_preprocessed_pens_"exposure".tiff	hdr_mosaiced_pens.hdr	hdr_demosaiced_pens.mat	hdr_xyz_pens.hdr	hdr_tonemapped_"method"_pens.png
Kerchief	78.8	raw_preprocessed_kerchief_"exposure".tiff	hdr_mosaiced_kerchief.hdr	hdr_demosaiced_kerchief.mat	hdr_xyz_kerchief.hdr	hdr_tonemapped_"method"_kerchief.png
Kiwi	216.1	raw_preprocessed_kiwi_"exposure".tiff	hdr_mosaiced_kiwi.hdr	hdr_demosaiced_kiwi.mat	hdr_xyz_kiwi.hdr	hdr_tonemapped_"method"_kiwi.png
Macbeth CC	153.3	raw_preprocessed_macbeth_"exposure".tiff	hdr_mosaiced_macbeth.hdr	hdr_demosaiced_macbeth.mat	hdr_xyz_macbeth.hdr	hdr_tonemapped_"method"_macbeth.png
Black swimsuit	231.4	raw_preprocessed_black_swimsuit_"exposure".tiff	hdr_mosaiced_black_swimsuit.hdr	hdr_demosaiced_black_swimsuit.mat	hdr_xyz_black_swimsuit.hdr	hdr_tonemapped_"method"_black_swimsuit.png
Origan	135.0	raw_preprocessed_origan_"exposure".tiff	hdr_mosaiced_origan.hdr	hdr_demosaiced_origan.mat	hdr_xyz_origan.hdr	hdr_tonemapped_"method"_origan.png
Orange object	42.5	raw_preprocessed_orange_object_"exposure".tiff	hdr_mosaiced_orange_object.hdr	hdr_demosaiced_orange_object.mat	hdr_xyz_orange_object.hdr	hdr_tonemapped_"method"_orange_object.png
Pastel	331.1	raw_preprocessed_pastel_"exposure".tiff	hdr_mosaiced_pastel.hdr	hdr_demosaiced_pastel.mat	hdr_xyz_pastel.hdr	hdr_tonemapped_"method"_pastel.png
Battery	274.7	raw_preprocessed_battery_"exposure".tiff	hdr_mosaiced_battery.hdr	hdr_demosaiced_battery.mat	hdr_xyz_battery.hdr	hdr_tonemapped_"method"_battery.png
Train side	296.6	raw_preprocessed_train_side_"exposure".tiff	hdr_mosaiced_train_side.hdr	hdr_demosaiced_train_side.mat	hdr_xyz_train_side.hdr	hdr_tonemapped_"method"_train_side.png
Raspberry	871.7	raw_preprocessed_raspberry_"exposure".tiff	hdr_mosaiced_raspberry.hdr	hdr_demosaiced_raspberry.mat	hdr_xyz_raspberry.hdr	hdr_tonemapped_"method"_raspberry.png
Ruler	145.6	raw_preprocessed_ruler_"exposure".tiff	hdr_mosaiced_ruler.hdr	hdr_demosaiced_ruler.mat	hdr_xyz_ruler.hdr	hdr_tonemapped_"method"_ruler.png
SD card	72.2	raw_preprocessed_sd_"exposure".tiff	hdr_mosaiced_sd.hdr	hdr_demosaiced_sd.mat	hdr_xyz_sd.hdr	hdr_tonemapped_"method"_sd.png
Painting	130.6	raw_preprocessed_painting_"exposure".tiff	hdr_mosaiced_painting.hdr	hdr_demosaiced_painting.mat	hdr_xyz_painting.hdr	hdr_tonemapped_"method"_painting.png

5. Analysis

5.1. Qualitative Evaluation

We provide an exhaustive example of images at each step of the pipeline for one scene in Figure 8.

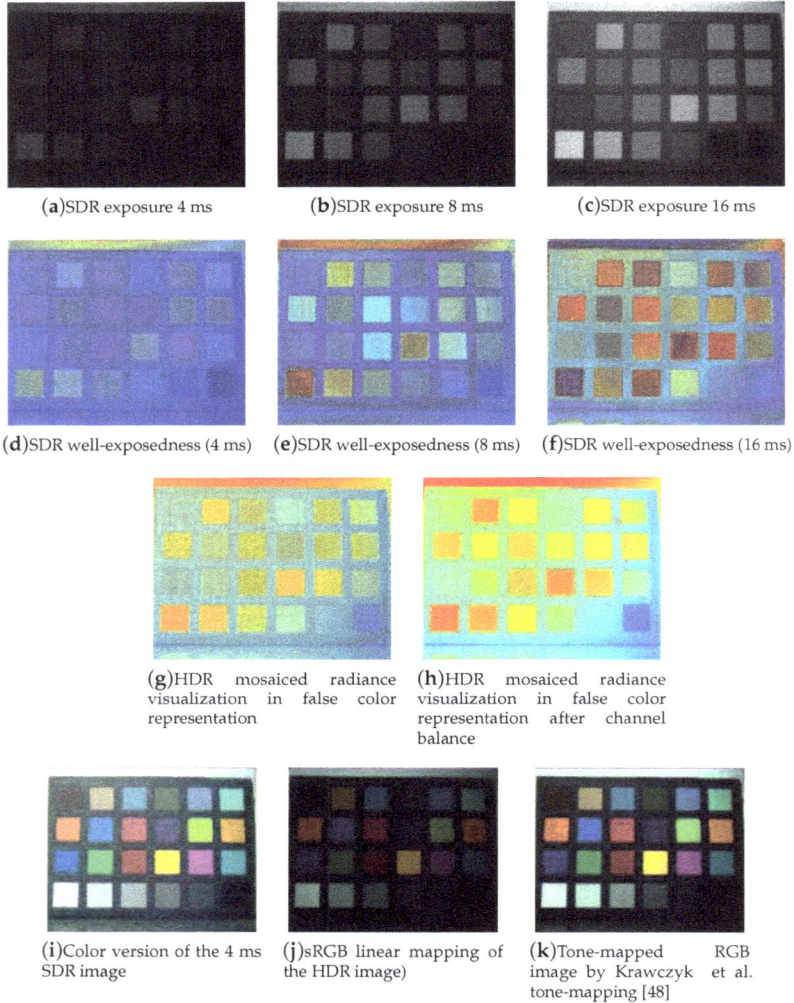

(**a**)SDR exposure 4 ms   (**b**)SDR exposure 8 ms   (**c**)SDR exposure 16 ms

(**d**)SDR well-exposedness (4 ms)   (**e**)SDR well-exposedness (8 ms)   (**f**)SDR well-exposedness (16 ms)

(**g**)HDR mosaiced radiance visualization in false color representation

(**h**)HDR mosaiced radiance visualization in false color representation after channel balance

(**i**)Color version of the 4 ms SDR image

(**j**)sRGB linear mapping of the HDR image)

(**k**)Tone-mapped RGB image by Krawczyk et al. tone-mapping [48]

Figure 8. Illustration of the pipeline results for the Macbeth ColorChecker image (a typical low dynamic range scene). (**a–c**) Raw images at different exposures; (**d–f**) false color well-exposedness representation that use the Jet colormap from MATLAB; (**g,h**) HDR radiance mosaiced image estimated from the three exposure set (**a–c**) and visualized before and after the channel balance using the Jet colormap from MATLAB; (**i-k**) color representation of the image based on the SDR single acquisition or after tone-mapping of the HDR images.

We observe that channels are unevenly affected by noise at the different exposures. This phenomenon is highlighted in Figure 8d–f, where a pixel position could hold a good intensity for a given exposure time (reddish colors of the Jet colormap), and a bad exposition in another (bluish colors). If we look at the raw images of Macbeth ColorChecker, at one neutral patch, we can clearly distinguish the inherent energy balance problems between pixel values through the 8 channels. In the white neutral patch, the NIR channel called S_8 is saturated in the middle integration time. These problems lead to visual noise when visualizing a single SDR reconstructed image (see Figure 8i). Our HDR-SFA imaging pipeline corrects the problem by a certain amount. On Figure 8g, we can observe the HDR mosaiced image of radiance, that exhibits unbalanced sensitivities by channel. After we applied the balance correction, we observe on Figure 8h that we have a more homogeneous representation of the achromatic patches through the different channels.

The global effect of applying our HDR pipeline can be visually appreciated on the color tone-mapped version of the image on Figure 8i,k.

In terms of scene dynamic range, we know that the Macbeth ColorChecker scene is a typical low dynamic range scene. We could capture the whole dynamic range of the scene with only one exposure (i.e., Figure 8i) and the HDR process used permits only to reduce the noise and to solve the energy balance issue. However, for a higher dynamic range scene (like the CD scene), in addition to balancing the exposure among the pixels, we also extend the dynamic range by a certain amount. This is evaluated below.

Figures 9–12 show the tone-mapped color images of the database processed with different algorithms. Namely, we applied a simple logarithmic mapping, Krawczyk et al. [48], Fattal et al. [49] and Banterle et al. [50].

Global impression is that noise is reduced and that the quality of the images is better than the best SDR versions of these images shown in Figure 13. Difference in the tone-mapping algorithms seems to impact mostly the global brightness of the images, which also depends on the scene. Highlights that are not handled by the shortest integration time are still not handled after the process. This is due to the selected exposure time, which may not be optimal for some scenes with high dynamic range of radiance. However, the quality of the scenes or part of the scenes of low dynamic have been greatly improved, such as in the Macbeth ColorChecker image. Spatial artifacts, such as seen on the SD card image, are due mostly to other optical and sampling effects, which we do not assess in this work (non-uniform illumination, optical effects, and aliasing in demosaicing).

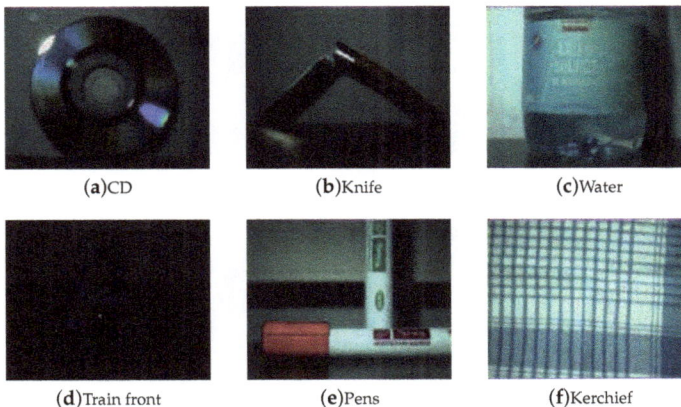

(**a**)CD (**b**)Knife (**c**)Water

(**d**)Train front (**e**)Pens (**f**)Kerchief

Figure 9. *Cont.*

(**g**)Kiwi

(**h**)Macbeth ColorChecker

(**i**)Black swimsuit

(**j**)Orange object

(**k**)Origan

(**l**)Pastel

(**m**)Battery

(**n**)Train side

(**o**)Raspberry

(**p**)Ruler

(**q**)SD card

(**r**)Painting

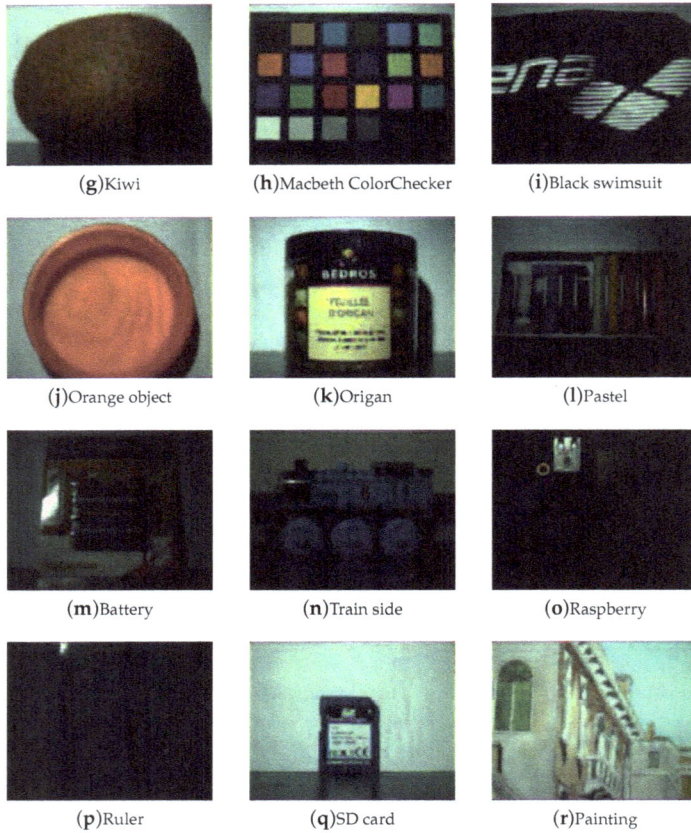

Figure 9. Database visualization of all scenes using a global logarithmic tone-mapping. In case of a high dynamic range scene with high specular reflection, good details are accomplished in specular regions, at the expense of a global image contrast reduction.

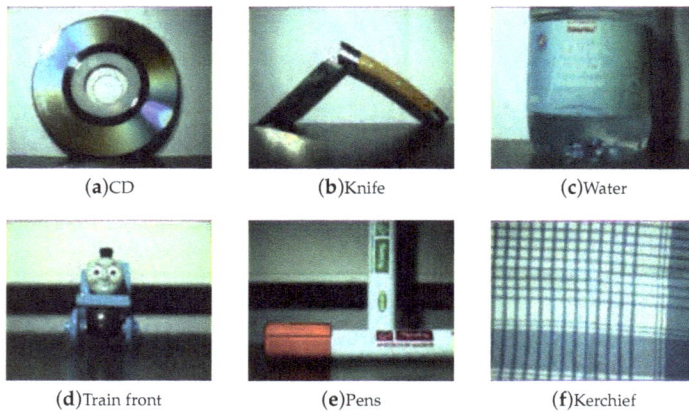

(**a**)CD

(**b**)Knife

(**c**)Water

(**d**)Train front

(**e**)Pens

(**f**)Kerchief

Figure 10. *Cont.*

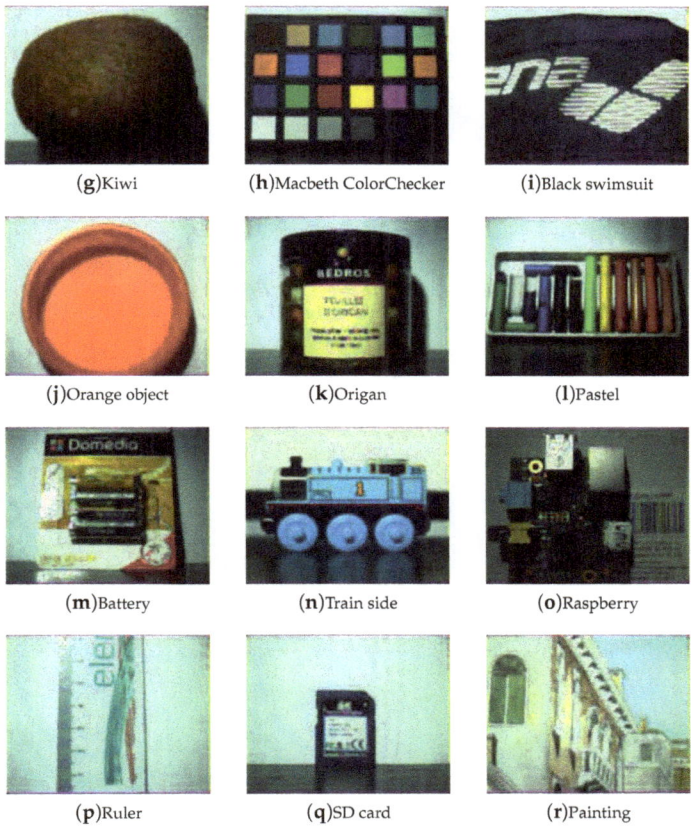

(**g**)Kiwi   (**h**)Macbeth ColorChecker   (**i**)Black swimsuit

(**j**)Orange object   (**k**)Origan   (**l**)Pastel

(**m**)Battery   (**n**)Train side   (**o**)Raspberry

(**p**)Ruler   (**q**)SD card   (**r**)Painting

Figure 10. Database visualization of all scenes using a tone-mapping that is a combination of local and global anchoring of brightness values; the Krawczyk et al. [48] tone-mapping. Global contrast is preserved even in the presence of high specular reflections.

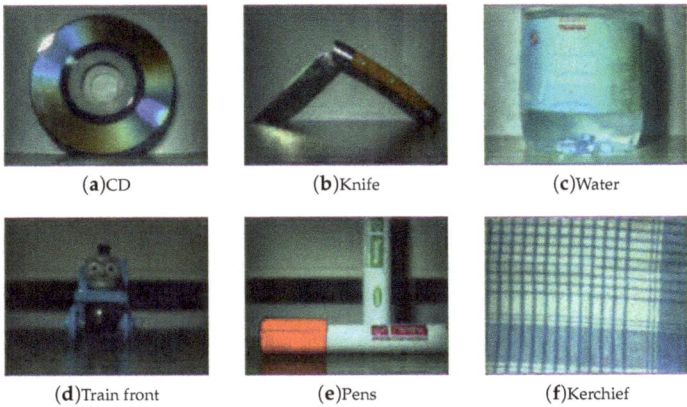

(**a**)CD   (**b**)Knife   (**c**)Water

(**d**)Train front   (**e**)Pens   (**f**)Kerchief

Figure 11. *Cont.*

(**g**)Kiwi

(**h**)Macbeth ColorChecker

(**i**)Black swimsuit

(**j**)Orange object

(**k**)Origan

(**l**)Pastel

(**m**)Battery

(**n**)Train side

(**o**)Raspberry

(**p**)Ruler

(**q**)SD card

(**r**)Painting

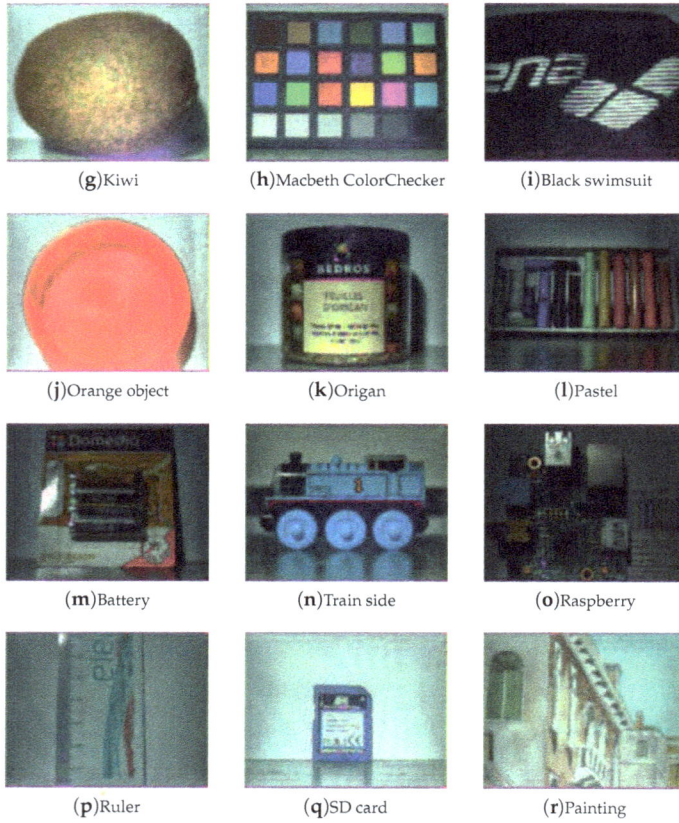

Figure 11. Database visualization of all scenes using the gradient domain compression tone-mapping by Fattal et al. [49]. We can see that this technique highlights details well in areas affected by shadows. It gives good details in specular regions, preserving a relatively good global contrast in the scene.

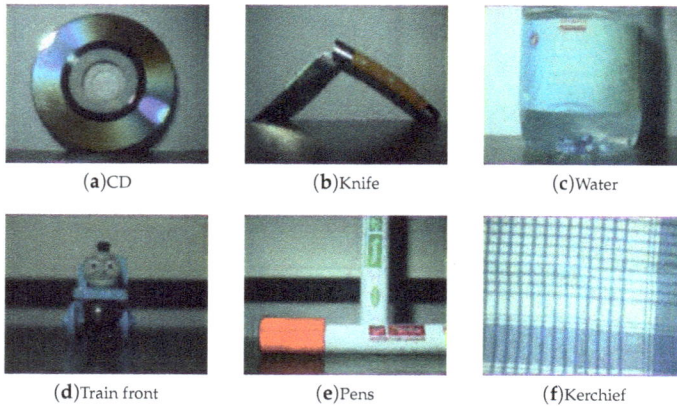

(**a**)CD

(**b**)Knife

(**c**)Water

(**d**)Train front

(**e**)Pens

(**f**)Kerchief

Figure 12. *Cont.*

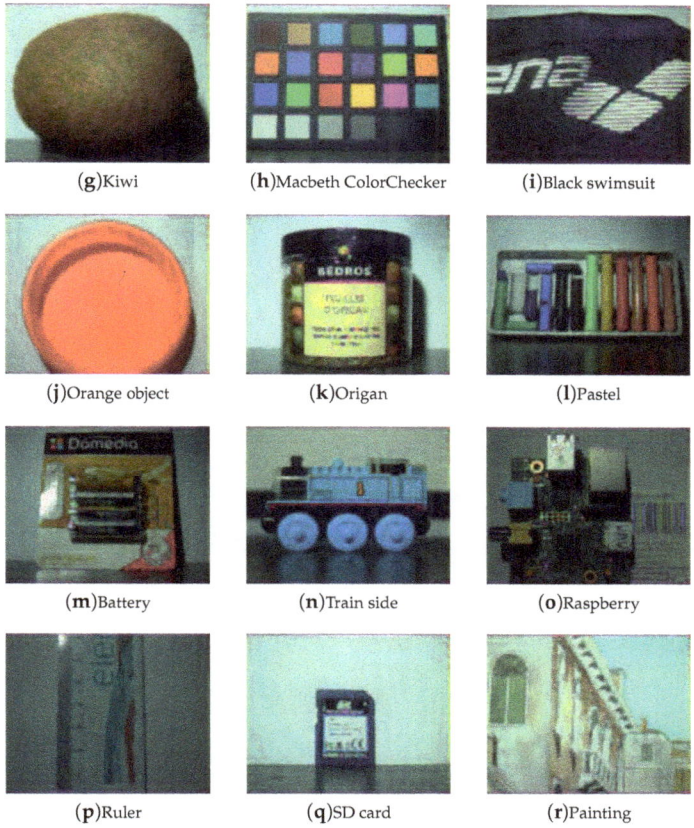

(**g**)Kiwi

(**h**)Macbeth ColorChecker

(**i**)Black swimsuit

(**j**)Orange object

(**k**)Origan

(**l**)Pastel

(**m**)Battery

(**n**)Train side

(**o**)Raspberry

(**p**)Ruler

(**q**)SD card

(**r**)Painting

Figure 12. Database visualization of all scenes using a tone-mapping that is a combination of local and global tone-mapping developed by Banterle et al. [50] tone-mapping. We observe that this technique achieves good rendering in term of local and global contrasts.

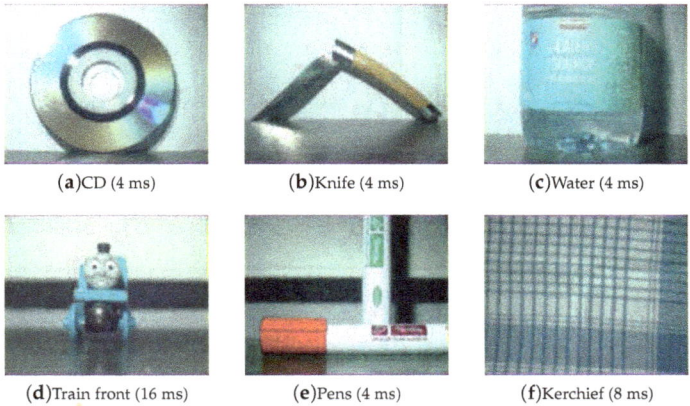

(**a**)CD (4 ms)

(**b**)Knife (4 ms)

(**c**)Water (4 ms)

(**d**)Train front (16 ms)

(**e**)Pens (4 ms)

(**f**)Kerchief (8 ms)

Figure 13. *Cont.*

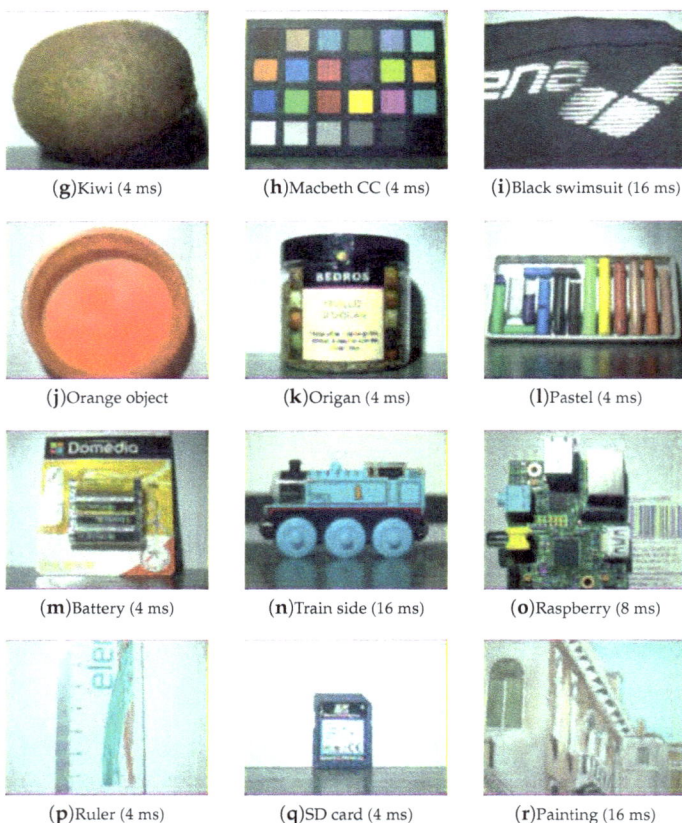

(**g**)Kiwi (4 ms)   (**h**)Macbeth CC (4 ms)   (**i**)Black swimsuit (16 ms)

(**j**)Orange object   (**k**)Origan (4 ms)   (**l**)Pastel (4 ms)

(**m**)Battery (4 ms)   (**n**)Train side (16 ms)   (**o**)Raspberry (8 ms)

(**p**)Ruler (4 ms)   (**q**)SD card (4 ms)   (**r**)Painting (16 ms)

Figure 13. Visualization of SDR color versions of the scenes without using any HDR processing. Integration times was selected to be the best exposure as described in [32]. Those SDR versions of scenes could be compared to the output images of HDR pipeline with tone-mapping.

5.2. Quantitative Evaluation

We propose to evaluate our pipeline quantitatively by several strategies. Difficulty in this task comes from the fact that no ground truth is available and from the fact that there are a great number of factors that affect the final image quality. We propose to evaluate only two aspects: The radiance estimation and the final tone-mapped color image.

5.2.1. Radiance Estimation

We are interested first in evaluating the estimation of the radiance of the scene. We propose three indicators:

- We first investigate the relation between the achromatic patches radiance of the Macbeth ColorChecker evaluated by the camera and computed theoretically by spectral simulation on Figure 14. The curves do not exhibit a very good linear behavior and show an offset. The CRF estimation may be impaired by the very high radiance of the lamp in the scene. However, issue with this evaluation is that it is quite affected by the spatial non-uniformity of the illumination, so it is not easy to draw strong conclusions from it. Indeed, the achromatic patches are distributed horizontally across the image, so vignetting and illumination shift impact the results. The curve of the channel sensitive to the NIR is showing an outlying behavior. This specific channel is not very well evaluated by this indicator due to lack of measurements of the illumination between 1000 and 1100 nm for material limitation (our measurement device did not reach beyond that limit).

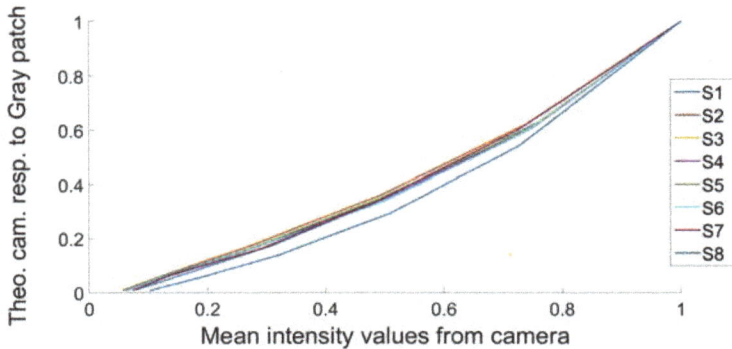

Figure 14. Study of the channel camera response according to the achromatic patches in the Macbeth ColorChecker chart. A theoretical response has been computed, taking into account the illuminant and the camera response (see Figure 7a,c).

- To produce a better evaluation and break the limitation of the above bullet point, we argue that the ratio of intensities, by channel, between patches in simulation and in practice should be the same if the radiance is well evaluated. In addition, if we compute only ratio between adjacent patches, the effect of the illumination and vignetting should be minimum. Difference in ratio r between a couple of horizontal adjacent patches is computed such as $\Delta R = \sqrt{(\hat{r}_i^j - r_i^j)^2}$, with $i \in \mathbb{N}$ and $i \in [1,8]$ being the channel considered and $j \in \mathbb{N}$ and $j \in [1,20]$, indexing the ratio between each pair considered.

  Results are shown on Table 3. It is shown that we have a minimum and maximum error in the range 0% to 46%. This evaluation demonstrates that we obtained a rather good radiance estimate with an average of 5% of error among patch couples and 5% of error among channels. Moreover, the error among channels is near to constant, which indicates that we have a good uniformity in radiance recovery over wavelengths. It appears that the maximum error is reached when couple of patches shows a large difference in intensity values like 4/5, 10/11, 13/14, 15/16. We investigated this point separately and found out that the signal to noise ratio is rather high for specific combinations of radiance and sensor spectral sensitivity. This is due to low radiance from the patch. When the sensor value is high for one patch and very small for another one, this noise is amplified. This creates those huge errors. It is yet to be investigated how this effect impacts the evaluation of HDR radiance accuracy.

- Image quality does not depend only on radiance accurate evaluation, so we also make a tentative to evaluate the process by means of established no-reference metric. This is an attempt to evaluate the global quality of the HDR images. BRISQUE [51] no-reference metric scores are computed per channel on the demosaiced HDR multispectral images. The scores are shown and averaged by channel on the right part of Table 4. The best scores of BRISQUE are close to 0 and the result on the HDR images are closer to 100, which should indicate a very bad quality. However, those results are difficult to interpret since we do not have very natural content, a consistent pixel intensity range, neither any data to compare with. In addition, we did not re-trained BRISQUE for those specific data, so that scores over 100 occurs, meaning that the quality of the image is worse than the training data set wost image quality, according to the measure. Also, HDR linear data may exhibit different statistics than usual gamma corrected SDR images. This is supported by the fact that the magnitude of the score seems not to correlate very well with the observation of the color images (see next paragraph). We observed similar difficulties with scores from NIQE [52] or BLIINDS-II [53] image quality measures. The scores of the two last metrics are not presented in the article because we consider that using those no-reference metrics for this purpose is not adequate. BRISQUE is shown to report an example of this attempt. We cannot state here how BLIINDS-II and BRISQUE would become efficient if they were trained on adequate images of radiance. Nevertheless, we could still observe that the quality of each channel for one image is of similar quality across channels, and since BRISQUE is computed based on image statistics, this indicates that we have an homogeneous quality, which is very good.

Table 3. Ratio difference between radiance computation and estimation between adjacent patches of the Macbeth ColorChecker. The index refers to the number of the patch on the chart. Except for some specific couple of patches, we could consider a good estimation at less than 5% in average.

-	S1	S2	S3	S4	S5	S6	S7	S8	Mean	STD
1/2	0.01	0.03	0.05	0.03	0.00	0.03	0.02	0.04	0.03	0.02
2/3	0.10	0.09	0.08	0.08	0.05	0.05	0.04	0.09	0.07	0.02
3/4	0.04	0.03	0.02	0.03	0.03	0.03	0.04	0.02	0.03	0.01
4/5	0.46	0.31	0.07	0.08	0.15	0.17	0.23	0.11	0.20	0.03
5/6	0.01	0.03	0.02	0.02	0.00	0.01	0.01	0.02	0.01	0.01
7/8	0.07	0.08	0.06	0.04	0.03	0.02	0.00	0.04	0.04	0.03
8/9	0.06	0.05	0.07	0.08	0.10	0.19	0.23	0.10	0.11	0.07
9/10	0.02	0.02	0.02	0.02	0.02	0.02	0.02	0.03	0.02	0.00
10/11	0.04	0.03	0.09	0.09	0.08	0.06	0.06	0.04	0.06	0.02
11/12	0.02	0.00	0.01	0.01	0.03	0.01	0.03	0.00	0.01	0.01
13/14	0.07	0.04	0.14	0.20	0.20	0.19	0.19	0.12	0.14	0.06
14/15	0.07	0.07	0.03	0.05	0.03	0.02	0.05	0.09	0.05	0.02
15/16	0.05	0.04	0.06	0.07	0.09	0.06	0.05	0.05	0.06	0.02
16/17	0.01	0.03	0.01	0.01	0.00	0.01	0.00	0.01	0.01	0.01
17/18	0.00	0.01	0.04	0.02	0.01	0.00	0.00	0.00	0.01	0.01
19/20	0.06	0.06	0.06	0.06	0.06	0.06	0.05	0.07	0.06	0.01
20/21	0.05	0.05	0.05	0.05	0.04	0.05	0.05	0.06	0.05	0.01
21/22	0.04	0.04	0.04	0.04	0.04	0.03	0.04	0.06	0.04	0.01
22/23	0.04	0.03	0.03	0.03	0.02	0.02	0.01	0.06	0.03	0.01
23/24	0.06	0.04	0.06	0.05	0.03	0.02	0.00	0.09	0.04	0.03
Mean	0.06	0.05	0.05	0.05	0.05	0.05	0.06	0.06	-	-
STD	0.10	0.06	0.03	0.04	0.05	0.06	0.07	0.04	-	-

Table 4. BRISQUE [51] no-reference metric computed on the SDR color images and also on each channel of the multispectral HDR images. Results estimate that the best exposure SDR color image is better than any of the tone-mapped. This is different to what is observed and we may discard BRISQUE to analyze such data. The results on the spectral channels shows very bad BRISQUE scores, but again, they are hardly comparable to anything we know. Nevertheless, they also show that scores are relatively similar across the channels, indicating stability. Red cells with the worst score are highlighted, whereas green cells mean best scores.

Image Scene	SDR	TM Banterle	TM Fattal	TM Log	TM Krawczyk	S1	S2	S3	S4	S5	S6	S7	S8
black_swimsuit	50.2	63.5	72.1	77.5	54.6	88.4	90.4	88.3	92.6	90.2	88.8	78.7	87.5
train_side	29.4	56.8	53.8	65.8	50.7	84.9	85.2	78.8	81.8	87.8	83.4	66.4	89.5
cd	34.0	53.5	53.9	59.7	46.2	54.0	52.5	48.9	47.7	53.5	50.1	44.4	44.2
kiwi	28.4	45.7	39.9	47.0	44.3	46.9	49.3	42.0	42.1	36.5	41.4	32.1	37.3
sd	36.4	50.8	52.0	51.6	47.6	51.7	51.4	49.1	50.8	53.1	56.9	49.5	45.3
pens	18.4	58.1	60.0	63.1	59.3	74.7	76.2	69.0	70.7	73.9	76.0	64.2	75.8
origan	26.9	52.0	50.2	53.1	50.4	55.3	60.7	49.4	57.3	53.1	52.8	49.3	50.3
painting	42.4	45.9	41.2	46.4	39.5	58.6	54.9	52.5	54.2	42.7	51.2	47.7	49.6
macbeth	29.6	61.2	46.1	69.1	66.2	62.0	67.1	63.8	67.5	69.9	69.9	54.4	65.1

Table 4. *Cont.*

Image Scene	SDR	TM Banterle	TM Fattal	TM Log	TM Krawczyk	S1	S2	S3	S4	S5	S6	S7	S8
knife	32.0	53.3	62.0	58.9	47.9	51.0	49.0	46.4	50.9	54.8	43.4	33.3	44.7
water	24.4	52.4	49.2	53.3	51.9	59.3	56.4	54.3	54.9	52.2	53.8	47.2	53.8
train_front	34.9	60.5	58.4	77.5	51.4	84.1	85.0	77.1	81.6	82.6	79.9	64.2	85.8
kerchief	87.4	102.6	97.5	105.0	105.1	93.6	96.5	98.8	99.0	102.3	99.4	52.9	45.9
pastel	52.1	71.9	59.6	69.7	67.8	70.4	73.5	71.3	69.9	73.9	73.8	68.0	68.3
orange_object	31.5	59.2	49.8	67.5	54.9	60.7	69.9	55.5	59.7	65.8	68.8	56.0	46.5
battery	42.7	56.7	48.4	60.3	49.8	55.0	57.4	55.8	51.3	52.6	49.8	48.7	50.5
raspberry	45.7	59.7	50.7	60.5	56.1	71.0	71.0	63.6	65.1	65.1	66.6	61.8	63.8
ruler	34.5	53.8	49.2	68.3	39.5	44.8	45.6	48.7	47.7	53.1	52.3	46.3	39.3
Mean	37.8	58.8	55.2	64.1	54.6	64.8	66.2	61.8	63.6	64.6	64.3	53.6	58.0
STD	15.2	12.6	13.0	13.7	14.7	15.0	15.6	15.8	16.3	17.7	16.6	12.0	17.0

5.2.2. Evaluation on Color Images

We are interested in the evaluation of the tone mapped color images to demonstrate that we improved the overall quality. For that, we used again the BRISQUE [51] no-reference metric, and evaluate the results on the left part of Table 4. Red cells with the worst score are highlighted, whereas green cells mean best scores. Results are somehow surprising, since the best observed visual quality is not indicated by the metric (cf. qualitative analysis). We still can explain the worst results by the dark low contrasted images tone-mapped by the global logarithm. Best results of BRISQUE indicates that SDR images are best, that probably satisfy more to the conditions evaluated by the metric. As mentioned above, scores higher than 100 mean that the quality of the image is worse than the set of training images. It seems however that BRISQUE is not the adequate metric to evaluate the quality of tone-mapped images. BLIINDS-II [53] exhibits generally similar results in giving generally the color SDR image as the best between the color images. NIQE [52] does not seem to provide any strong tendency. The scores of the two last metrics are not presented in the article because we estimate that using those no-reference metrics for this purpose is not adequate. BRISQUE is shown to report an example of this attempt.

New metrics are specially designed to evaluate tone-mapped HDR images. We evaluated the different color images by using a state-of-the-art method called HIGRADE [54]. HIGRADE is a dedicated method for non-reference image quality assessment. It evaluates the SDR images obtained by algorithms such as tone-mapping, multi-exposure fusion, or other dedicated processing. This method is dedicated to perceptual evaluation of HDR images and has been considered as the most efficient algorithm on a large database of HDR images, called ESPL-LIVE HDR Image Quality database [55]. Table 5 shows the output of the algorithm for SDR images. This evaluation clearly demonstrates that the HDR versions of the images are better than the SDR one, especially when using the Krawczyk et al. [48] technique. This technique is based on both local and global contrast enhancements, that seems to give a good general rendering, as the metric suggests. Other tone-mapping techniques have relatively good scores, and all provide better scores than the SDR one after averaging. Standard deviation remains quite stable, except for Banterle tone mapping.

Table 5. HIGRADE [54] results for evaluation of color images. Scores indicate that the HDR tone-mapped color images are always better than the SDR single exposure version. Scores indicate also that Krawczyk et al. tone-mapping provide best results amongst the tested algorithms, which is also supported qualitatively by visualization of the images.

Image	SDR	TM Banterle	TM Fattal	TM Log	TM Krawczyk
black_swimsuit	−1.33	−0.84	−0.92	−0.91	−0.57
train_side	−0.79	−0.94	−0.96	−1.19	−0.71
cd	−0.72	−0.12	−0.09	−0.93	−0.08
kiwi	−1.07	−0.86	−1.02	−0.62	−0.45
sd	−1.31	−0.39	−0.13	−0.45	−0.47

Table 5. *Cont.*

Image	SDR	TM Banterle	TM Fattal	TM Log	TM Krawczyk
pens	−1.23	−0.54	−0.70	−0.56	−0.46
origan	−0.90	−0.41	−0.44	−0.13	−0.14
painting	−0.60	−0.64	−0.84	−0.47	−0.43
macbeth	−0.90	−0.50	−0.42	−0.45	0.02
knife	−0.79	−0.39	−0.49	−0.87	−0.07
water	−1.01	−0.92	−0.80	−0.86	−0.70
train_front	−0.95	−0.64	−0.66	−1.18	−0.72
kerchief	−2.02	−2.05	−1.53	−1.77	−1.46
pastel	−1.13	−0.59	−0.55	−0.73	−0.32
orange_object	−0.99	−0.96	−0.42	−0.61	−0.73
battery	−0.92	−0.34	−0.38	−0.79	−0.14
raspberry	−0.95	−0.81	−0.89	−0.93	−0.24
ruler	−0.40	−0.74	−0.77	−1.21	0.02
Mean	−1.00	−0.70	−0.67	−0.81	−0.42
STD	0.34	0.40	0.34	0.36	0.36

5.3. Discussion

The implementation of the HDR-SFA pipelines produce images of better visual quality that its SDR counterpart. In the previous section, we made a tentative to evaluate the quality of data which are not easy to handle by quality procedures. First because we do not have references, neither groundtruth or good knowledge on HDR scenes. Indeed, in general HDR images are evaluated visually after being tone-mapped. In our context, it is a little more complex since we also add some spectral dimension to the problem. We try in the following to discuss this according to our evaluation.

One major aspect of SFA images is that we do not have ground truth, for both SDR and HDR domains. For spatial information, we are limited to the result of the demosaicing method or by the fact that the information is sparsely distributed over the whole sensor. For spectral information, we are limited to visual evaluation of content projected in a color space. Evaluation may however be produce on usability on a specific vision task, which is out of the scope of that article.

In our tentative to produce an evaluation on data quality, it appears that we cannot directly compare SDR, HDR and HDR tone-mapped results using a SDR image quality metric (such as BRISQUE). Generally, it is understood that a tone-mapped HDR result tends to compress a high dynamic range of radiance into a displayable range of intensity (e.g., 8-bits), as the logarithmic tone-mapping do. It often leads to an image of rather "washed and gray" appearance and devoid of local and global contrast. In other words, the quantity of visual information is enhanced at the expanse of the naturalness. So evaluation fails when trying to use SDR metrics to evaluate the image quality. That is why dedicated HDR quality metrics, such as HIGRADE [54], have been introduced in the literature, along with emerging perceptual-based tone-mapping techniques [48].

Although we hardly evaluate the radiance estimation, we validated the pipeline for the visualization output by examples and by objective metrics.

Based on this pipeline, we consider two typical types of potential applications. SFA camera are being used in several applications, with promising recent results in, e.g., medical imaging, where the snapshot aspect is rather important to have stability over physiological parameter changes [56]. However, the dynamic of the scene may be very large in such applications (for instance specular reflections on wet tissues or difference of intensity light between outside body and inside an opening in the body). In this case, such pipeline can be useful in situations that are well controlled, including knowledge about the illumination. Evaluation on how three consecutive image captures generate noise for a specific application is yet to be investigated. Temporal corrections of the time-dependent artifacts introduced are well understood in the literature [28]. On the other hand, general computer vision tasks, e.g., background subtraction [57], within uncontrolled illumination, where the dynamic range of the scenes could also be great can be targeted (for instance, automotive car getting out of a tunnel). One of the major constraint in this case is to handle the changes of illumination within the pipeline. Recent works on illuminant estimation from multispectral images [41,42,45] shall permit to implement correction in real time as for camera white balancing. Further works are required to investigate the impact of illumination in those applications. In any application, the HDR imaging pipeline permits to solve issues with energy balance of the sensor, i.e., when two bands have very different sensitivities for a similar integration time.

6. Conclusions

We defined a generalized imaging pipeline for HDR-SFA cameras. This is a similar pipeline to CFA architecture, adding spectral processing blocks and HDR enhancement for the channel balance correction and noise reduction. By demonstrating the pipeline, we enable the use of SFA camera in computer vision systems at reduce modification of the existing CFA pipeline.

Further works include the evaluation of the impact of each of the imaging pipeline components with respect to either visualization or usability of the HDR spectral data. We presented one instantiation, while many are possible. Further works also include standardization of camera and pipeline as well as file format and transmission line mixing multiple channel and HDR radiance data.

Acknowledgments: The authors thank the NTNU open access publishing policies for funding the cost of publication. The research was mostly funded by two projects: The Open Food System project. Open Food System is a research project supported by Vitagora, Cap Digital, Imaginove, Aquimer, Microtechnique and Agrimip, funded by the French State and the Franche-Comté Region as part of The Investments for the Future Programme managed by BPIfrance, www.openfoodsystem.fr. And the EU-project, H2020-662222, EXIST. http://cordis.europa.eu/project/rcn/198017_en.html.

Author Contributions: Main contributions can be distributed like in the following. We put the percentage as a rough indication of contributions only. Design and conception: J.-B.T. (40%), P.-J.L. (40%), P.G. (20%); Methodology: J.-B.T. (40%), P.-J.L. (40%), P.G. (20%); Acquisitions and implementations: J.-B.T. (20%), P.-J.L. (80%); State of the art: J.-B.T. (40%), P.-J.L. (40%), P.G. (20%); Analysis: J.-B.T. (50%), P.-J.L. (30%), P.G. (20%).

Conflicts of Interest: The authors declare no conflict of interest.

References

1. Lapray, P.J.; Wang, X.; Thomas, J.B.; Gouton, P. Multispectral Filter Arrays: Recent Advances and Practical Implementation. *Sensors* **2014**, *14*, 21626–21659.
2. Park, H.; Dan, Y.; Seo, K.; Yu, Y.J.; Duane, P.K.; Wober, M.; Crozier, K.B. Vertical Silicon Nanowire Photodetectors: Spectral Sensitivity via Nanowire Radius. In Proceedings of the 2013 CLEO: Science and Innovations, San Jose, CA, USA, 9–14 June 2013.
3. Yi, D.; Kong, L.; Wang, J.; Zhao, F. Fabrication of multispectral imaging technology driven MEMS-based micro-arrayed multichannel optical filter mosaic. *Proc. SPIE* **2011**, *7927*, 792711.
4. Eichenholz, J.M.; Dougherty, J. Ultracompact fully integrated megapixel multispectral imager. *Proc. SPIE* **2009**, *7218*, 721814.
5. Kiku, D.; Monno, Y.; Tanaka, M.; Okutomi, M. Simultaneous capturing of RGB and additional band images using hybrid color filter array. *Proc. SPIE* **2014**, *9023*, 90230V-9.
6. Geelen, B.; Blanch, C.; Gonzalez, P.; Tack, N.; Lambrechts, A. A tiny VIS-NIR snapshot multispectral camera. *Proc. SPIE* **2015**, *9374*, 937414.
7. Lapray, P.J.; Thomas, J.B.; Gouton, P. A Multispectral Acquisition System using MSFAs. In Proceedings of the Color and Imaging Conference, Boston, MA, USA, 3–7 November 2014; pp. 97–102.
8. Thomas, J.B.; Lapray, P.J.; Gouton, P.; Clerc, C. Spectral Characterization of a Prototype SFA Camera for Joint Visible and NIR Acquisition. *Sensors* **2016**, *16*, 993.
9. IMEC. Hyperspectral-Imaging. Available online: https://www.imec-int.com (accessed on 5 May 2017).
10. SILIOS Technologies. MICRO-OPTICS Supplier. Available online: http://www.silios.com/ (accessed on 5 May 2017).
11. PIXELTEQ. Micro-Patterned Optical Filters. Available online: https://pixelteq.com/ (accessed on 5 May 2017).
12. Jia, J.; Barnard, K.J.; Hirakawa, K. Fourier Spectral Filter Array for Optimal Multispectral Imaging. *IEEE Trans. Image Process.* **2016**, *25*, 1530–1543.

13. Monno, Y.; Kikuchi, S.; Tanaka, M.; Okutomi, M. A Practical One-Shot Multispectral Imaging System Using a Single Image Sensor. *IEEE Trans. Image Process.* **2015**, *24*, 3048–3059.

14. Ewerlöf, M.; Larsson, M.; Salerud, E.G. Spatial and temporal skin blood volume and saturation estimation using a multispectral snapshot imaging camera. *Proc. SPIE* **2017**, *10068*, 1006814.

15. Constantin, D.; Rehak, M.; Akhtman, Y.; Liebisch, F. Detection of crop properties by means of hyperspectral remote sensing from a micro UAV. In Proceedings of the 20. und 21. Workshop Computer-Bildanalyse in der Landwirtschaft - 3. Workshop Unbemannte autonom fliegende Systeme (UAS) in der Landwirtschaft, Braunschweig, Germany, 7 May 2015.

16. Péguillet, H.; Thomas, J.B.; Gouton, P.; Ruichek, Y. Energy balance in single exposure multispectral sensors. In Proceedings of the 2013 Colour and Visual Computing Symposium (CVCS), Gjøvik, Norway, 5–6 September 2013; pp. 1–6.

17. Lapray, P.J.; Thomas, J.B.; Gouton, P.; Ruichek, Y. Energy balance in Spectral Filter Array camera design. *J. Eur. Opt. Soc.* **2017**, doi:10.1186/s41476-016-0031-7.

18. Knickerbocker, J.U.; Andry, P.; Dang, B.; Horton, R.; Patel, C.S.; Polastre, R.; Sakuma, K.; Sprogis, E.; Tsang, C.; Webb, B.; et al. 3D silicon integration. In Proceedings of the 58th Electronic Components and Technology Conference, Lake Buena Vista, FL, USA, 27–30 May 2008; pp. 538–543.

19. Brochard, N.; Nebhen, J.; Ginhac, D. 3D-IC: New Perspectives for a Digital Pixel Sensor. In Proceedings of the 10th International Conference on Distributed Smart Camera, (ICDSC '16), Paris, France, 12–15 September 2016; pp. 92–97.

20. Mann, S.; Picard, R. *On Being 'undigital' with Digital Cameras: Extending Dynamic Range by Combining Differently Exposed Pictures*; Massachusetts Institute of Technology: Cambridge, MA, USA, 1995; pp. 422–428.

21. Debevec, P.E.; Malik, J. Recovering High Dynamic Range Radiance Maps from Photographs. In Proceedings of the 24th Annual Conference on Computer Graphics and (SIGGRAPH '97), Los Angeles, CA, USA, 3–8 August 1997; pp. 369–378.

22. Mitsunaga, T.; Nayar, S.K. Radiometric self calibration. In Proceedings of the 1999 IEEE Computer Society Conference on Computer Vision and Pattern Recognition (CVPR), Fort Collins, CO, USA, 23–25 June 1999; Volume 1, p. 380.

23. Robertson, M.A.; Borman, S.; Stevenson, R.L. Dynamic range improvement through multiple exposures. In Proceedings of the 1999 International Conference on Image Processing, (ICIP 99), Kobe, Japan, 24–28 October 1999; Volume 3, pp. 159–163.

24. Mann, S.; Lo, R.C.H.; Ovtcharov, K.; Gu, S.; Dai, D.; Ngan, C.; Ai, T. Realtime HDR (high dynamic range) video for eyetap wearable computers, FPGA-based seeing aids, and glasseyes (eyetaps). In Proceedings of the 25th IEEE Canadian Conference on Electrical & Computer Engineering (CCECE), Montreal, QC, Canada, 29 April–2 May 2012; pp. 1–6.

25. Lapray, P.J.; Heyrman, B.; Ginhac, D. HDR-ARtiSt: An adaptive real-time smart camera for high dynamic range imaging. *J. Real Time Image Process.* **2014**, *12*, 747.

26. Lapray, P.J.; Heyrman, B.; Ginhac, D. Hardware-based smart camera for recovering high dynamic range video from multiple exposures. *Opt. Eng.* **2014**, *53*, 102110.

27. An, J.; Ha, S.J.; Cho, N.I. Probabilistic motion pixel detection for the reduction of ghost artifacts in high dynamic range images from multiple exposures. *EURASIP J. Image Video Process.* **2014**, *2014*, 42.

28. Bouderbane, M.; Lapray, P.J.; Dubois, J.; Heyrman, B.; Ginhac, D. Real-time Ghost Free HDR Video Stream Generation Using Weight Adaptation Based Method. In Proceedings of the 10th International Conference on Distributed Smart Camera, (ICDSC '16), Paris, France, 12–15 September 2016; pp. 116–120.

29. Brauers, J.; Schulte, N.; Bell, A.; Aach, T. Color Accuracy and Noise Analysis in Multispectral HDR Imaging. In Proceedings of the 14th Workshop Farbbildverarbeitung 2008, Aachen, Germany,1–2 October 2008; pp. 33–42.

30. Simon, P.M. Single Shot High Dynamic Range and Multispectral Imaging Based on Properties of Color Filter Arrays. Ph.D. Thesis, University of Dayton, Dayton, OH, USA, 2011.

31. Thomas, J.B.; Lapray, P.J.; Gouton, P. HDR imaging pipeline for spectral filter array cameras. Presented at the Scandinavian Conference on Image Analysis (SCIA), Tromsø, Norway, 12–14 June 2017.

32. Lapray, P.J.; Thomas, J.B.; Gouton, P. A Database of Spectral Filter Array Images that Combine Visible and NIR. In Proceedings of the 6th International Workshop on Computational Color Imaging (CCIW 2017), Milan, Italy, 29–31 March 2017; pp. 187–196.

33. Tsin, Y.; Ramesh, V.; Kanade, T. Statistical calibration of CCD imaging process. In Proceedings of the Eighth IEEE International Conference on Computer Vision, (ICCV 2001), Vancouver, BC, Canada, 7–14 July 2001; Volume 1, pp. 480–487.

34. Zhou, J. *Getting the Most out of Your Image-Processing Pipeline*; White Paper; Texas Instruments: Dallas, Texas, United States, 2007.

35. Rani, K.S.; Hans, W.J. FPGA implementation of bilinear interpolation algorithm for CFA demosaicing. In Proceedings of the 2013 International Conference on Communications and Signal Processing (ICCSP), Melmaruvathur, India, 3–5 April 2013 ; pp. 857–863.

36. Sharma, G.; Trussell, H.J. Digital color imaging. *IEEE Trans. Image Process.* **1997**, *6*, 901–932.

37. Ramanath, R.; Snyder, W.E.; Yoo, Y.; Drew, M.S. Color image processing pipeline. *IEEE Signal Process. Mag.* **2005**, *22*, 34–43.

38. Kao, W.C.; Wang, S.H.; Chen, L.Y.; Lin, S.Y. Design considerations of color image processing pipeline for digital cameras. *IEEE Trans. Consum. Electron.* **2006**, *52*, 1144–1152.

39. Reinhard, E.; Stark, M.; Shirley, P.; Ferwerda, J. Photographic tone reproduction for digital images. *ACM Trans. Graph.* **2002**, *21*, 267–276.

40. Tamburrino, D.; Alleysson, D.; Meylan, L.; Süsstrunk, S. Digital camera workflow for high dynamic range images using a model of retinal processing. In Proceedings of the Electronic Imaging 2008, International Society for Optics and Photonics. *Proc. SPIE* **2008**, *6817*, 68170J.

41. Thomas, J.B. Illuminant estimation from uncalibrated multispectral images. In Proceedings of the 2015 Colour and Visual Computing Symposium (CVCS), Gjøvik, Norway, 25–26 August 2015; pp. 1–6.

42. Ahmad, H.; Thomas, J.B.; Hardeberg, J.Y. Multispectral constancy based on spectral adaptation transform. Presented at the Scandinavian Conference on Image Analysis (SCIA), Tromsø, Norway, 12–14 June 2017.

43. Miao, L.; Qi, H.; Ramanath, R.; Snyder, W.E. Binary Tree-based Generic Demosaicking Algorithm for Multispectral Filter Arrays. *IEEE Trans. Image Process.* **2006**, *15*, 3550–3558.

44. Banterle, F.; Artusi, A.; Debattista, K.; Chalmers, A. *Advanced High Dynamic Range Imaging: Theory and Practice*; AK Peters (CRC Press): Natick, MA, USA, 2011.

45. Khan, H.A.; Thomas, J.B.; Hardeberg, J.Y.; Laligant, O. Illuminant estimation in multispectral imaging. *J. Opt. Soc. Am. A* **2017**, *34*, in press.

46. Mantiuk, R.; Krawczyk, G.; Mantiuk, R.; Seidel, H.-P. High dynamic range imaging pipeline: Perception-motivated representation of visual content. *Proc. SPIE* **2007**, *6492*, 649212.

47. CHIC website. Website at Université de Bourgogne. Available online: http://chic.u-bourgogne.fr (accessed on 5 May 2017).

48. Krawczyk, G.; Myszkowski, K.; Seidel, H.P. Lightness perception in tone reproduction for high dynamic range images. *Comput. Graphics Forum* **2005**, *24*, 635–645.

49. Fattal, R.; Lischinski, D.; Werman, M. Gradient Domain High Dynamic Range Compression. In Proceedings of the 29th Annual Conference on Computer Graphics and Interactive Techniques (SIGGRAPH '02), San Antonio, TX, USA, 23–26 July 2002; pp. 249–256.

50. Banterle, F.; Artusi, A.; Sikudova, E.; Bashford-Rogers, T.; Ledda, P.; Bloj, M.; Chalmers, A. Dynamic Range Compression by Differential Zone Mapping Based on Psychophysical Experiments. In Proceedings of the ACM Symposium on Applied Perception (SAP '12), Los Angeles, CA, USA,3–4 August 2012; pp. 39–46.

51. Mittal, A.; Moorthy, A.K.; Bovik, A.C. No-Reference Image Quality Assessment in the Spatial Domain. *IEEE Trans. Image Process.* **2012**, *21*, 4695–4708.

52. Mittal, A.; Soundararajan, R.; Bovik, A.C. Making a "completely blind" image quality analyzer. *IEEE Signal Process. Lett.* **2013**, *20*, 209–212.

53. Saad, M.A.; Bovik, A.C.; Charrier, C. Blind Image Quality Assessment: A Natural Scene Statistics Approach in the DCT Domain. *IEEE Trans. Image Process.* **2012**, *21*, 3339–3352.

54. Kundu, D.; Ghadiyaram, D.; Bovik, A.; Evans, B. No-Reference Quality Assessment of Tone-Mapped HDR Pictures. *IEEE Trans. Image Process.* **2017**, *26*, 2957–2971.

55. Kundu, D.; Ghadiyaram, D.; Bovik, A.C.; Evans, B.L. ESPL-LIVE HDR Image Quality Database. Available online: http://live.ece.utexas.edu/research/HDRDB/hdr_index.html (accessed on 1 May 2017).

Sensors **2017**, *17*, 1281

56. Luthman, A.S.; Dumitru, S.; Quiros-Gonzalez, I.; Joseph, J.; Bohndiek, S.E. Fluorescence hyperspectral imaging (fHSI) using a spectrally resolved detector array. *J. Biophotonics* **2017**, doi:10.1002/jbio.201600304.

57. Benezeth, Y.; Sidibé, D.; Thomas, J.B. Background subtraction with multispectral video sequences. In Proceedings of the IEEE ICRA Workshops on Non-classical Cameras, Camera Networks and Omnidirectional Vision (OMNIVIS), Hong Kong, China, 31 May–7 June 2014.

sensors

MDPI

Article

Diazonium Salt-Based Surface-Enhanced Raman Spectroscopy Nanosensor: Detection and Quantitation of Aromatic Hydrocarbons in Water Samples

Inga Tijunelyte [1], Stéphanie Betelu [2,*], Jonathan Moreau [3], Ioannis Ignatiadis [2], Catherine Berho [2], Nathalie Lidgi-Guigui [1], Erwann Guénin [4], Catalina David [5], Sébastien Vergnole [5], Emmanuel Rinnert [3] and Marc Lamy de la Chapelle [1,*]

[1] CSPBAT Laboratory, UMR 7244, UFR SMBH, University of Paris 13, Sorbonne Paris Cite, 93017 Bobigny, France; inga.tijunelyte@univ-paris13.fr (I.T.); nathalie.lidgi-guigui@univ-paris13.fr (N.L.-G.)

[2] BRGM, F-45060 Orléans CEDEX 02, France; i.ignatiadis@brgm.fr (I.I.); C.Berho@brgm.fr (C.B.)

[3] IFREMER, Brittany Center, Detection, Sensors and Measurements Laboratory, CS10070, 29280 Plouzané, France; Jonathan.Moreau@ifremer.fr (J.M.); Emmanuel.Rinnert@ifremer.fr (E.R.)

[4] Laboratoire TIMR, EA4297, Sorbonne Universités—Université de Technologie de Compiègne, Centre de recherche de Royallieu, rue du docteur Schweitzer, CS 60319, 60203 Compiègne CEDEX, France; erwann.guenin@utc.fr

[5] HORIBA Jobin Yvon SAS, 59650 Villeneuve d'Ascq, France; Catalina.DAVID@horiba.com (C.D.); sebastien.vergnole@horiba.com (S.V.)

[*] Correspondence: s.betelu@brgm.fr (S.B.); marc.lamydelachapelle@univ-paris13.fr (M.L.d.l.C.); Tel.: +33-2-38-64-3268 (S.B.); +33-1-48-38-7691 (M.L.d.l.C.)

Academic Editors: Nicole Jaffrezic-Renault and Gaelle Lissorgues
Received: 31 March 2017; Accepted: 16 May 2017; Published: 24 May 2017

Abstract: Here, we present a surface-enhanced Raman spectroscopy (SERS) nanosensor for environmental pollutants detection. This study was conducted on three polycyclic aromatic hydrocarbons (PAHs): benzo[a]pyrene (BaP), fluoranthene (FL), and naphthalene (NAP). SERS substrates were chemically functionalized using 4-dodecyl benzenediazonium-tetrafluoroborate and SERS analyses were conducted to detect the pollutants alone and in mixtures. Compounds were first measured in water-methanol (9:1 volume ratio) samples. Investigation on solutions containing concentrations ranging from 10^{-6} g L^{-1} to 10^{-3} g L^{-1} provided data to plot calibration curves and to determine the performance of the sensor. The calculated limit of detection (LOD) was 0.026 mg L^{-1} (10^{-7} mol L^{-1}) for BaP, 0.064 mg L^{-1} (3.2×10^{-7} mol L^{-1}) for FL, and 3.94 mg L^{-1} (3.1×10^{-5} mol L^{-1}) for NAP, respectively. The correlation between the calculated LOD values and the octanol-water partition coefficient (K_{ow}) of the investigated PAHs suggests that the developed nanosensor is particularly suitable for detecting highly non-polar PAH compounds. Measurements conducted on a mixture of the three analytes (i) demonstrated the ability of the developed technology to detect and identify the three analytes in the mixture; (ii) provided the exact quantitation of pollutants in a mixture. Moreover, we optimized the surface regeneration step for the nanosensor.

Keywords: polynuclear aromatic hydrocarbon (PAH); surface-enhanced Raman spectroscopy (SERS); nanosensor; diazonium salt; surface functionalization; detection

1. Introduction

As part of the latest European Water Framework Directives (Directives 2000/60/EC, 2006/118/EC, and 2006/11/EC), the development of analytical tools allowing on-site, accurate, and sensitive detection of pollutants in environmental waters is of primary importance. Although extensive efforts have been devoted to developing highly sensitive, reproducible, accurate, and robust analytical sensors for on-site or in situ monitoring of trace elements (i.e., qualitative and quantitative analysis) [1–7], organic contaminants have received less attention.

Among the strategies investigated to meet the need for detecting organic contaminants in water, novel extraction techniques have been widely reported. Solid phase microextraction (SPME) [8–11] or stir-bar sorptive extraction (SBSE) [12] combined with gas chromatography (GC) [13–15], high performance liquid chromatography (HPLC) [15,16], mass spectrometry (MS) [17], Raman spectroscopy [10], or capillary electrophoresis (CE) have led to accurate results down to pg L^{-1} concentrations for chlorinated solvents [16], benzene, toluene, ethylbenzene, and xylenes (BTEXs) [10], and polycyclic aromatic hydrocarbons (PAHs) [9,15]. Despite their high sensitivity, procedures using these passive sensors are ill-suited for on-site monitoring due to their limitations regarding the co-injection of solvents [18,19] and the desorption procedure under high temperature and/or pressure [20].

New alternative methodologies with similar efficiencies are thus needed for detecting organic contaminants. Given their low limit of detection on the order of ng L^{-1}, piezoelectric chemical sensors (quartz crystal microbalance, QCM) [21–31] have attracted considerable attention. Although QCM is suitable for investigating the preconcentration of organic contaminants using coated polymers or calixarenes [22–29,31,32], the use of QCM-based sensors for the detection of various organic compounds in natural samples has not been reported as far as we know, probably because they cannot identify individual compounds.

In comparison, issues related to compound identification in complex environmental conditions can be overcome by exploiting surface-enhanced Raman spectroscopy (SERS). SERS is a powerful technique based on nanostructured metallic surfaces that greatly enhance the Raman signal via both electromagnetic and chemical effects. The first effect arises from the interaction between the incident light and the metallic nanostructures, inducing local enhancement of the electromagnetic field through the excitation of localized surface plasmons [33–37]. As a result, the Raman signal of any molecule located in close vicinity to the nanostructured surface can be enhanced up to 10^8-fold [38–42]. The second effect contributing to Raman signal enhancement by up to 10^2-fold is due to the electronic interaction between the molecules and the metallic nanostructures (i.e., charge transfer between surface and chemisorbed molecules) [43]. SERS nanosensors can detect very low concentrations of analytes and thus, attain high sensitivity and low limit of detection (LOD) values, down to the individual molecule [44–46]. Most importantly, SERS provides a molecular fingerprint, thus molecules can be individually identified and deciphered in complex mixtures [47,48].

Finally, due to the improvements in Raman spectrometer miniaturization [11], the innovations in nanoscale technologies applied to sensors (essentially based on colloid systems [49,50]), and the implemented surface chemistry strategies (i.e., surface functionalization by using Self-Assembled Monolayers (SAMs) for analyte preconcentration), SERS has already been recognized as a powerful tool for on-site monitoring of organic contaminants at the ng L^{-1} level [50–57]. However, despite great improvements made to produce colloidal nanostructures with controlled sizes and shapes [49,50], SERS nanosensors have a major drawback related to poor SERS signal reproducibility due to the uncontrollable aggregation of metal colloids. To overcome the lack of reproducibility and capitalize on the increasing number of SERS substrates available commercially, we selected substrates (Wavelet) based on supported gold nanorod arrays produced and distributed by the S.T. Japan company. These SERS substrates are devoted for highly reproducible and very sensitive measurements.

Regarding surface functionalization, SAMs are widely recognized as excellent systems for sensing applications because they offer well-defined organization and densely-packed structures [58]. However, there are limitations with regard to the stability of the bond between the thiol and the metallic surface and hence the stability of the resulting SAMs [59,60]. As a powerful, innovative alternative to SAMs, we adopted a diazonium salt-based surface functionalization strategy. Immobilization of aryldiazonium has been described as a versatile method, providing surface properties that can be fine-tuned given the wide variety of available diazonium salts and stable grafted organic layers [61–68], thereby ensuring sensor robustness. Therefore, we studied a diazonium-salt-functionalized SERS sensor for the detection and the quantitation of three PAHs. Among the 130 PAHs released into the environment, benzo[a]pyrene (BaP), fluoranthene (FL), and naphthalene (NAP) belong to the priority substances under the European Water Framework Directive (2000/60/EC) and the United States Environment Protection Agency. The investigation of selected pollutants was conducted to demonstrate the proof-of-concept of this novel SERS nanosensor for compound sensing. PAH compounds were analyzed alone and in mixture. Investigations of the targeted compounds were carried out with different concentrations prepared in water-methanol (9:1 volume ratio) solutions. Calibration curves and detection limits were established for each compound. Moreover, given that the preconcentration of targeted molecules at the nanosensor surface is essentially based on weak molecular forces, we tested the possibility of regeneration and reuse of the developed nanosensor.

2. Materials and Methods

2.1. Reagents

Diethyl ether ($\geq$98%), tetrafluoroboric acid (49.5–50.5%), sodium nitrite ($\geq$97%), 4-dodecylaniline (97%), benzo[a]pyrene ($\geq$96% HPLC), fluoranthene (98%), and naphthalene (99%) were purchased from Sigma-Aldrich and used without further purification. Ethanol (96%) and methanol (100%) were purchased from VWR; sulfuric acid (95–98%) was purchased from JT Baker. Milli-Q water (resistivity of 18.2 MΩ cm^{-1}) was used in all experiments.

2.2. Diazonium Salt Chemical Synthesis

4-dodecyl benzenediazonium-tetrafluoroborate (DS-C$_{10}$H$_{21}$), was synthesized regarding the reaction between aryl amines and sodium nitrite at 0 °C [69]. Briefly, the primary amine (4 $\times$ 10^{-3} mol L^{-1}) was dissolved in a tetrafluoroboric acid (HBF$_4$) solution and the mixture was then cooled for 15 min. A precooled aqueous sodium nitrite (4.3 $\times$ 10^{-3} mol L^{-1}) solution was added dropwise to the acid amine mixture under stirring. The mixture was then allowed to react for 40 min. The obtained precipitate was filtered through a 0.2 µm cellulose ester filter (Whatman) and generously washed with milli-Q water. The diazonium salt was purified by re-crystallization in diethyl ether for 48 h at 6 °C. Afterwards, the prepared salt was dried under vacuum and preserved at −20 °C.

The synthesized diazonium salt was characterized by nuclear magnetic resonance (NMR). Spectra were recorded using a Bruker Avance III 400 MHz instrument in a d^6-DMSO solvent. Tetramethylsilane (TMS) was used as an internal standard. DS-C$_{10}$H$_{21}$:^{1}H NMR (400 MHz, DMSO-d^6): d (ppm) 0.84–0.87 (t, *J* = 6.4 Hz, 3H); 1.25 (s, 14H); 1.61 (s, 2H); 2.81–2.85 (t, *J* = 7.6 Hz, 2H); 7.81–7.83 (d, *J* = 8.4 Hz, 2H); 8.55–8.57 (d, *J* = 8.4 Hz, 2H).

2.3. SERS Substrate

Commercially available gold nanorod arrays (Wavelet) (hereafter referred to as gold nanostructures, GNSs) were purchased from S.T. Japan and were used in this study as SERS active substrates.

2.4. Surface Functionalization

Chemical (spontaneous) grafting of the diazonium salt was performed by immersing SERS substrates in a 30 mL solution containing the diazonium salt (10^{-3} mol L^{-1}) dissolved in H$_2$SO$_4$ (10^{-3} mol L^{-1}). Substrates were then incubated for 12 h at +4 °C. They were then rinsed under mild stirring by immersion in (i) milli-Q water (3 times) and (ii) pure methanol (3 times), for 10 min for each immersion. Afterwards, the functionalized substrates were dried under a mild nitrogen flux.

2.5. Solutions of the Targeted Compounds

Benzo[a]pyrene (BaP), fluoranthene (FL), and naphthalene (NAP) were used in this study to test the ability of the diazonium-salt-based SERS sensor to detect the PAH compounds. These PAHs were selected for their different solubility values related to their polarity defined by the octanol-water partition coefficient, K_{OW}. The K_{OW} of each compound is summarized in Table 1.

Table 1. Physical properties of the selected targeted pollutants [70].

Polycyclic Aromatic Hydrocarbon	Structure	Solubility in Water (mg L^{-1})	K$_{ow}$
Benzo[a]pyrene (BaP)		0.0038 at 25 °C	10^6
Fluoranthene (FL)		0.26 at 25 °C	3.4×10^5
Naphthalene (NAP)		32	2.3×10^3

Due to solubility issues, stock solutions were prepared by dissolving the targeted pollutants in pure methanol (MeOH). Stock solutions of BaP, FL, and NAP were prepared at different concentrations as follows: 10 mg L^{-1} for BaP, 200 mg L^{-1} for FL, and 1000 mg L^{-1} for NAP, and were stored at -20 °C. Stock solutions were diluted daily with milli-Q water (10% of pollutant stock solution in 90% of water % v/v) to obtain stock solutions which were then stored at +4 °C and used for further dilution, keeping a constant volume ratio (9:1) between milli-Q water and MeOH.

2.6. Nanosensor Surface Regeneration

The preconcentration of the targeted pollutants on the SERS nanosensors was driven by weak interactions with the DS-C$_{10}$H$_{21}$ layer, such as hydrophobic interactions and/or $\pi-\pi$ stacking (Scheme 1). To regenerate the surface, MeOH was found to be an efficient solution. Thus, after performing the detection of the targeted pollutants, the SERS nanosensors were regenerated with MeOH for 30 min.

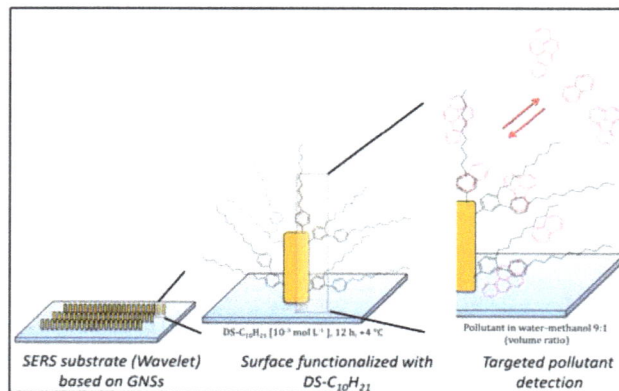

SERS substrate (Wavelet) based on GNSs	**Surface functionalized with DS-C$_{10}$H$_{21}$**	**Targeted pollutant detection**

Scheme 1. General illustration of nanosensor surface functionalization and preconcentration of the targeted compounds via weak interactions. Gold Nanostructures GNSs, DS-C$_{10}$H$_{21}$, 4-dodecyl benzenediazonium-tetrafluoroborate diazonium salt.

2.7. Raman and SERS Measurements

Raman and SERS measurements were performed using a transportable micro-Raman spectrometer prototype designed by HORIBA Scientific for on-site applications. The instrumental set-up was equipped with a 691 nm laser diode (Ondax) and a 60× magnification objective (0.7 N.A.) with a collar for glass correction (Olympus). The spectral range was recorded from 400 to 2100 cm^{-1} with a spectral resolution better than 4 cm^{-1}. For SERS experiments, the laser power was set to 4 mW to avoid molecular degradation induced by photochemical or thermal effects. The integration time for SERS measurements was set to 20 s with three accumulations to reach a relevant signal-to-noise ratio. Spectral calibration was performed daily on a crystalline silicon sample (peak position at 520 cm^{-1}). A fluidic cell located above the microscope objective allowed the sample solutions to flow onto the SERS substrate.

2.8. Chemometrics

To remove the Raman and SERS spectral background in a similar way for all spectra, leaving the analytical signal intact, an algorithm was programmed in MatLab 7.0.1 based on [71,72]. Background correction was possible by minimizing the following S function:

$$S = \sum_{(i)} \kappa_i (y_i - z_i)^2 + \lambda \sum_{(i)} (\Delta^2 z_i)^2 \qquad (1)$$

where y is the signal intensity for each i wavenumber, z is the baseline, λ is the smoothing parameter, and *p* is the asymmetric parameter as $\kappa_i = p$ if $y_i > z_i$ and $\kappa_i = 1 - p$ otherwise. The last term was defined as follows:

$$\Delta^2 z_i = (z_i - z_{i-1}) - (z_{i-1} - z_{i-2}) \qquad (2)$$

The two parameters needed for the calculations, *p* and λ, were set to 10^3 and 10^{-3}, respectively. However for the diazonium-salt-based system during the detection of BaP, the λ value was set to 10^4 to remove fluorescence induced by impurities. Figure S1 illustrates the background removal process.

3. Results and Discussion

3.1. Diazonium Salt Based Surface Functionalization

Surface functionalization by aryldiazonium is a convenient method leading to a robust, grafted organic coating. Grafting can be accomplished by either chemical (spontaneous grafting) [65–68], electrochemical [62–64], or photochemical methods [73,74]. A spontaneous grafting strategy was used in this study to functionalize SERS-active substrates. The substrates were immersed in a diazonium salt (DS-C$_{10}$H$_{21}$) solution and incubated at +4 °C for 12 h. The temperature control during the functionalization step is important to drive the covalent grafting and at the same time reduce the rate of spontaneous polymerization [75]. Functionalized substrates were then investigated using SERS. The SERS spectra for DS-C$_{10}$H$_{21}$-GNSs are shown in Figure 1. These spectra are presented in real intensities with the SERS spectral background removed using the algorithm detailed in the Material and Methods section. Because the complete characterization of DS-C$_{10}$H$_{21}$-GNSs can be found in the literature [61], here we just briefly mention that the recorded spectra have a number of features in the spectral ranges that represent the vibrations related to the aromatic ring (see Figure S2 for direct comparison between the SERS spectra of the grafted phenyl derivative and the Raman spectrum of the synthesized diazonium salt). The peak at 1079 cm^{-1} for DS-C$_{10}$H$_{21}$ assigned to the C-H in-plane bending mode coupled with the C-N stretching mode [76] constitutes the signature of the coating. Thus, this peak was used as an internal reference during the quantitation of the detected pollutants. Moreover, this peak does not overlap with the Raman signature of the targeted pollutants (see Figure 1), providing further support for selecting this internal reference.

3.2. Detection of PAHs

Diazonium-salt-based SERS substrates were then evaluated on the sensing of the targeted compounds. Thus, functionalized SERS substrates were incubated in pollutant solutions for 30 min. Investigated concentrations for the initial test were 0.75 mg L^{-1} for BaP, 5 mg L^{-1} for FL, and 50 mg L^{-1} for NAP. After incubation, SERS spectra were measured in six different randomly selected areas on the substrates, covering the entire substrate surface. Averaged spectra for BaP, FL, and NAP detection are shown in Figure 1A–C, respectively. All SERS spectra are presented with removed background.

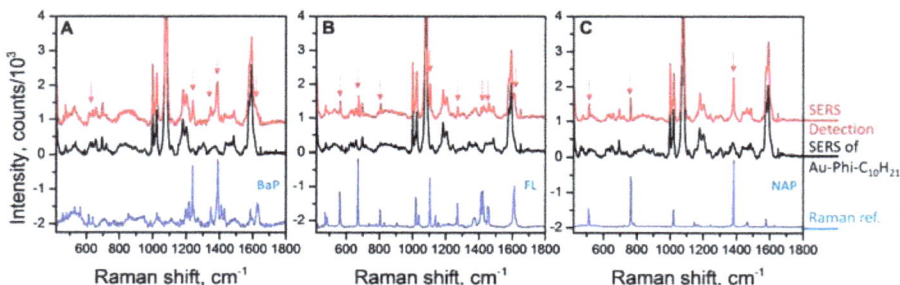

Figure 1. Detection of benzo[a]pyrene (BaP) at 0.75 mg L^{-1} (**A**), fluoranthene (FL) at 5 mg L^{-1} concentration (**B**) and naphthalene (NAP) at 50 mg L^{-1} (**C**) using gold nanostructures (GNSs) functionalized with diazonium salt (Au-Phi-C$_{10}$H$_{21}$). The black and red spectra correspond to the SERS signals before and after incubation in the pollutants solutions, respectively. Blue spectra in (**A**–**C**) correspond to the Raman reference spectra for BaP, FL, and NAP, respectively, acquired on powder. Red and blue spectra are shifted vertically for better visualization. Red arrows indicate the peaks of detected pollutants.

The successful preconcentration and sensing of the targets by the DS-C$_{10}$H$_{21}$-based SERS sensor was validated after the observation of characteristic fingerprint peaks of each pollutant as summarized in Table 2.

Table 2. Positions in cm^{-1} of the main polycyclic aromatic hydrocarbon (PAH) bands observed in the Raman and SERS spectra (see [77,78] for assignments).

Benzo[a]Pyrene (BaP)		Fluoranthene (FL)		Naphthalene (NAP)	
Raman	SERS	Raman	SERS	Raman	SERS
615	615	562	562	509	511
1215	1215	671	672	762	762
1236	1239	803	803	1021	1021
1345	1345	1018	1018	1382	1382
1389	1385	1104	1104	1577	1577
1623	1623	1269	1270		
		1612	1612		

The interaction between the analyte and the coating layer did not induce important shifts in the position of the pollutant peaks. Hence, most of the peak positions were nearly identical to those of the Raman reference spectra. However, in the case of the preconcentration of BaP (Figure 1A), the peak at 1236 cm^{-1} corresponding to the ring stretching and the C-H stretching modes was found shifted 3 cm^{-1} upwards towards higher wavenumbers after adsorption on the nanosensor surface, whereas the peak observed at 1389 cm^{-1} assigned to C-C ring stretching shifted 4 cm^{-1} downwards to lower wavenumbers. Spectral differences mostly related to the vibrational mode intensities were observed for FL (Figure 1B). For instance, the peaks around 1400 cm^{-1} were very weakly enhanced in the SERS spectrum. Such weak spectral modifications may be due to weak interactions between the pollutants and the functionalization layer.

3.3. Sensing Performance

After having first proven the ability of the developed DS-C$_{10}$H$_{21}$-based nanosensor to preconcentrate (the targeted) compounds, calibration curves were established to compare the analytical performance of the nanosensor for each pollutant. To do so, the substrate was fixed on the microfluidic cell of the Raman set-up and solutions with increasing concentrations of pollutants were injected onto the substrate at a flow rate of 3 μL min^{-1} for 30 min for each solution.

3.4. Benzo[a]pyrene (BaP) Calibration Curve

BaP solutions with concentrations increasing from 0.1 mg L^{-1} to 0.75 mg L^{-1} (0.4–3.0 μmol L^{-1}) were tested to establish the calibration curve. Between each concentration, SERS measurements were carried out on six different areas on the surface. Averaged spectra are shown in Figure 2A, focusing on the peak of BaP at 1239 cm^{-1} (the full spectral range of averaged SERS spectra can be seen in Figure S3). The calibration curve was plotted using the relative intensity of this peak compared with the intensity of the characteristic peak of the coating (i.e., the peak at 1079 cm^{-1}) as a function of the analyte concentration (Figure 2B).

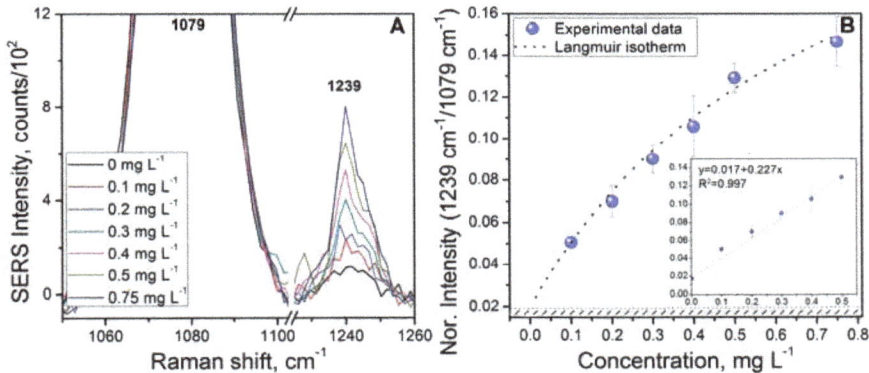

Figure 2. (**A**) Surface-enhanced Raman spectroscopy (SERS) spectra of benzo[a]pyrene (BaP) detection for concentrations from 0 up to 0.75 mg L^{-1}; (**B**) calibration curve of BaP detection (blue circles: experimental data, dotted line: Langmuir adsorption isotherm fit). Insert: zoom on the concentration range from 0 to 0.5 mg L^{-1} (blue circles: experimental data, dotted line: linear fit). The hatched area corresponds to the noise level.

The calibration curve was then fitted using the Langmuir adsorption isotherm expressed as follows:

$$I = \frac{I_{max}Kc}{1 + Kc} \tag{3}$$

where I is the normalized SERS intensity, K is the adsorption constant, and c is the analyte concentration.

The calibration plot was found to be linear over the range of 0.1–0.5 mg L^{-1} (0.4–2.0 μmol L^{-1}) but surface saturation was not reached. The limiting factor was the solubility of BaP in the milli-Q water-MeOH (9:1 volume ratio) solution.

The slope of the linear regression on the low concentrations (insert in Figure 3B) was used for the calculation of the LOD and the LOQ. LOD is defined as the lowest concentration of analyte in the sample that can be detected [79] and can be calculated using the following formula [80]: LOD = 3 σ/s, where s is the slope of calibration curve and σ is the standard deviation of the blank response. LOQ is the lowest concentration of an analyte that can be determined with acceptable precision and accuracy. LOQ was calculated using the same formula that was used for the calculation of LOD with a confidence level increased to 10: LOQ = 10 σ/s. The calculated LOD and LOQ for BaP were 0.026 mg L^{-1} (0.10 μmol L^{-1}) and 0.087 mg L^{-1} (0.34 μmol L^{-1}), respectively (Table 3).

3.5. Fluoranthene (FL) Calibration Curve

The calibration curve for FL was performed using the same protocol as that for BaP. In comparison with BaP, the investigated concentrations of FL ranged from 0.1 mg L^{-1} to 5 mg L^{-1} (0.5–25 μmol L^{-1}) due to its higher solubility (see Table 1). As before, between each concentration, SERS was measured on six different areas on the surface. Averaged spectra are shown in Figure 3A, focusing on the peak at 1104 cm^{-1} (the full spectral range of the averaged SERS spectra is given in Figure S4). Assigned to the FL C-C in-plane stretching mode, this peak was the most intense band observed in the SERS spectrum (Figure 1B). It was thus selected to establish the calibration curve (Figure 3B), in which the relative intensity of this peak compared with the intensity of the peak at 1079 cm^{-1}—characteristic of the diazonium salt DS-C$_{10}$H$_{21}$—was plotted as a function of the FL concentration.

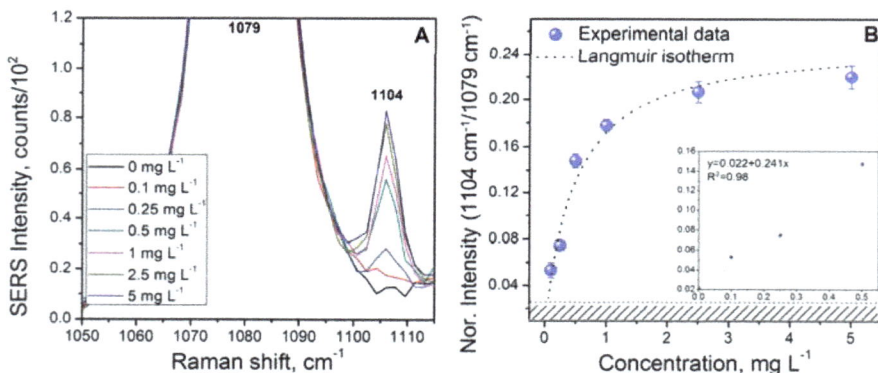

Figure 3. (**A**) Surface-enhanced Raman spectroscopy (SERS) spectra of fluoranthene (FL) detection for concentrations ranging from 0 to 5 mg L^{-1}; (**B**) calibration curve of FL detection (blue circles: experimental data, dotted line: Langmuir adsorption isotherm fit). Insert: zoom on the concentration in the range from 0 to 0.5 mg L^{-1} (blue circles: experimental data, dotted line: linear fit). The hatched area corresponds to the noise level.

At values greater than 3 mg L^{-1}, there is a plateau, indicating that the adsorption equilibrium was reached. The calibration curve can be considered linear in the range of 0.1–0.5 mg L^{-1}. The LOD of FL was calculated to be 0.064 mg L^{-1} (0.32 μmol L^{-1}). The calculated LOQ value was 0.214 mg L^{-1} (1.06 μmol L^{-1}).

3.6. Naphthalene (NAP) Calibration Curve

For the study of NAP pre-concentration and detection, SERS measurements were carried out as previously described in milli-Q/MeOH (9:1 v/v) media containing different concentrations of NAP ranging from 1 mg L^{-1} up to 50 mg L^{-1} (7.8–390 μmol L^{-1}). NAP peaks observed in the SERS spectra are shown in Figure 1C and summarized in Table 2. Figure 4A focuses on the increase in the intensity of the peak located at 1382 cm^{-1}, assigned to the C=C stretching mode (the full spectral range averaged

SERS spectra are shown in Figure S5). This peak was chosen to establish the calibration curve shown in Figure 4B since it is the only one that does not overlap with the vibrational signature of the grafted diazonium salt.

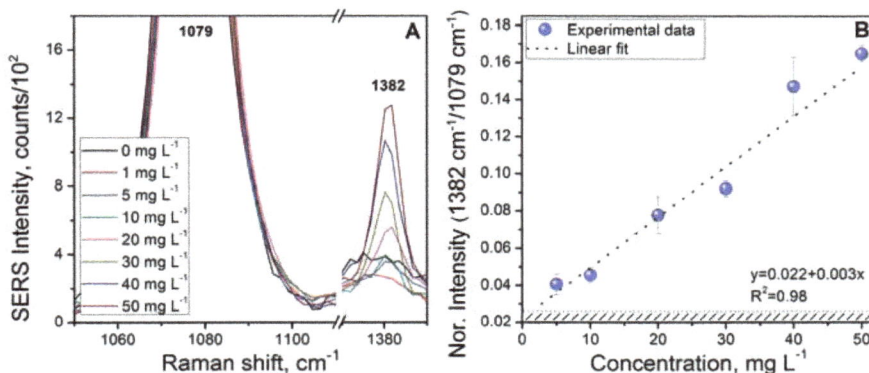

Figure 4. (**A**) Surface-enhanced Raman spectroscopy (SERS) spectra of naphthalene (NAP) detection for concentrations ranging from 0 to 50 mg L^{-1}; (**B**) calibration curve of NAP detection (blue circles: experimental data, dotted line: linear fit). The hatched area corresponds to the noise level.

The calibration curve was found to be linear over the investigated range. The calculated LOD and LOQ values were 3.9 mg L^{-1} (30 µmol L^{-1}) and 13 mg L^{-1} (0.1 mmol L^{-1}), respectively (Table 3).

Table 3. Summarized limit of detection (LOD) and limit of quantitation (LOQ) values for all targeted pollutants.

Polycyclic Aromatic Hydrocarbon	LOD		LOQ	
	(mg L^{-1})	(mol L^{-1})	(mg L^{-1})	(mol L^{-1})
Benzo[a]pyrene (BaP)	0.026	10^{-7}	0.087	3.4×10^{-7}
Fluoranthene (FL)	0.064	3.2×10^{-7}	0.214	1.1×10^{-6}
Naphthalene (NAP)	3.9	3.0×10^{-5}	13	1.0×10^{-4}

3.7. Analysis in a Mixture of Analytes

An experiment using a mixture of the three investigated pollutants was performed to further assess the feasibility of the designed nanosensor for PAHs sensing in water samples and to demonstrate the SERS spectra suitability for on-site screening purposes. For this experiment, a mixed solution of BaP (0.75 mg L^{-1}, 3.0 µmol L^{-1}), FL (5 mg L^{-1}, 25 µmol L^{-1}), and NAP (5 mg L^{-1}, 39 µmol L^{-1}) was prepared in pure water-MeOH (9:1 volume ratio) media. The concentration of BaP was notably lower due to the solubility issue discussed above. The SERS substrate was exposed to the prepared mixture for 30 min before starting SERS measurements. The averaged SERS spectra obtained before and after substrate exposure to the pollutants are shown in Figure 5. At the selected concentrations, BaP and FL were detected and well identified. However, most NAP peaks overlapped either the signal of the coating (i.e., at 1022 cm^{-1}) or the signal of the other analytes (i.e., the peak used for establishing the calibration curve for NAP at 1382 cm^{-1} was found combined with the signal of BaP). Nonetheless, it showed one weak, but characteristic peak at 762 cm^{-1}. Together, these results demonstrate that the SERS nanosensor can detect the different pollutants in a mixed solution and that we can clearly identify them by using their spectral signatures. However, the quantitation of BaP, FL, and NAP in the mixture is more problematic, as discussed below.

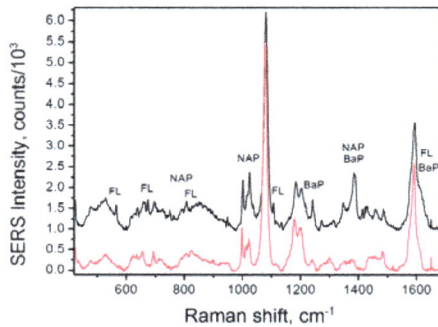

Figure 5. Black spectrum corresponds to the surface-enhanced Raman spectroscopy (SERS) spectrum obtained for the analysis of a mixed solution composed of benzo[a]pyrene BaP (0.75 mg L^{-1}, 3 µmol L^{-1}), fluoranthene FL (5 mg L^{-1}, 25 µmol L^{-1}), and naphthalene (NAP) (5 mg L^{-1}, 39 µmol L^{-1}). The red spectrum shows the SERS signal of the DS-$C_{10}H_{21}$ diazonium-salt-based gold nanostructures (GNSs) for comparison. The black spectrum is vertically shifted for clarity.

3.8. Feasibility of the Nanosensor for Sensing Polycyclic Aromatic Hydrocarbons

The results obtained on targeted pollutants in their individual solutions and in mixture provide important information on (i) the sensor effectiveness for PAH sensing and (ii) the sensor selectivity for BaP, FL, or NAP. First, the LOD concentrations for the investigated PAH compounds varied by two orders of magnitude (Table 3) between BaP and NAP. The correlation of the observed LOD values versus the log of the K_{ow} factor of the investigated analytes is given in Figure 6. This curve suggests that the hydrophobic coating of the sensor is more suitable for the preconcentration of highly non-polar PAH compounds.

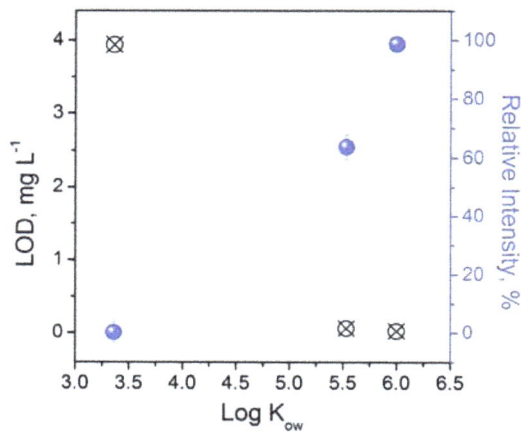

Figure 6. Correlation of the limit of detection (LOD) concentration (crossed circles) and the observed related surface-enhance Raman spectroscopy (SERS) intensities of the analytes in the mixed solution (blue circles) plotted as a function of the K_{ow} factor of the investigated compounds.

This relationship is confirmed by the results obtained from the detection of pollutants in the mixed solution. Figure 6 shows the relative SERS intensity of the pollutant peaks calculated as follows:

$$Relative\ Intensity(\%) = 100\% * \frac{I'_{Peak\ of\ Analyte}}{I'_{Peak\ of\ DS-C_{10}H_{21}}} * \frac{I_{Peak\ of\ DS-C_{10}H_{21}}}{I_{Peak\ of\ Analyte}} \tag{4}$$

where I' is the SERS intensity of the selected peaks measured in the mixed solution and I is the SERS intensity of the peaks measured for each individual PAH at the same concentration as that used in the mixed solution (0.75 mg L^{-1} for BaP and 5 mg L^{-1} for FL and NAP). The peaks used for this calculation are those used for the establishment of the calibration curves (the peak of DS-$C_{10}H_{21}$ at 1079 cm^{-1} and the peaks at 1239 cm^{-1}, 1104 cm^{-1}, and 1382 cm^{-1} for BaP, FL, and NAP, respectively). With this formula, the relative SERS intensity of the PAH signal obtained in the mixed solution was directly compared with the relative SERS intensity given by the calibration curve for the same pollutant concentration and it was then expressed as a percentage. For instance, for BaP, at a concentration of 0.75 mg L^{-1}, the relative intensity of the peak at 1239 cm^{-1} measured in the calibration curve and in the mixed solution were found to be almost identical, leading to a calculated relative intensity of 98.7%. This value shows that the adsorption of BaP onto the nanosensor surface was not restricted by the presence of competitive analytes. The FL calibration curve indicates that at a concentration of 5 mg L^{-1}, the adsorption equilibrium was reached with a maximum relative intensity of 0.22 ± 0.01. At the same concentration, FL in the mixed solution exhibited a relative intensity of 0.140 ± 0.012. This corresponds to 64% of the relative intensity reached with FL alone. The calculation of the NAP signal recorded in the mixed solution was more challenging. Due to the low intensity of the non-overlapping peak, it was not possible to quantify the intensity with the same formula. The approximate contribution of NAP was calculated indirectly. Since NAP and BaP both have peaks around 1382 cm^{-1}, we subtracted the SERS spectra of BaP detection at 0.75 mg L^{-1} concentration from that measured in the mixed solution. The relative intensity of NAP was then found to be 0.14%.

The calculated relative intensities of the PAHs peaks in the SERS spectra obtained in the mixed solution confirmed that the strength of the interaction between the coating layer and the analyte have a strong influence on the preconcentration and thus on the detection of the analytes. Highly hydrophobic molecules having a high K_{ow} value will be easier to detect and will exhibit lower LOD. Moreover, the SERS nanosensor allowed the detection of BaP with a sensitivity comparable to the detection carried out using colloidal nanoparticles [81,82].

3.9. Surface Regeneration

To demonstrate the reusability of the developed nanosensor, we performed regeneration experiments using a pure MeOH solution to remove the adsorbed PAHs from the sensor surface. For each step of regeneration, MeOH was injected and circulated in the fluidic cell for 30 min. To determine the impact of the surface regeneration step on the nanosensor performance, we screened for BaP at a concentration of 0.75 mg L^{-1}. For this demonstration, BaP was selected for its strong interaction with the coated layers (as discussed above). Figure 7 summarizes the results of (i) SERS signal intensities for the peak at 1079 cm^{-1} assigned to the DS-$C_{10}H_{21}$ coating and (ii) the SERS intensity ratio between the peak of BaP at 1236 cm^{-1} and the peak of the DS-$C_{10}H_{21}$ coating for each experimental step (detection and washing). The surface regeneration process did not affect the stability of the functionalization layer, and the average intensity of the peak was found to be nearly constant with a variation of 11% between the different experimental steps. Moreover, several cycles of BaP screening show highly reproducible SERS signals of the analyte with an average intensity ratio of 0.152 with a standard deviation lower than 5%. This result demonstrates that the nanosensor surface can be washed and therefore reused several times without affecting the detection efficiency.

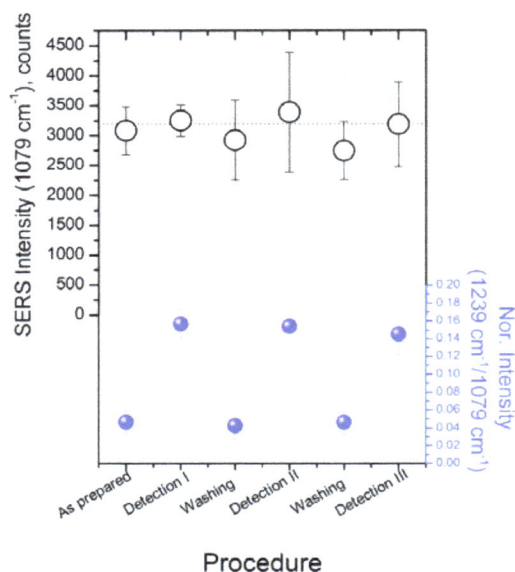

Figure 7. Surface-enhanced Raman spectroscopy (SERS) intensity variations for the peak at 1079 cm^{-1} of the diazonium salt (DS-C$_{10}$H$_{21}$) coating measured after each procedure (open black circles); SERS intensity ratios between the peaks at 1236 cm^{-1} and 1079 cm^{-1} for benzo[a]pyrene (BaP) and the DS-C$_{10}$H$_{21}$ coating, respectively (filled blue circles).

3.10. Detection Reproducibility and Repeatability

The high reproducibility and repeatability of the SERS signal of the developed nanosensor is of primary importance for its application to quantitative analyte detection. Statistical analysis was carried out to estimate the repeatability of the SERS signal intensity between various spots on the substrate as well as to define reproducibility between two functionalized nanosensors. For this, SERS spectra of BaP at a concentration of 0.75 mg L^{-1} were recorded on six individual spots for two substrates. By comparing the calculated relative intensity ratios between the signal of BaP and the signal of the coating ($I_{1236\,cm^{-1}}/I_{1079\,cm^{-1}}$), we attained very reproducible targeted pollutant detection. Calculated average values were 0.147 and 0.152 for different substrates with a coefficient of variation of 15% and 13%, respectively. These results indicate that commercially available Wavelet SERS substrates functionalized with a diazonium salt can be considered as a repeatable and reproducible SERS solution. To further demonstrate the suitability of our diazonium salt-SERS solution, we compared it with a popular commercial SERS substrate, Klarite, for which a coefficient of variation of 45% has been reported from the measurement of benzenethiol grafted to the surface [83]. Moreover, the low coefficient of variation observed from spot-to-spot measurements suggests that DS-C$_{10}$H$_{21}$ diazonium-salt-based surface functionalization offers robust and homogeneous surface coverage.

4. Conclusions

Here, we demonstrated a novel approach for the decoration of a SERS nanosensor for the reversible and reproducible detection of non-polar pollutants. Diazonium salt (DS-C$_{10}$H$_{21}$) was tested for substrate surface functionalization and for subsequent preconcentration of the targeted compounds. Spontaneous DS-C$_{10}$H$_{21}$ layer formation was found to be highly reproducible under the operating conditions. During this conceptual study, we demonstrated the detection of BaP, FL, and NAP with LOD values of 0.026 mg L^{-1} (0.1 µmol L^{-1}), 0.064 mg L^{-1} (0.32 µmol L^{-1}), and 3.94 mg L^{-1} (31 µmol L^{-1}), respectively. The LOD concentrations of the detected compounds were strongly correlated with their physicochemical properties. The observed selectivity of the coated nanosensor favoring highly non-polar molecules such as BaP suggests that the preconcentration of aromatic pollutants can be optimized by varying the nature of the aryldiazonium salt used for surface functionalization.

The level of BaP detection obtained in this study is comparable with that documented in the literature using colloidal nanoparticles. Given that the major drawback of colloidal nanoparticles' application for pollutant monitoring is signal reproducibility, our proposed nanosensor functionalized with diazonium salt holds great promise for applications to in situ measurements due to its demonstrated sensitivity, reproducibility, repeatability, and reusability.

Supplementary Materials: The Supplementary Materials containing Figures S1–S5 are available online at http://www.mdpi.com/1424-8220/17/6/1198/s1.

Acknowledgments: The authors would like to acknowledge the REMANTAS project: Enhanced Raman scattering for aquatic media: a new technology for on-site analysis 2011–2014 (REMANTAS project; ANR-11-ECOT-0010) for financial support. They also thank Carolyn Engel-Gautier for providing a proofreading of the English text. The authors thank Ifremer for covering the costs to publish in open access.

Author Contributions: I.T. and S.B. contributed equally to this work: they synthesized diazonium salt and performed the experiments on surface enhanced Raman scattering (SERS) at BRGM, treated the obtained data, and wrote the paper; J.M. adapted the algorithm for chemometrics; E.R. provided information on the pollution of marine waters; C.D. and S.V. developed a prototype of a transportable micro-Raman spectrometer; I.I., C.B., E.G., N.L.-G. and M.L.d.l.C. followed the experiments and advised on the questions related with nanosensors and SERS applications.

Conflicts of Interest: The authors declare no conflict of interest.

References

1. Bessbousse, H.; Nandhakumar, I.; Decker, M.; Barsbay, M.; Cuscito, O.; Lairez, D.; Clochard, M.C.; Wade, T.L. Functionalized nanoporous track-etched β-pvdf membrane electrodes for lead (ii) determination by square wave anodic stripping voltammetry. *Anal. Methods* **2011**, *3*, 1351–1359. [CrossRef]

2. Parat, C.; Authier, L.; Betelu, S.; Petrucciani, N.; Potin-Gautier, M. Determination of labile cadmium using a screen-printed electrode modified by a microwell. *Electroanalysis* **2007**, *19*, 403–406. [CrossRef]

3. Betelu, S.; Parat, C.; Petrucciani, N.; Castetbon, A.; Authier, L.; Potin-Gautier, M. Semicontinuous monitoring of cadmium and lead with a screen-printed sensor modified by a membrane. *Electroanalysis* **2007**, *19*, 399–402. [CrossRef]

4. Parat, C.; Betelu, S.; Authier, L.; Potin-Gautier, M. Determination of labile trace metals with screen-printed electrode modified by a crown-ether based membrane. *Anal. Chim. Acta* **2006**, *573*, 573–574. [CrossRef] [PubMed]

5. Buffle, J.; Tercier-Waeber, M.L. Trends in analytical chemistry. Voltammetric environmental trace-metal analysis and speciation: From laboratory to in situ measurements. *Trends Anal. Chem.* **2005**, *24*, 172–191. [CrossRef]

6. Finch, M.S.; Hydesa, D.J.; Claysona, C.H.; Weiglb, B.; Dakinc, J.; Gwilliama, P. A low power ultra violet spectrophotometer for measurement of nitrate in seawater: Introduction, calibration and initial sea trials. *Anal. Chim. Acta* **1998**, *377*, 167–177. [CrossRef]

7. Mikkelsen, Ø.; Skogvold, S.M.; Schrøder, K.H. Continuous heavy metal monitoring system for application in river and seawater. *Electroanalysis* **2005**, *17*, 431–439. [CrossRef]

8. Heglund, D.L.; Tilotta, D.C. Determination of volatile organic compounds in water by solid phase microextraction and infrared spectroscopy. *Environ. Sci. Technol.* **1996**, *30*, 1212–1219. [CrossRef]

9. Potter, D.W.; Pawliszyn, J. Rapid determination of polyaromatic hydrocarbons and polychlorinated biphenyls in water using solid-phase. *Environ. Sci. Technol.* **1994**, *28*, 298–305. [CrossRef] [PubMed]

10. Wittkamp, B.L.; Tillota, D.C. Determination of btex compounds in water by solid-phase microextraction and raman spectroscopy. *Anal. Chem.* **1995**, *67*, 600–605. [CrossRef]

11. Young, M.A.; Stuart, D.A.; Lyandres, O.; Glucksberg, M.R.; Van Duyne, R.P. Surface-enhanced raman spectroscopy with a laser pointer light source and miniature spectrometer. *Can. J. Chem.* **2004**, *82*, 1435–1441. [CrossRef]
12. Demeestere, K.; Dewulf, J.; De Witte, B.; Van Langenhove, H. Review: Sample preparation for the analysis of volatile organic compounds in air and water matrices. *J. Chromatogr. A* **2007**, *1153*, 130–144. [CrossRef] [PubMed]
13. Arthur, C.L.; Killiam, L.M.; Motlagh, S.; Lim, M.; Potter, D.W.; Pawliszyn, J. Analysis of substituted benzene compounds in groundwater using solid-phase microextraction. *Environ. Sci. Technol.* **1992**, *26*, 979–983. [CrossRef]
14. Buchholz, K.D.; Pawliszyn, J. Determination of phenols by solid-phase microextraction and gas chromatographic analysis. *Environ. Sci. Technol.* **1993**, *27*, 2844–2848. [CrossRef]
15. Zhang, Z.; Pawliszyn, J. Quantitative extraction using an internally cooled solid phase microextraction device. *Anal. Chem.* **1995**, *67*, 34–43. [CrossRef]
16. Arthur, C.L.; Pawliszyn, J. Solid phase microextraction with thermal desorption using fused silica optical fibers. *Anal. Chem.* **1990**, *62*, 2145–2148. [CrossRef]
17. Qiaoa, M.; Qia, W.; Liua, H.; Qua, J. Simultaneous determination of typical substituted and parent polycyclic aromatic hydrocarbons in water and solid matrix by gas chromatography–mass spectrometry. *J. Chromatogr. A* **2013**, *1291*, 129–136. [CrossRef] [PubMed]
18. Eisert, R.; Levsen, K. Determination of pesticides in aqueous samples by solid-phase microextraction in-line coupled to gas chromatography-mass spectrometry. *J. Am. Soc. Mass. Spectrom.* **1995**, *6*, 1119–1130. [CrossRef]
19. Roy, G.; Vuillemin, R.; Guyomarch, J. On-site determination of polynuclear aromatic hydrocarbons in seawater by stir bar sorptive extraction (SBSE) and thermal desorption GC-MS. *Talanta* **2005**, *66*, 540–546. [CrossRef] [PubMed]
20. Huang, X.; Qiu, N.; Yuan, D.; Huang, B. A novel stir bar sorptive extraction coating based on monolithic material for apolar, polar organic compounds and heavy metal ions. *Talanta* **2009**, *78*, 101–106. [CrossRef] [PubMed]
21. Brunink, J.A.J.; Di Natale, C.; Bungaro, F.; Davide, F.A.M.; D'Amico, A.; Paolesse, R.; Boschi, T.; Faccio, M.; Ferri, G. The application of metalloporphyrins as coating material for quartz microbalance-based chemical sensors. *Anal. Chim. Acta* **1995**, *325*, 53–64. [CrossRef]
22. Cygan, M.T.; Collins, G.E.; Dunbar, T.D.; Allara, D.L.; Gibbs, C.G.; Gutsche, C.D. Calixarene monolayers as quartz crystal microbalance sensing elements in aqueous solution. *Anal. Chem.* **1999**, *71*, 142–148. [CrossRef] [PubMed]
23. Harbeck, M.; Erbahar, D.D.; Gürol, I.; Musluoglu, E.; Ahsen, V.; Öztürk, Z.Z. Phthalocyanines as sensitive coatings for qcm sensors operating in liquids for the detection of organic compounds. *Sens. Actuators B* **2010**, *150*, 346–354. [CrossRef]
24. Lieberzeit, P.; Halikias, K.; Afzal, A.; Dickert, F. Polymers imprinted with pah mixtures-comparing fluorescence and qcm sensors. *Anal. Bioanal. Chem.* **2008**, *392*, 1405–1410. [CrossRef] [PubMed]
25. Lucklum, R.; Rösler, S.; Hartmann, J.; Hauptmann, P. On-line detection of organic pollutants in water by thickness shear mode resonators. *Sens. Actuators B* **1996**, *35–36*, 103–111. [CrossRef]
26. Menon, A.; Zhou, R.; Josse, F. Coated-quartz crystal resonator (qcr) sensors for on-line detection of organic contaminants in water. *IEEE Trans. Ultrason. Ferroelect. Freq. Control* **1998**, *45*, 1416–1426. [CrossRef]
27. Patel, R.; Zhou, R.; Zinszer, K.; Josse, F. Real-time detection of organic compounds in liquid environments using polymer-coated thickness shear mode quartz resonator. *Anal. Chem.* **2000**, *72*, 4888–4898. [CrossRef] [PubMed]
28. Pejcic, B.; Barton, C.; Crooke, E.; Eadington, P.; Jee, E.; Ross, A. Hydrocarbon sensing. Part 1: Some important aspects about sensitivity of a polymer-coated quartz crystal microbalance in the aqueous phase. *Sens. Actuators B* **2009**, *135*, 436–443. [CrossRef]
29. Rösler, S.; Lucklum, R.; Borngräber, R.; Hartmann, J.; Hauptmann, P. Sensor system for the detection of organic pollutants in water by thickness shear mode resonators. *Sens. Actuators B* **1998**, *48*, 415–424. [CrossRef]
30. Schierbaum, K.D.; Weimar, U.; Göpel, W. Comparison of ceramic, thick-film and thin-film chemical sensors based upon sno2. *Sens. Actuators B* **1992**, *7*, 709–716. [CrossRef]

31. Zhou, R.; Haimbodi, M.; Everhart, D.; Josse, F. Polymer-coated QRC sensors for the detection of organic solvents in water. *Sens. Actuators B* **1996**, *35*, 176–182. [CrossRef]

32. Dickert, F.L.; Lieberzeit, P.; Miarecka, S.G.; Mann, K.J.; Hayden, O.; Palfinger, C. Synthetic receptors for chemical sensors-subnano- and micrometre patterning by imprinting techniques. *Biosens. Bioelectron.* **2004**, *20*, 1040–1044. [CrossRef] [PubMed]

33. Colas, F.J.; Cottat, M.; Gillibert, R.; Guillot, N.; Djaker, N.; Lidgi-Guigui, N.; Toury, T.; Barchiesi, D.; Toma, A.; Di Fabrizio, E.; et al. Red-shift effects in surface enhanced raman spectroscopy: Spectral or intensity dependence of the near-field? *J. Phys. Chem. C* **2016**, *120*, 13675–13683. [CrossRef]

34. Grand, J.; de la Chapelle, M.L.; Bijeon, J.L.; Adam, P.M.; Vial, A.; Royer, P. Role of localized surface plasmons in surface-enhanced raman scattering of shape-controlled metallic particles in regular arrays. *Phys. Rev. B* **2005**, *72*, 033407. [CrossRef]

35. Schatz, G.C.; Van Duyne, R.P. Electromagnetic mechanism of surface enhanced raman spectroscopy. In *Handbook of Vibrational Spectroscopy*; Chalmers, J.M., Griffiths, P.R., Eds.; Wiley: Hoboken, NJ, USA, 2002; Volume 1, p. 759.

36. Wokaun, A. Surface-enhanced electromagnetic processes. *Solid State Phys.* **1984**, *38*, 223–294.

37. Guillot, N.; de la Chapelle, M.L. The electromagnetic effect in surface enhanced raman scattering: Enhancement optimization using precisely controlled nanostructures. *J. Quant. Spectrosc. Radiat. Transf.* **2012**, *113*, 2321–2333. [CrossRef]

38. Cottat, M.; Lidgi-Guigui, N.; Hamouda, F.; Bartenlian, B.; Venkataraman, D.; Marks, R.S.; Steele, T.W.S.; Lamy de la Chapelle, M. Highly sensitive detection of paclitaxel by surface-enhanced raman scattering. *J. Opt.* **2015**, *17*, 114019. [CrossRef]

39. Cottat, M.; D'Andrea, C.; Yasukuni, R.; Malashikhina, N.; Grinyte, R.; Lidgi-Guigui, N.; Fazio, B.; Sutton, A.; Oudar, O.; Charnaux, N.; et al. High sensitivity, high selectivity sers detection of mnsod using optical nanoantennas functionalized with aptamers. *J. Phys. Chem. C* **2015**, *119*, 15532–15540. [CrossRef]

40. Albrecht, M.G.; Creighton, J.A. Intense raman spectra of pyridine at a silver electrode. *J. Am. Chem. Soc.* **1977**, *99*, 5215–5217. [CrossRef]

41. Fleischmann, M.; Hendra, P.J.; McQuillan, A.J. Raman spectra of pyridine adsorbed at a silver electrode. *Chem. Phys. Lett.* **1974**, *26*, 163–166. [CrossRef]

42. Jeanmaire, D.L.; Van Duyne, R.P. Surface raman spectroelectrochemistry: Part I. Heterocyclic, aromatic, and aliphatic amines adsorbed on the anodized silver electrode. *J. Electroanal. Chem. Interfacial Electrochem.* **1977**, *84*, 1–20. [CrossRef]

43. Kambhampati, P.; Child, C.M.; Foster, M.C.; Campion, A. On the chemical mechanism of surface enhanced raman scattering: Experiment and theory. *J. Chem. Phys.* **1998**, *108*, 5013–5026. [CrossRef]

44. Guillot, N.; Lamy de la Chapelle, M. Lithographied nanostructures as nanosensors. *J. Nanophotonics* **2012**, *6*, 064506–064528. [CrossRef]

45. Kneipp, K.; Wang, Y.; Kneipp, H.; Perelman, L.T.; Itzkan, I.; Dasari, R.R.; Feld, M.S. Single molecule detection using surface-enhanced raman scattering (sers). *Phys. Rev. Lett.* **1997**, *78*, 1667–1670. [CrossRef]

46. Nie, S.; Emory, S.R. Probing single molecules and single nanoparticles by surface-enhanced raman scattering. *Science* **1997**, *275*, 1102–1106. [CrossRef] [PubMed]

47. Schmidt, H.; Bich Ha, N.; Pfannkuche, J.; Amann, H.; Kronfeldt, H.-D.; Kowalewska, G. Detection of pahs in seawater using surface-enhanced raman scattering (sers). *Mar. Pollut. Bull.* **2004**, *49*, 229–234. [CrossRef] [PubMed]

48. Xie, Y.; Wang, X.; Han, X.; Song, W.; Ruan, W.; Liu, J.; Zhao, B.; Ozaki, Y. Selective sers detection of each polycyclic aromatic hydrocarbon (PAH) in a mixture of five kinds of pahs. *J. Raman Spectrosc.* **2011**, *42*, 945–950. [CrossRef]

49. Otto, A.; Mrozek, I.; Grabhorn, H.; Akemann, W. Surface-enhanced raman scattering. *J. Phys. Condens. Matter* **1992**, *4*, 1143. [CrossRef]

50. Peron, O.; Rinnert, E.; Lehaitre, M.; Crassous, P.; Compere, C. Detection of polycyclic aromatic hydrocarbon (pah) compounds in artificial sea-water using surface-enhanced raman scattering (sers). *Talanta* **2009**, *79*, 199–204. [CrossRef] [PubMed]

51. Allan, I.J.; Vrana, B.; Greenwood, R.; Mills, G.A.; Roig, B.; Gonzales, C. A "toolbox" for biological and chemical monitoring requirements for the european union's water framework directive. *Talanta* **2006**, *69*, 302–322. [CrossRef] [PubMed]

52. Peron, O.; Rinnert, E.; Florent, C.; Lehaitre, M.; Compere, C. First steps of in situ surface-enhanced raman scattering during shipboard experiments. *Appl. Spectrosc.* **2010**, *64*, 1086–1093. [CrossRef] [PubMed]

53. Peron, O.; Rinnert, E.; Toury, T.; Lamy de la Chapelle, M.; Compère, C. Quantitative sers sensors for environmental analysis of naphthalene. *Analyst* **2011**, *136*, 1018–1022. [CrossRef] [PubMed]

54. Prien, R.D. The future of chemical in situ sensors. *Mar. Chem.* **2007**, *107*, 422–432. [CrossRef]

55. Rogers, J.A.; Jackman, R.J.; Whitesides, G.M. Constructing single- and multiple-helical microcoils and characterizing their performance as components of microinductors and microelectromagnets. *J. Microelectromech. Syst.* **1997**, *6*, 184–192. [CrossRef]

56. Ulman, A. Formation and structure of self-assembled monolayers. *Chem. Rev.* **1996**, *96*, 1533–1554. [CrossRef] [PubMed]

57. Zielinski, O.; Busch, J.A.; Cembella, A.D.; Daly, K.L.; Engelbrektsson, J.; Hannides, A.K.; Schmidt, H. Detecting marine hazardous substances and organisms: Sensors for pollutants, toxins, and pathogens. *Ocean Sci. Discuss.* **2009**, *6*, 953–1005. [CrossRef]

58. Chow, E.; Hibbert, D.B.; Gooding, J.J. Voltammetric detection of cadmium ions at glutathione-modified gold electrodes. *Analyst* **2005**, *130*, 831–837. [CrossRef] [PubMed]

59. Delamarche, E.; Michel, B.; Kang, H.; Gerber, C. Thermal stability of self-assembled monolayers. *Langmuir* **1994**, *10*, 4103–4108. [CrossRef]

60. Horn, A.B.; Russell, D.A.; Shorthouse, L.J.; Simpson, T.R.E. Ageing of alkanethiol self-assembled monolayers. *J. Chem. Soc.* **1996**, *92*, 4759–4762. [CrossRef]

61. Betelu, S.; Tijunelyte, I.; Boubekeur-Lecaque, L.; Ignatiadis, I.; Ibrahim, J.; Gaboreau, S.; Berho, C.; Toury, T.; Guenin, E.; Lidgi-Guigui, N.; et al. Evidence of the grafting mechanisms of diazonium salts on gold nanostructures. *J. Phys. Chem. C* **2016**, *120*, 18158–18166. [CrossRef]

62. Betelu, S.; Vautrin-Ul, C.; Chaussé, A. Novel 4-carboxyphenyl-grafted screen-printed electrode for trace cu(ii) determination. *Electrochem. Commun.* **2009**, *11*, 383–386. [CrossRef]

63. Betelu, S.; Vautrin-Ul, C.; Ly, J.; Chaussé, A. Screen-printed electrografted electrode for trace uranium analysis. *Talanta* **2009**, *80*, 372–376. [CrossRef] [PubMed]

64. Delamar, M.; Hitmi, R.; Pison, J.; Savéan, J. Covalent modification of carbon surfaces by grafting of functionalized aryl radicals produced from electrochemical reduction of diazonium salts. *J. Am. Chem. Soc.* **1992**, *114*, 5883–5884. [CrossRef]

65. Adenier, A.; Cabet-Deliry, E.; Chaussé, A.; Griveau, S.; Mercier, F.; Pinson, J.; Vautrin-Ul, C. Grafting of nitrophenyl groups on carbon without electrochemical induction. *Chem. Mater.* **2005**, *17*, 491–501. [CrossRef]

66. Bekyarova, E.; Itkis, M.E.; Ramesh, P.; Berger, C.; Sprinkle, M.; de Heer, W.A.; Haddon, R.C. Chemical modification of epitaxial graphene: Spontaneous grafting of aryl groups. *J. Am. Chem. Soc.* **2009**, *131*, 1336–1337. [CrossRef] [PubMed]

67. Combellas, C.M.; Delamar, F.; Kanoufi, F.; Pinson, J.; Podvorica, F.I. Spontaneous grafting of iron surfaces by reduction of aryldiazonium salts in acidic or neutral aqueous solution. Application to the protection of iron against corrosion. *Chem. Mater.* **2005**, *17*, 3968–3975. [CrossRef]

68. Laurentius, L.; Stoyanov, S.R.; Gusarov, S.; Kovalenko, A.; Du, R.; Lopinski, G.P.; McDermott, M.T. Diazonium-derived aryl films on gold nanoparticles: Evidence for a carbon–gold covalent bond. *ACS Nano* **2011**, *5*, 4219–4227. [CrossRef] [PubMed]

69. Griveau, S.; Mercier, D.; Vautrin-Ul, C.; Chaussé, A. Electrochemical grafting by reduction of 4-aminoethylbenzenediazonium salt: Application to the immobilization of (bio)molecules. *Electrochem. Commun.* **2007**, *9*, 2768–2773. [CrossRef]

70. Manoli, E.; Samara, C. Polycyclic aromatic hydrocarbons in natural waters: Sources, occurrence and analysis. *Trends Anal. Chem.* **1999**, *18*, 417–428. [CrossRef]

71. Eilers, P.H.C. A perfect smoother. *Anal. Chem.* **2003**, *75*, 3631–3636. [CrossRef] [PubMed]

72. Moreau, J.; Rinnert, E. Fast identification and quantification of btex coupling by raman spectrometry and chemometrics. *Analyst* **2015**, *140*, 3535–3542. [CrossRef] [PubMed]

73. Busson, M.; Berisha, A.; Combellas, C.; Kanoufi, F.; Pinson, J. Photochemical grafting of diazonium salts on metals. *Chem. Commun.* **2011**, *47*, 12631–12633. [CrossRef] [PubMed]

74. Bouriga, M.; Chehimi, M.M.; Combellas, C.; Decorse, P.; Kanoufi, F.; Deronzier, A.; Pinson, J. Sensitized photografting of diazonium salts by visible light. *Chem. Mater.* **2013**, *25*, 90–97. [CrossRef]

75. Abiman, P.; Wildgoose, G.G.; Compton, R.G. Investigating the mechanism for the covalent chemical modification of multiwalled carbon nanotubes using aryl diazonium salts. *Int. J. Electrochem. Sci.* **2008**, *3*, 104–117.

76. Ahmad, R.; Boubekeur-Lecaque, L.; Nguyen, M.; Lau-Truong, S.; Lamouri, A.; Decorse, P.; Galtayries, A.; Pinson, J.; Felidj, N.; Mangeney, C. Tailoring the surface chemistry of gold nanorods through au–c/ag–c covalent bonds using aryl diazonium salts. *J. Phys. Chem. C* **2014**, *118*, 19098–19105. [CrossRef]

77. Qu, L.-L.; Li, Y.-T.; Li, D.-W.; Xue, J.-Q.; Fossey, J.S.; Long, Y.-T. Humic acids-based one-step fabrication of sers substrates for detection of polycyclic aromatic hydrocarbons. *Analyst* **2013**, *138*, 1523–1528. [CrossRef] [PubMed]

78. Tijunelyte, I.; Dupont, N.; Milosevic, I.; Barbey, C.; Rinnert, E.; Lidgi-Guigui, N.; Guenin, E.; de la Chapelle, M. Investigation of aromatic hydrocarbon inclusion into cyclodextrins by raman spectroscopy and thermal analysis. *Environ. Sci. Pollut. Res.* **2015**, 1–13. [CrossRef] [PubMed]

79. International Union of Pure and Applied Chemistry. *Compendium of Chemical Terminology*; Blackwell Science: Oxford, UK, 1997.

80. Guideline, I.H.T. Validation of Analytical Procedures: Text and Methodology. Available online: https://www.ich.org/fileadmin/Public_Web_Site/ICH_Products/Guidelines/Quality/Q2_R1/Step4/Q2_R1__Guideline.pdf (accessed on 24 May 2017).

81. Na, W.; Hai-Feng, Y.; Xuan, Z.; Rui, Z.; Yao, W.; Guan-Feng, H.; Zong-Rang, Z. Synthesis of anti-aggregation silver nanoparticles based on inositol hexakisphosphoric micelles for a stable surface enhanced raman scattering substrate. *Nanotechnology* **2009**, *20*, 315603.

82. Fu, S.; Guo, X.; Wang, H.; Yang, T.; Wen, Y.; Yang, H. Functionalized au nanoparticles for label-free raman determination of ppb level benzopyrene in edible oil. *Sens. Actuators B* **2015**, *212*, 200–206. [CrossRef]

83. Zheng, Y.; Thai, T.; Reineck, P.; Qiu, L.; Guo, Y.; Bach, U. DNA-directed self-assembly of core-satellite plasmonic nanostructures: A highly sensitive and reproducible near-ir sers sensor. *Adv. Funct. Mater.* **2013**, *23*, 1519–1526. [CrossRef]

![sensors logo] *sensors* MDPI

Article

Development of Diamond and Silicon MEMS Sensor Arrays with Integrated Readout for Vapor Detection

Maira Possas-Abreu [1,*], Farbod Ghassemi [1], Lionel Rousseau [1], Emmanuel Scorsone [2], Emilie Descours [3] and Gaelle Lissorgues [1]

[1] ESYCOM, ESIEE-Paris, Cité Descartes BP99, 93162 Noisy-le-Grand, France; farbod.ghassemi@esiee.fr (F.G.); lionel.rousseau@esiee.fr (L.R.); gaelle.lissorgues@esiee.fr (G.L.)
[2] CEA, LIST, Diamond Sensor Laboratory, 91191 Gif-sur-Yvette, France; emmanuel.scorsone@cea.fr
[3] ISIPCA, 34–36 Rue du Parc de Clagny, 78000 Versailles, France; edescours@isipca.fr
* Correspondence: maira.possasabreu@esiee.fr; Tel.: +33-1-45-92-66-96

Academic Editor: Stefano Mariani
Received: 31 March 2017; Accepted: 12 May 2017; Published: 24 May 2017

Abstract: This paper reports on the development of an autonomous instrument based on an array of eight resonant microcantilevers for vapor detection. The fabricated sensors are label-free devices, allowing chemical and biological functionalization. In this work, sensors based on an array of silicon and synthetic diamond microcantilevers are sensitized with polymeric films for the detection of analytes. The main advantage of the proposed system is that sensors can be easily changed for another application or for cleaning since the developed gas cell presents removable electrical connections. We report the successful application of our electronic nose approach to detect 12 volatile organic compounds. Moreover, the response pattern of the cantilever arrays is interpreted via principal component analysis (PCA) techniques in order to identify samples.

Keywords: microcantilevers; electronic nose; VOC discrimination; gas sensors; sensor arrays; synthetic diamond

1. Introduction

A recent study revealed that humans can discriminate among more than a trillion different smells [1] and the mammalian nose remains the primary "apparatus" used in many applications to evaluate the smell of products. Despite the recent progress in research in the field, the mammalian olfactory system is complex and mechanisms of olfaction are still not fully understood [2]. Driven by the needs of odour detection for medical applications, environmental monitoring, security or food monitoring, the development of electronic noses has increased over the years [3–9]. These sensing technologies operate by mimicking the manner that mammalian noses proceed to discriminate odorant volatile compounds. The first study in this field [10] reported that a system aiming to mimic the mammalian olfactory system may be composed of two main elements: roughly tuned receptor cells, not selective toward specific odorant molecules and a system capable of performing parallel processing of the output signals. The processing may include qualitative analysis of sensor signal reports, by using, for example, pattern recognition techniques. Since then, various electronic noses have been developed based on different sensor technologies and different identification and classification methods [11].

Multi-gas detection using portable systems requires the use of sensors adapted for this specific application, especially in terms of sensitivity, response time and recovery time, selectivity, application to a wide range of gases, simplicity and convenience with respect to the use and replacement of sensors, and equally important, the sensor life-time. To meet this demand, a large number of technologies, based on optical, mechanical and electrical techniques, have been developed for chemical transduction.

Regarding the choice of these sensor techniques, solutions are distinguished by the type of sensitive layer and the principle of transduction. The most used for electronic nose applications are: semiconductor metal oxides gas sensors, conducting polymer sensors [12], Surface Acoustic Wave (SAW) sensors [13], Quartz Crystal Microbalance (QCM) [14] and optical fibre sensors [15]. Gas sensors differ in size, sensitivity, operating temperature, response and recovery times.

Electronic noses technologies offer a cheaper alternative to existing analytical instruments such as gas chromatography, mass spectrometry or ion mobility spectrometry [16]. They are supposed to be an alternative mobile or transportable and easy to use. In this context, the size and the number of sensors are important parameters, not only to obtain a smaller instrument but also to promote the use of smaller gas volumes, smaller detection surfaces and shorter detection times.

In the last two decades, advances in the micro-electromechanical systems (MEMS) field have promoted the development of miniaturized sensors that are able to transduce mechanical energy (e.g., gravitational potential energy) to electrical energy. The operation principle of MEMS sensors is that chemical, physical or biological stimuli can be transduced to mechanical stimuli and affect mechanical characteristics of the sensor structure in a manner that these changes can be measured by electrical or optical means. In this context, bio-chemical detection is possible by measuring mass or surface stress changes. In particular, microcantilevers, the simplest MEMS structures, offer the possibility of label-free biochemical detection with very high sensitivity [17]. The sensitivity of resonant microcantilevers is related to the dimensional scale of these devices.

In this perspective, a microcantilever-based electronic nose can offer highly desirable characteristics, including fast responses, height sensitivity and being able to accommodate a large number of sensors in a small volume. Moreover, they are suitable for mass production, taking advantage of micro-machining techniques and circuit integration.

In this work, we report the fabrication and the development of a silicon and synthetic diamond microcantilever array-based electronic nose. In the MEMS field, silicon is widely used. In fact, processing methods such as etching and photolithography have been thoroughly developed by the electronics industry, and have been easily adapted for MEMS production. As a consequence, the development of new techniques has not been favoured. Nonetheless, diamond is expected to be a very promising alternative in the micro-sensors field. In fact, diamond is a highly suitable material for the manufacture of resonant microcantilevers because of its exceptional mechanical and thermal properties, biocompatibility as well as excellent hardness and robustness. Polycrystalline diamond can be an excellent choice for the fabrication of resonant sensors due to its high elasticity modulus (in the order of 10^3 GPa [18]). Moreover, because of its carbon nature, this material is convenient for stable grafting of a wide range of bio-receptors by covalent C–C binding [19].

The originality of the study reported in this paper is related to the development of a complete modular and autonomous system which is designed from sensors (silicon and diamond) to signal processing to be low noise, sensitive and easy-to-operate. In order to increase the sensitivity of our microcantilever sensors to volatile organic compounds detection, a variety of polymer coatings has been used to coat microcantilever surfaces. These sensors present mass resolution down to the ng range. Finally, we report the successful application of this electronic nose approach to discriminate some volatile organic compounds.

2. Material and Methods

2.1. Microcantilevers

The MEMS sensor presented in this paper consists on an array of independent microcantilevers attached to a chip (2 cm × 5 cm) in which we can find electrical pads. We have fabricated silicon and polycrystalline diamond cantilevers using same geometry (same masks) in order to use in the functionalization step, a priori, silicon cantilevers for polymer coatings and diamond cantilevers for direct proteins binding. As diamond sensors have presented higher mass sensitivity than silicon ones, they have also been used with polymer coatings. The fabrication process of silicon and diamond cantilevers takes advantage of thin film technology and surface micro-machining techniques. In both cases, three metallic contact pads allow electrical connection to the integrated pair of poly-silicon strain gauges that serve as transducer elements. The piezoresistors in the cantilevers are connected to form a Wheatstone half-bridge circuit per chip as shown in Figure 1a. A silicon-based sensor is illustrated in Figure 1b.

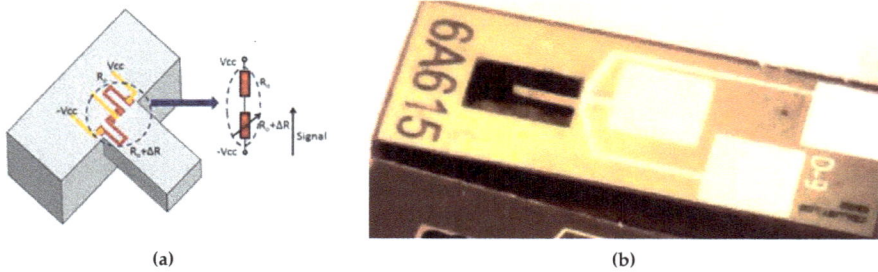

(a)　　　　　　　　　　　　　　　　　　　　　　　(b)

Figure 1. (**a**) Schematic of the placement of the piezoresistive gauges at the microcantilever anchorage and its equivalent circuit (**b**) A microscope picture of a silicon microcantilever with the three contact pads.

The fabrication of silicon cantilevers follows well-known techniques using a silicon-on-insulator wafer (SOI) as substrate [20]. In order to produce diamond microcantilevers, a novel polycrystalline diamond structuration method was developed and sensors were fabricated following a process previously described in [18]. Briefly, the fabrication process shown in Figure 2 starts with a four-inch single-side-polished silicon substrate. The wafer is thermally oxidized in order to create an electrical insulation layer (step 1). The strain gauges are created by sputter-deposition of polysilicon which is then patterned by etching (step 2). In order to prepare the wafer-to-diamond growth steps, a layer of tungsten has been deposited and patterned serving as an etch stop layer for a later etching step of synthetic diamond (step 3). In this work, diamond layer is structured by chemical vapor deposition (CVD). This technique consists of depositing carbon atoms over a substrate from methane gas using specific concentration, pressure and temperature conditions. Since polycrystalline diamond does not grow spontaneously on non-diamond materials, we seeded diamond nanoparticles over the substrate before proceeding to CVD steps. Nano-seeding of diamond is realized by incorporating a solution of diamond nanoparticles in Poly-vinyl alcohol (PVA) and by spin coating the solution over the substrate (step 4) using the process described in [21]. The wafer is then submitted to a Microwave Plasma Enhanced Chemical Vapor Deposition (MPECVD) reactor to synthesize diamond (step 5). After step 5, one can note that the diamond film is not homogeneous. The thickness varies over the wafer and on the zones protected by tungsten, the structure of diamond is fragile and thinner. In order to remove this layer, aluminium is sputter-deposited and structured using photolithography as a masking layer in opposition to the tungsten layer to generate the microcantilever shape and openings to access strain gauges (step 6). Subsequently, exposed diamond is removed by the Deep Reactive Ion Etching (DRIE) process (step 7). The aluminium mask and tungsten etch-stop layer are chemically removed (step 8). Immediately thereafter, the electrical contacts (chromium and gold) are structured by standard photolithography and etching techniques (step 9). Another aluminium masking layer is used to release the microcantilevers by performing DRIE etching of silicon on the wafer back side (steps 10 and 11). Diamond cantilevers have been designed to resonate at similar frequencies than silicon cantilevers of same length and width which means that diamond microcantilevers are thinner than the silicon ones.

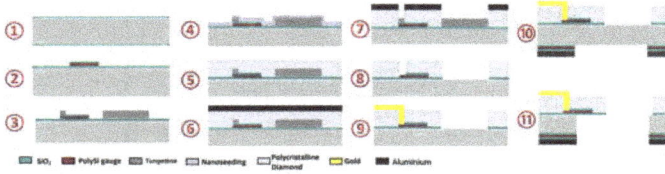

Figure 2. Polycrystalline diamond cantilever fabrication process. Fabrication steps start by structuring polysilicon gauges in order to place them at the bottom of the cantilever structure. Such a configuration is necessary to prevent the crystalline diamond layer from oxidising, which could occur during a subsequent step of depositing polysilicon. A layer of tungsten has been deposited to avoid diamond growth in the protected zones, since the diamond nanoparticles are spread on the whole wafer.

Our sensors have been conceived to operate in dynamic mode to detect changes of the device's mass due to adsorption of analytes on the microcantilever surface. For a microcantilever uniformly loaded on one side, as is the case of cantilevers coated with a sensitive layer, the mass change can be calculated from the measured resonance frequency shift using the following equation (model of harmonic oscillator):

$$\Delta m = \frac{k}{4\pi^2} \left(\frac{1}{f_2^2} - \frac{1}{f_1^2} \right) \tag{1}$$

where Δm is the mass variation of the sensor, or the adsorbed mass, k is its spring constant, and f_2 and f_1 are the final and initial frequency, respectively. The equation is valid when coating layers does not significantly change the cantilever spring constant. In order to characterize our MEMS devices, resonance frequency measurements have been performed using a Micro Scanning Laser Doppler vibrometer (Polytec). Grain size and morphology of the fabricated cantilevers have been verified using scanning electron microscopy (SEM). The average size of the diamond grain is 1 μm. Regarding frequency profiles characterizations, sensor responses are comprised between 20 kHz and 150 kHz for all geometries. This range of values corresponds to cantilevers of different lengths (five geometries). The thickness of diamond films can also vary over the wafer, producing sensors with different resonance frequency. After fabrication and measurement tests, cantilevers were selected regarding their quality factor (greater or equal to 600) and mass sensitivity (in the range of hundreds of Hz/ng). Figure 3 shows some examples of SEM images of diamond microcantilevers of different geometries (A,B,C) and a photo of a diamond cantilever during one of the rising steps (D).

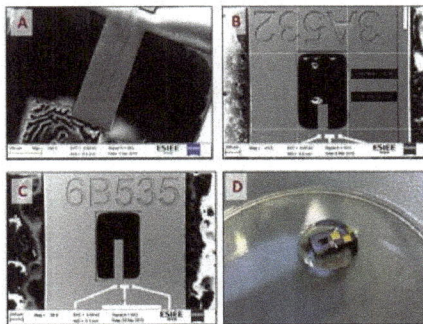

Figure 3. Pictures of some of manufactured devices: (**a**) SEM image of a beam Diamond L = 360 μm, (**b**) Same beam (**a**) with lower magnification (**c**) SEM image of a diamond beam L = 660 μm (**d**) Photography of a diamond cantilever during one of the rinsing steps.

2.2. Microcantilever Functionalization

Silicon microcantilevers conceived to be used as chemical sensors must be adapted so that their surface acquires high affinity to the target analyte. Many types of coating layers can be used to increase cantilevers chemical sensitivity and selectivity. Noble metal coating layers have been used because they can provide surfaces which can be modified to bind biological or synthetic receptors [22,23]. These materials can also be used for chemical detections, particularly for gases such as hydrogen and mercury, for which we can use palladium and gold coatings respectively [24,25]. In the field of vapor detection, gas sensors are conceived to detect complex volatile organic compound (VOC) mixtures. Many studies have demonstrated that the use of polymeric materials in the form of thick films can increase the sensitivity of mechanical sensors to several volatile organic compounds [26–28]. Inorganic coatings such as zeolithes have also been used for VOC detection [29].

In terms of diamond chemical sensors, several possibilities can be envisaged to increase sensor sensitivity. Because of their carbon nature, diamond sensors are good candidates for stable grafting of a wide range of biological receptors via covalent C--C binding [30–33]. In addition to present excellent chemical properties, diamond is well rated for sensors development due to its very high hardness and its inertness. As a bulk material, diamond sensors can also be used with other types of coatings as well as silicon sensors.

In this study, polymeric materials have been used as thin film coatings for silicon and diamond microcantilevers. Polymer solutions were deposited on diamond and silicon microcantilevers by a spray coating and in some cases, by spreading a droplet. A commercial airbrush (Evolution Silverline FPC, Harder & Steenbeck GMBH & CO., Norderstedt, Germany) was the equipment used for spray coating. The devices were cleaned prior to the polymer solution deposition (deionized water). Moreover, distance between sample and nozzle together with the pressure have been optimized to allow reproducible layer deposition and avoid the formation of droplets. Freshly coated microcantilevers were brought to 40 °C in an oven for 30 min in order to evaporate solvents. Film thickness was controlled by varying the concentration of the polymer solution and the number of depositions steps. Figure 4a presents the results for the characterization of the spray coating process. A shadow mask was fabricated to determine the area to be exposed (Figure 4b).

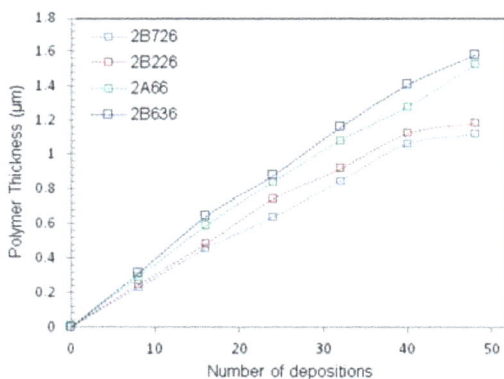

Figure 4. (a) Polymer film thickness as a function of the number of depositions for four microcantilevers of width = 140 μm and length = 260 μm (b) Schematic of the spray coating technique using a shadow mask to limit the area to be exposed over the microcantilever.

Film thickness was measured using a mechanical profilometer. Because microcantilevers are flexible, it is not possible to measure polymer thickness on the microcantilever surface. Therefore, film thickness can be measured over the immobile part of the device thanks to the shadow mask corners.

Sensors **2017**, *17*, 1163

The validity of thickness measurements was confirmed by estimations of the deposit mass of polymer via the measurements of resonance frequency shifts of bare and coated microcantilevers. The thickness of the layer can be calculated from the estimated added mass, the density of the polymer and the area of the coated surface. Figure 5 presents a SEM image of a silicon microcantilever coated with poly(epichlorohydrin) (PECH) polymer.

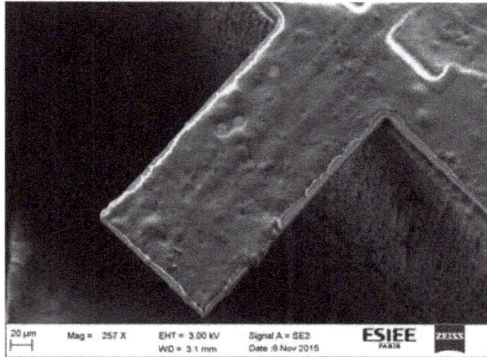

Figure 5. Silicon microcantilever coated with a film of polyepichlorohydrin (PECH) polymer by spray coating. The polymer solution was prepared by dissolving PECH on tetrahydrofuran (THF) at less than 5% of the mass of THF. Final polymer thickness = 2.42 μm.

The impact of coating on microcantilever sensors regarding energy losses depends on the geometry and the material of the sensor. We have observed that, for polymer thin layers (low added mass), the quality factor does not present significant changes for silicon cantilevers as can be observed in Figure 6. In the case of diamond sensors, the quality factor seems to increase when we add mass. In fact, this behaviour remains unexplained. One possible explanation is that as diamond cantilevers present a higher spring constant (due to the high elastic modulus), the added polysilicon acts like an added mass to a resonant system and does not degrade the equivalent elastic modulus. Thus, the quality factor of the systems may increase with increasing of mass.

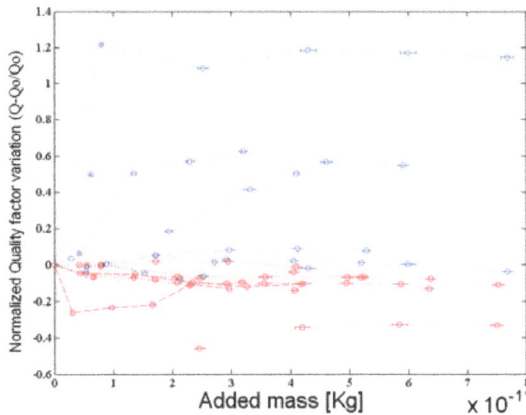

Figure 6. Relative changes in quality factor for diamond (**blue**) and silicon (**red**) for microcantilevers of different geometries.

2.3. Gas Cell and Electronic System

Detection of vapors using MEMS sensors relies on real-time measurements of cantilever deflection or cantilever resonance properties with high accuracy. Therefore, actuation and readout systems are a very important part of the development of resonant microcantilever-based sensors. The first aspect to consider is a practical aspect: microcantilevers must be placed inside a hermetically sealed gas cell which may also allow the direct actuation and read-out of the sensors. Another important characteristic of the cell is its volume size and geometry. In order to reduce time of detection, it is necessary to work with small volumes of gas. Internal geometry may also be optimized to avoid dead corners and provide a homogeneous gas flow.

In this context, a customized gas analysis cell has been conceived to accommodate up to eight sensors. The cell consists of a stainless steel chamber which presents eight cavities to accommodate sensors. A 2 mm rigid conduit is attached to the chamber serving as a gas inlet/outlet (Figure 7). Sensor chips are organized radially on a removable part (shown in green in Figure 7). A piezoelectric cell placed under the removable part is used to excite each microcantilever to its resonance frequency (first transverse mode). The design and volume of the chamber was optimized to create a homogeneous flow over all sensors and to avoid the dead corners using a minimum sample volume of gas. The cell is hermetically sealed and electrical contacts to the sensors are provided by using 24 spring-loaded pins (eight times three pads per cantilever). These pins are soldered to a PCB board screwed on the chamber cover. In this configuration, no wire-bounding is needed and sensors can be easily and individually changed for maintenance or for use in another application.

Figure 7. Three-dimensional drawing of the gas analysis cell showing the MEMS cantilever sensors placed inside (**left**) and a photo of the gas cell with eight diamond sensors placed inside (**right**). The internal volume of the cell is 1 cm^3, small enough to ensure homogeneity of the gas inside. Furthermore, the main advantage of this gas cell is that it is easily adaptable to other applications, since it is possible to easily exchange one or more sensors in order to take account of the gases detected (patented).

In order to properly interface our microcantilevers, we need to have an autonomous and stable electronic circuit that can detect very small changes in sensor resonance frequency (typically a few hertz in tens of kilohertz) over very long periods of time (minutes to hours). In recent years, the need to design superior instruments for electronic nose applications has drawn researchers attention to the development of new solutions [34,35]. In this context, we propose a low-noise and reconfigurable system as well as a dedicated human–machine interface. The major requirements in the design of this system are:

- Small dimensions: for electronic nose applications, it is not possible to use a measuring instrument such as gain-phase analyzers;
- A human–machine interface: an interface is necessary for the visualization of the data in real time and for the intervention on the measurement configurations;
- An analog processing interface to detect the responses of the sensors with sufficient resolution;

- A memory to store a certain quantity of measurements and calibration data of the sensors;
- An autonomous architecture: calibration and diagnostics executed autonomously, without user intervention;
- Communication interfaces: to communicate with other modules (pump control system, human–machine interface, etc).

With regard to the operation principle, in dynamic mode, the electronic system can operate following two methods of frequency shift detection: oscillator configuration or frequency sweep mode. In this work, the variations caused by dispersions during sensors manufacturing and the fact that we use an array of sensors make the sweep mode the best choice since it allows modularity and is easy to implement. In this operation mode, a sine-sweep signal is used to excite sensors near their resonance frequency while measuring sensors' response for each frequency. The resonance frequency (maximum amplitude response) is recorded while sensors are submitted to a reference gas in order to generate a baseline. As the sensor response is continuously monitored, when gas samples are sent to the sensors, shifts on the frequency response can be detected.

Figure 8 provides a functional diagram of the electronic signal processing architecture. The left hand-side of the figure shows the interface circuit to the cantilever sensors. The piezoresistive gauges integrated in the microcantilevers allow an electronic reading of the resonance frequency. To obtain a compensation for the effects of temperature, another gauge is integrated in the substrate (fixed part) and is used as a reference gauge. Under the operating conditions used, our microcantilevers can vibrate at amplitudes of the order of a few tens of nanometers to a few hundred nanometers. For small oscillations, it is preferable to work in Wheatstone bridge configuration. This front-end interface allows the differential reading between the voltages of each half bridge. Low noise reference voltages have been used in order to maximize the signal-to-noise ratio. The pre-amplification stage is based on eight fixed gain instrumentation amplifiers (Analog Devices - AD8428) connected to the Wheatstone bridge. In order to balance each Wheatstone bridge and optimize the dynamic range of the output signal, digitally controlled potentiometers are connected in series with sensors resistances. In order to optimize the signal-to-noise ratio, this analog front-end is mounted straight on top of the gas cell as shown in Figure 9.

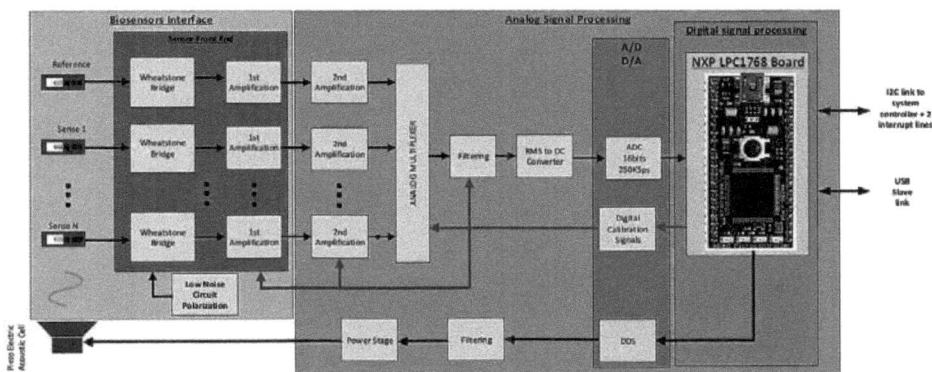

Figure 8. Functional architecture of the electronic processing circuit. The left part shows the biosensor interface corresponding to the board in Figure 9 (left). The analog signal processing circuit (right) corresponds to the board in Figure 9 (right).

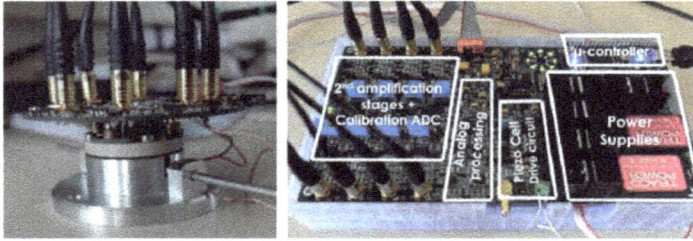

Figure 9. Gas analysis cell and electronic system—On the left: Bio-sensors interface circuit. On the right: Analog signal processing circuit.

The second part (Figure 8, right) of the electronic system starts with a second amplification stage (Analog Devices—AD8429) with digitally programmable gain. Sensor signals are then time-multiplexed in order to use a single chain circuit. The signal is then treated through a digitally programmable low-pass filter (Linear Technology—LTC1565-31) which presents a cut-off frequency ranging from 10 kHz up to 150 kHz. To detect the maximum amplitude (related to the resonance frequency), the Root Mean Square (RMS) value of the filtered signal is computed using a delta-sigma RMS-to-DC converter (Linear Technology—LTC1968). The amplitude information is converted from analog to digital and sent to an embedded microcontroller (NXP LPC1768) which correlates each sensor amplitude response to an excitation frequency and builds up the eight corresponding frequency profiles. The sensors actuation signal sent to the piezo-electrical cell is generated through a Direct Digital Synthesizer (DDS—Analog Devices—AD5932). The sensors interface and processing board have been built apart in order to simplify mechanical assembly and reduce noise injection on the sensors interface board.

A dedicated LABVIEW application was developed to ensure communication between the micro-controller and the user. Figure 10 presents a picture of the LABVIEW interface showing the real-time measurement of the frequency response profile of eight cantilevers. This interface also allows the control of filters' cut-off frequency and gain and the set-up of the frequency step and frequency span.

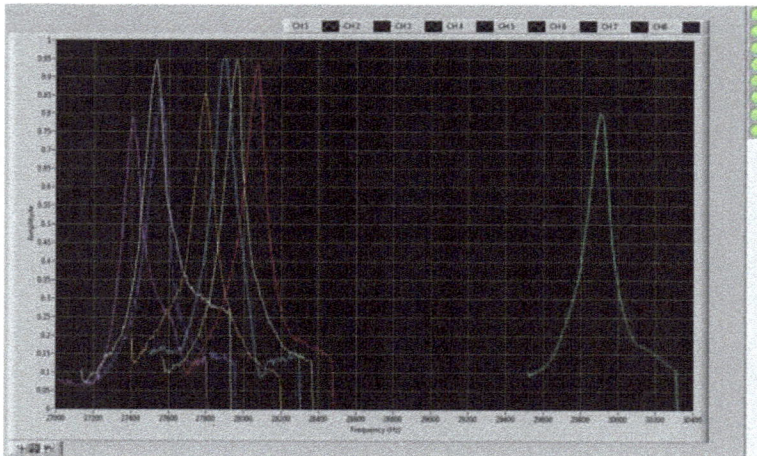

Figure 10. A picture of one of the screens of the Labview interface showing the real-time measurements of eight resonant microcantilevers. The curves are the frequency response profile of each sensor.

3. Experimental

3.1. Sensor Preparations

Eight cantilever sensors were functionalized by coating individual microcantilevers with six different polymer layers using the techniques presented in the previous section. Two diamond cantilevers have not been coated, but their surfaces were treated in order to change their hydrophobic properties. After the functionalization steps, we verified the mass changes and thickness of each cantilever in order to estimate the minimum concentration of volatile organic compounds to be used in this experiment. Table 1 summarizes the cantilever coatings used for this experiment.

Table 1. Cantilever coatings, surface treatment and resonance properties.

Cantilever	Material	Sensitive Layer (Solvent > 90%)		Resonant Frequency [Hz]
1	Silicon	PDMS	Polydimethylsiloxane	113,954
2	Diamond	Hydrophile treatment	-	28,026
3	Diamond	Hydrophobic treatment	-	31,549
4	Silicon	PMMA	Polymethylmethacrylate	127,136
5	Silicon	PAC	Poly(acetylene)	59,769
6	Silicon	PECH	Polyepichlorohydrin	126,627
7	Silicon	P(EM)MA	Poly(ethylene-co-methyl methacrylate)	124,236
8	Silicon	PIB	Polyisobutylene	125,388

3.2. Measurement Set-Up

The eight microcantilevers are placed on the gas analysis chamber in which gases and vapors can be introduced. The exposure to different VOCs was carried out by using a gas generator calibrated using a photon ionization detector (PID). Analytes in liquid phase are placed in a temperature controlled enclosure and headspace above the sample is carried in a stream of dry nitrogen gas to the measurement chamber via flow controllers and pumps. Detection set-up is completely independent on the system as can be seen in Figure 11.

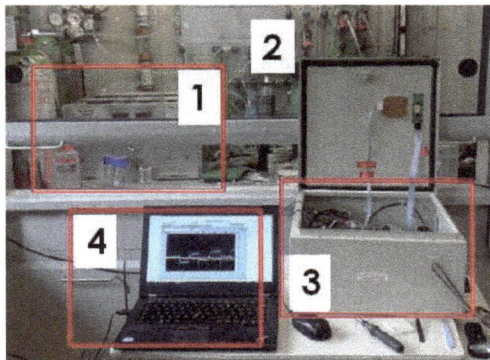

Figure 11. Picture of detection test set-up—1: volatile organic compound (VOC) generator; 2: photon ionization detector (PID) for real-time calibration; 3: the system under test; 4: PC to read-out of detection results.

Before using the sensors for the detection of vapors, we first checked the possible variations of original resonance frequency that are not related to the detection of analytes. Microcantilevers (bare or coated) are sensitive to pressure, temperature and humidity. We ensure the validity of the experiment by controlling the temperature and setting a constant flow of nitrogen at 0% RH.

Test conditions may also vary depending on the acquisition configurations of the electronic system. Indeed, the number of points (sampling step) and the frequency span of measurement are configured for each case so as to optimize the response time without deteriorating the accuracy of the measurement. This consists of reducing the frequency band to the minimum possible and also the sampling step in order to keep the maximum number of points focused on the resonance peak.

4. Results

In this experiment, the detection of vapors is achieved by the diffusion of the analyte into the polymer layer and also by surface interaction with diamond sensors (bare cantilevers). As the operation mode for this application is the dynamic mode, changes in the polymer layers lead to an increase of sensor total mass, resulting in a negative frequency shift. Shifts of resonance frequency of each microcantilever are specific to the interaction between vapor molecules and the polymer. When we change to a flow of pure nitrogen in the same temperature and flow rate conditions, the trapped analyte starts to diffuse out of the sensitive layer, back to the environment, causing a decrease of mass and a positive resonant frequency shift. Figure 12 presents some examples of cantilever-array responses in dynamic mode. Considering the frequency shift as the system output signal, the signal-to-noise ratio will depend on the sensitive layer and the substance to be detected. However, in any case, the estimated noise level (fluctuation of the output signal when the system receives reference gas) is 4 Hz.

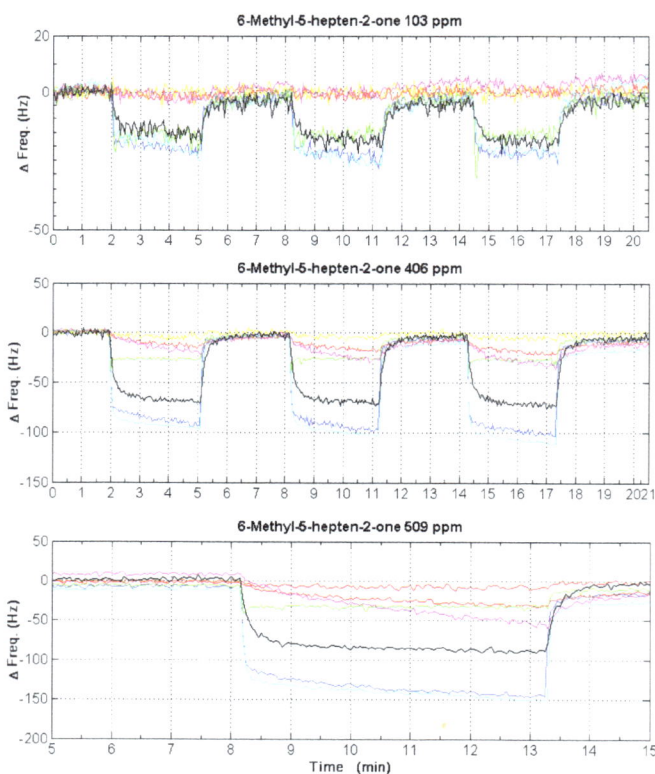

Figure 12. Cantilever responses for 6-Methyl-5-hepten-2-one at 103 ppm, 406 ppm and 509 ppm. Each concentration was generated many times in order to verify the repeatability and stability of our sensor array. Each colour in the graph represents the response of one sensor.

For each vapor tested, we have generated different concentrations in order to verify the linearity of the microcantilever sensors. Each concentration was also repeated at least three times in order to ensure that the response is reversible as shown in Figure 12.

This procedure was repeated for 13 substances : Toluene; Styrene; Pentanal; Octanal; Hexanal; Ethanol; 2-Methyl-1-propanol; Butanol; Benzadehyde; Acetone; 6-Methyl-5-hepten-2-one; Phenyl acetate; Isopropanol. Each vapor was generated at 500 ppm and sensor response (resonance frequency shift) was measured. Relative response patterns for 12 of 13 vapors are shown in Figure 13. The best sensitivity was estimated for the couple "Ppy-Phenyl Acetate" and is about 1.62 Hz/ppm. Considering the noise level of 4 Hz, the limit-of-detection in this case is 7.5 ppm. The relative response of a sensor (coating) is equal to the frequency shift of the sensor divided by the sum of the shifts for all eight sensor coatings. The sum of these scaled responses is the unity for each vapor, which facilitates comparisons. As sensor 2 was a diamond cantilever without coating, it was used as reference. Therefore, the response of this sensor is close to 0 Hz and was not represented in Figure 13. As we can observe, patterns of vapors are all different from each other. From Figure 13, we have an indication of the discriminating capability of our sensor array for this group of solvents. Some similarities can be found among the patterns from the same chemical class. For example, response patterns for ethanol and acetone are similar. One can also notice the complexity of the problem of recognizing and discriminating among more than a few vapors.

In order to identify the tested samples and evaluate the selectivity of the system, cantilever-array responses of different concentrations have been used to generate a set of "time-shift" vectors that corresponds to the dynamic development of the detection curves for each analyte. The data set was evaluated using principal component analysis (PCA) techniques, allowing to extract the most dominant deviations in the responses for the various sample vapors. In order to reduce the complexity of the analysis, we have only kept eight analytes, the best represented on the PCA first plane. As shown in Figure 14, each VOC is comprised of a cluster without any overlap. This result demonstrates the ability of the system to discriminate a large number of VOCs in dynamic mode.

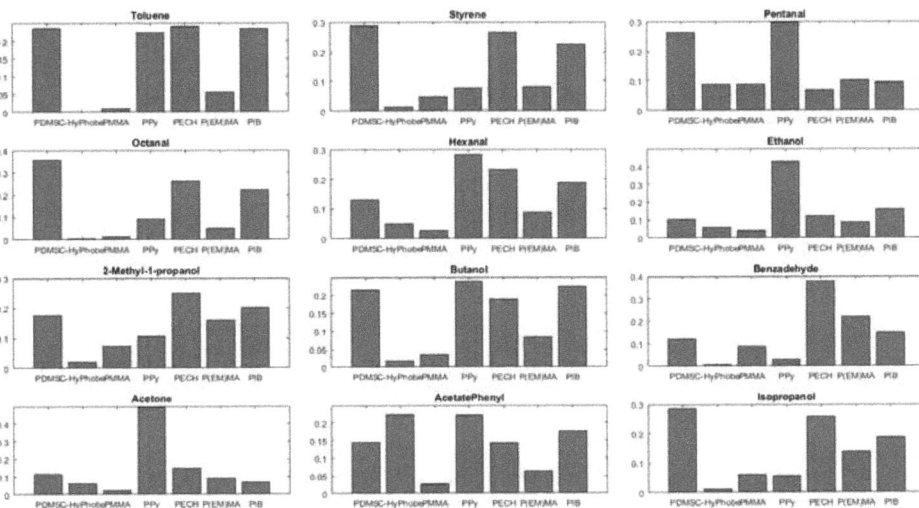

Figure 13. Relative response patterns for 12 vapors on seven microcantilevers with different coatings. The eighth one was used as reference and is not shown. The highest sensitivity was estimated for the couple "Ppy-Phenyl Acetate" and is about 1.62 Hz/ppm. For this case, LOD is estimated at 7.5 ppm.

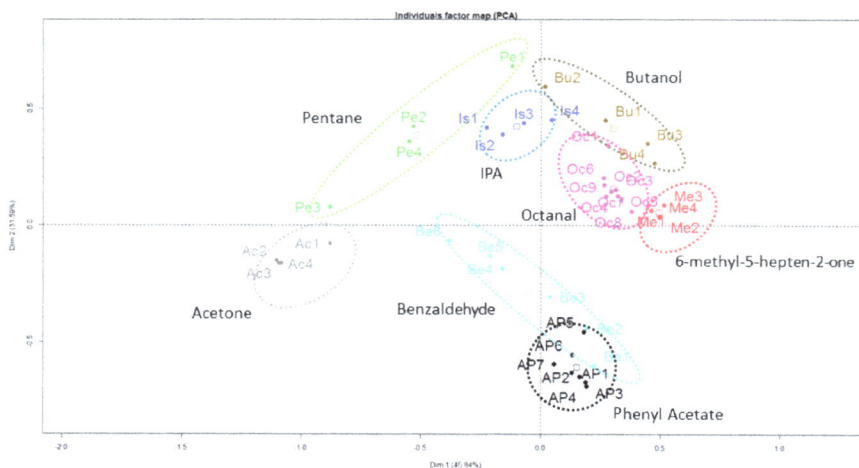

Figure 14. Principal Component Analysis (PCA) performed on results of some detected VOCs in a two-dimensional principal component space—Components 1 and 2. PCA shows good discrimination between species.

5. Conclusions

In this work, we have reported the development of a system for vapor detection. Firstly, development results address the fabrication and characterization of silicon and diamond microcantilever-based sensors for VOC detection. In this study, microcantilevers were coated with different polymer layers in order to improve the sensitivity of the sensor array. A complete three-element electronic nose system (sensors, electronics and data treatment) was developed and we have demonstrated that an array of resonant micromechanical cantilevers can be used as chemical sensors for electronic noses. The system developed is completely autonomous and modular. One of the most important results of the present work is the development of a gas analysis cell (patented) able to hold up to eight sensors which are exposed to a homogeneous gas flow while providing electrical read-out of sensors. Moreover, the gas cell also provides an easy and reliable actuation method employing a piezo-electrical cell. A dedicated electronic read-out circuit has been conceived in two different boards which provides low-noise detection of sensor output signals and accurately tracks frequency shifts. Sensor frequency profiles are updated to the user interface every second, which allows detections to be followed in real-time, during field operation. The system was used to detect 13 volatile organic compounds in the range of hundreds of ppm and PCA techniques were applied to identify samples. In conclusion, we have developed a microcantilever sensor array-based system that exhibits significant potential as a tool for vapor analysis. Improvements in the sensor quality factor, sensitivity and specificity of sensitive layers may be considered in future work in order to improve the identification and separation of samples.

Acknowledgments: This work was supported by European Community's Seventh Framework Program (FP7) under grant agreement n 285203 (FP7 SNIFFER PROJECT) and is continuing within the European project n 653323 (H2020 C-BORD PROJECT). Special thanks are due to Nadine Vallet of ISIPCA for her assistance in gas analysis and for having welcomed some experiments in her laboratory.

Author Contributions: Maira Pôssas Abreu and Lionel Rousseau conceived and fabricated the sensors. Gaelle Lissorgues, Lionel Rousseau and Maira Pôssas Abreu designed the experiments. Farbod Ghassemi and Maira Pôssas Abreu designed the electronic interface and performed experiments. Emmanuel Scorsone and Maira Pôssas Abreu designed gas related experiments. Emilie Descours was involved in the gas chromatography preliminary tests for calibration of the device and provided chemicals for gas experiments and for sensors functionalization. Farbod Ghassemi, Lionel Rousseau, Emmanuel Scorsone and Maira Pôssas Abreu designed the gas cell. Maira Pôssas Abreu and Emmanuel Scorsone analyzed data. Maïra Pôssas Abreu and Gaelle Lissorgues wrote the paper.

Conflicts of Interest: The authors declare no conflict of interest.

References

1. Bushdid, C.; Magnasco, M.O.; Vosshall, L.B.; Keller, A. Humans can discriminate more than 1 trillion olfactory stimuli. *Science* **2014**, *343*, 1370–1372.
2. Sharma, R.; Matsunami, H. Mechanisms of olfaction. In *Bioelectronic Nose*; Springer: Berlin, Germany, 2014; pp. 23–45.
3. Schaller, E.; Bosset, J.O.; Escher, F. 'Electronic noses' and their application to food. *LWT Food Sci. Technol.* **1998**, *31*, 305–316.
4. Delpha, C.; Siadat, M.; Lumbreras, M. An electronic nose for the discrimination of forane 134a and carbon dioxide in a humidity controlled atmosphere. *Sens. Actuators B Chem.* **2001**, *78*, 49–56.
5. Yinon, J. Peer reviewed: detection of explosives by electronic noses. *Anal. Chem.* **2003**, *75*, 98A.
6. Dragonieri, S.; Schot, R.; Mertens, B.J.; Le Cessie, S.; Gauw, S.A.; Spanevello, A.; Resta, O.; Willard, N.P.; Vink, T.J.; Rabe, K.F.; et al. An electronic nose in the discrimination of patients with asthma and controls. *J. Allergy Clin. Immunol.* **2007**, *120*, 856–862.
7. Lim, S.H.; Feng, L.; Kemling, J.W.; Musto, C.J.; Suslick, K.S. An optoelectronic nose for the detection of toxic gases. *Nat. Chem.* **2009**, *1*, 562–567.
8. Montuschi, P.; Santonico, M.; Mondino, C.; Pennazza, G.; Mantini, G.; Martinelli, E.; Capuano, R.; Ciabattoni, G.; Paolesse, R.; Di Natale, C.; et al. Diagnostic performance of an electronic nose, fractional exhaled nitric oxide, and lung function testing in asthma. *CHEST J.* **2010**, *137*, 790–796.
9. Wilson, A.D.; Baietto, M. Advances in electronic-nose technologies developed for biomedical applications. *Sensors* **2011**, *11*, 1105–1176.
10. Persaud, K.; Dodd, G. Analysis of discrimination mechanisms in the mammalian olfactory system using a model nose. *Nature* **1982**, *299*, 352–355.
11. Gardner, J.W.; Bartlett, P.N. *Sensors and Sensory Systems for an Electronic Nose*; Springer: Berlin, Germany, 1992.
12. Hatfield, J.; Neaves, P.; Hicks, P.; Persaud, K.; Travers, P. Towards an integrated electronic nose using conducting polymer sensors. *Sens. Actuators B Chem.* **1994**, *18*, 221–228.
13. Yang, Y.M.; Yang, P.Y.; Wang, X.R. Electronic nose based on SAWS array and its odor identification capability. *Sens. Actuators B Chem.* **2000**, *66*, 167–170.
14. El Barbri, N.; Mirhisse, J.; Ionescu, R.; El Bari, N.; Correig, X.; Bouchikhi, B.; Llobet, E. An electronic nose system based on a micro-machined gas sensor array to assess the freshness of sardines. *Sens. Actuators B Chem.* **2009**, *141*, 538–543.
15. McDonagh, C.; Burke, C.S.; MacCraith, B.D. Optical chemical sensors. *Chem. Rev.* **2008**, *108*, 400–422.
16. Lang, H.P.; Hegner, M.; Gerber, C. Cantilever array sensors. *Mater. Today* **2005**, *8*, 30–36.
17. Lavrik, N.V.; Sepaniak, M.J.; Datskos, P.G. Cantilever transducers as a platform for chemical and biological sensors. *Rev. Sci. Instrum.* **2004**, *75*, 2229–2253.
18. Possas, M.; Rousseau, L.; Ghassemi, F.; Lissorgues, G.; Scorsone, E.; Bergonzo, P. Fabrication and micromechanical characterization of polycrystalline diamond microcantilevers. In Proceedings of the 2014 Symposium on Design, Test, Integration and Packaging of MEMS/MOEMS (DTIP), Cannes, France, 1–4 April 2014; pp. 1–5.
19. Manai, R.; Scorsone, E.; Rousseau, L.; Ghassemi, F.; Abreu, M.P.; Lissorgues, G.; Tremillon, N.; Ginisty, H.; Arnault, J.; Tuccori, E.; et al. Grafting odorant binding proteins on diamond bio-MEMS. *Biosens. Bioelectron.* **2014**, *60*, 311–317.
20. Yu, Q.; Qin, G.; Darne, C.; Cai, C.; Wosik, W.; Pei, S.S. Fabrication of short and thin silicon cantilevers for AFM with SOI wafers. *Sens. Actuators A Phys.* **2006**, *126*, 369–374.
21. Scorsone, E.; Saada, S.; Arnault, J.; Bergonzo, P. Enhanced control of diamond nanoparticle seeding using a polymer matrix. *J. Appl. Phys.* **2009**, *106*, 014908.
22. Fritz, J.; Baller, M.; Lang, H.; Rothuizen, H.; Vettiger, P.; Meyer, E.; Güntherodt, H.J.; Gerber, C.; Gimzewski, J. Translating biomolecular recognition into nanomechanics. *Science* **2000**, *288*, 316–318.

23. Arntz, Y.; Seelig, J.D.; Lang, H.; Zhang, J.; Hunziker, P.; Ramseyer, J.; Meyer, E.; Hegner, M.; Gerber, C. Label-free protein assay based on a nanomechanical cantilever array. *Nanotechnology* **2002**, *14*, 86.

24. Thundat, T.; Wachter, E.; Sharp, S.; Warmack, R. Detection of mercury vapor using resonating microcantilevers. *Appl. Phys. Lett.* **1995**, *66*, 1695–1697.

25. Baselt, D.; Fruhberger, B.; Klaassen, E.; Cemalovic, S.; Britton, C.; Patel, S.; Mlsna, T.; McCorkle, D.; Warmack, B. Design and performance of a microcantilever-based hydrogen sensor. *Sens. Actuators B Chem.* **2003**, *88*, 120–131.

26. Maute, M.; Raible, S.; Prins, F.; Kern, D.; Ulmer, H.; Weimar, U.; Göpel, W. Detection of volatile organic compounds (VOCs) with polymer-coated cantilevers. *Sens. Actuators B Chem.* **1999**, *58*, 505–511.

27. Zimmermann, C.; Rebière, D.; Dejous, C.; Pistré, J.; Chastaing, E.; Planade, R. A love-wave gas sensor coated with functionalized polysiloxane for sensing organophosphorus compounds. *Sens. Actuators B Chem.* **2001**, *76*, 86–94.

28. Bietsch. A.; Zhang, J.; Hegner, M.; Lang, H.P.; Gerber, C. Rapid functionalization of cantilever array sensors by inkjet printing. *Nanotechnology* **2004**, *15*, 873.

29. Urbiztondo, M.; Peralta, A.; Pellejero, I.; Sesé, J.; Pina, M.; Dufour, I.; Santamaria, J. Detection of organic vapours with Si cantilevers coated with inorganic (zeolites) or organic (polymer) layers. *Sens. Actuators B Chem.* **2012**, *171*, 822–831.

30. Härtl, A.; Schmich, E.; Garrido, J.A.; Hernando, J.; Catharino, S.C.; Walter, S.; Feulner, P.; Kromka, A.; Steinmüller, D.; Stutzmann, M. Protein-modified nanocrystalline diamond thin films for biosensor applications. *Nat. Mater.* **2004**, *3*, 736–742.

31. Wang, J.; Carlisle, J.A. Covalent immobilization of glucose oxidase on conducting ultrananocrystalline diamond thin films. *Diamond Relat. Mater.* **2006**, *15*, 279–284.

32. Salmi, Z.; Lamouri, A.; Decorse, P.; Jouini, M.; Boussadi, A.; Achard, J.; Gicquel, A.; Mahouche-Chergui, S.; Carbonnier, B.; Chehimi, M.M. Grafting polymer–protein bioconjugate to boron-doped diamond using aryl diazonium coupling agents. *Diamond Relat. Mater.* **2013**, *40*, 60–68.

33. Manai, R.; Habchi, M.; Kamouni-Belghiti, D.; Persuy, M.; Rousseau, L.; Abreu, M.P.; Grebert, D.; Badonnel, K.; Bergonzo, P.; Pajot-Augy, E.; et al. Diamond micro-cantilevers as transducers for olfactory receptors-based biosensors: Application to the receptors M71 and OR7D4. *Sens. Actuators B Chem.* **2017**, *238*, 1199–1206.

34. Dyer, D.C.; Gardner, J.W. High-precision intelligent interface for a hybrid electronic nose. *Sens. Actuators A Phys.* **1997**, *62*, 724–728.

35. Depari, A.; Ferrari, P.; Flammini, A.; Marioli, D.; Rosa, S.; Taroni, A. A new modular approach for low-cost electronic noses. In Proceedings of the Instrumentation and Measurement Technology Conference, Sorrento, Italy, 24–27 April 2006; pp. 578–583.

Sample Availability: Samples of the fabricated silicon and polycristalline diamond sensors are available from the authors.

![sensors](sensors logo)

MDPI

Article

Highly Sensitive Sputtered ZnO:Ga Thin Films Integrated by a Simple Stencil Mask Process on Microsensor Platforms for Sub-ppm Acetaldehyde Detection

Lionel Presmanes [1,*], Yohann Thimont [1], Audrey Chapelle [2], Frédéric Blanc [2], Chabane Talhi [2], Corine Bonningue [1], Antoine Barnabé [1], Philippe Menini [2] and Philippe Tailhades [1]

[1]   CIRIMAT, Université de Toulouse, CNRS, INPT, UPS, 118 Route de Narbonne,
      F-31062 Toulouse Cedex 9, France; thimont@chimie.ups-tlse.fr (Y.T.); bonning@chimie.ups-tlse.fr (C.B.);
      barnabe@chimie.ups-tlse.fr (A.B.); tailhades@chimie.ups-tlse.fr (P.T.)
[2]   LAAS-CNRS, Université de Toulouse, UPS, INSA, 7 avenue du colonel Roche, F-31031 Toulouse, France;
      chapelle@laas.fr (A.C.); frederic.blanc@laas.fr (F.B.); chabane.talhi@laas.fr (C.T.);
      Philippe.Menini@laas.fr (P.M.)
*   Correspondence: presmane@chimie.ups-tlse.fr; Tel.: 05-6155-7751; Fax: 05-6155-6163

Academic Editors: Nicole Jaffrezic-Renault and Gaelle Lissorgues
Received: 31 March 2017; Accepted: 4 May 2017; Published: 6 May 2017

Abstract: The integration of a 50-nm-thick layer of an innovative sensitive material on microsensors has been developed based on silicon micro-hotplates. In this study, integration of ZnO:Ga via radio-frequency (RF) sputtering has been successfully combined with a low cost and reliable stencil mask technique to obtain repeatable sensing layers on top of interdigitated electrodes. The variation of the resistance of this n-type Ga-doped ZnO has been measured under sub-ppm traces (500 ppb) of acetaldehyde (C_2H_4O). Thanks to the microheater designed into a thin membrane, the generation of very rapid temperature variations (from room temperature to 550 °C in 25 ms) is possible, and a rapid cycled pulsed-temperature operating mode can be applied to the sensor. This approach reveals a strong improvement of sensing performances with a huge sensitivity between 10 and 1000, depending on the working pulsed-temperature level.

Keywords: gas sensors; ZnO:Ga; RF sputtering; stencil mask; metal-oxide microsensor; acetaldehyde; pulsed temperature

1. Introduction

In 1988, Demarne et al. [1] patented the first metal-oxide semiconductor (MOS) gas sensors based on a micromachined silicon substrate. It was a groundbreaking development that has since led to mature and robust technology [2] with few examples of devices on the market, notably based on SnO_2 and WO_3 metal oxides. To decrease the resistivity of the gas sensitive film, as well as to improve the kinetics of the chemical reactions, commercial MOS-type gas sensors are operated in constant high temperature mode (isothermal), knowing that the interactions between the sensitive material and the surrounding gases are temperature-dependent. The most important disadvantage of MOS-type sensors is their well-known poor selectivity [3]. Functionalization of sensitive materials with suitable catalytic elements including noble metals or metal oxides can be used to improve the selectivity [4]. Recently, it has been shown that the well-defined pore structure of metal-organic frameworks was able to provide molecular sieving at the surface of ZnO nanowires [5]. Another common method to enhance selectivity is to use sensor arrays based on two or more sensing elements in order to detect gas with data of higher dimensions [6]. On the other hand, because the temperature dependence is not similar for all gases, operating a sensor at different temperatures can provide pertinent information about the gas matrix composition, or the concentration of a specific gas in a background of other gases [7]. A cycled temperature mode allowed by the low thermal mass of micro-hotplates (the thermal time constant ranges from a few to tens of milliseconds) was first introduced by Sears et al. [8] in 1989 in an attempt to avoid the interference of humidity and to enable the discrimination of several gases with a single sensor. The sensitivity can be further improved by changing the number of oxygen species at the surface of the metal-oxide when its temperature is changing. Thus, Llobet et al. [9] showed that the transient response of thermally cycled metal oxide sensors decreases the influence of humidity on sensor response and the drift in the resistance of the gas sensitive layer. In this approach, it has been shown that, with very short temperature pulses, transient sensor responses are strongly dependent on the ambient mixture of gases, so this approach can enhance sensor selectivity [10].

Because the heater operates at a relatively high temperature, the reliability of micromachined hotplates is important for MOS-type gas sensors. Since the end of the 1980s, the technology has evolved significantly and has led to performing devices at operational temperatures up to 500 °C, with a homogeneous temperature distribution over the sensing area and with minimum power consumption [2,11]. Power consumption for continuous operation is in the order of a few tens of mW, but sub-mW consumption can be reached only by using a pulsed operating temperature. These micro-hotplates can now be elaborated in an array configuration with different types of semiconducting sensitive layers and with the very interesting possibility of modulating independently of their own operating temperature [12]. Current technologies allow temperature cycling up to several millions of cycles without failure. Finally, the sensors and the near electronics can be integrated into a small substrate to obtain an autonomous embedded system.

Microsensors have many advantages as, for example, high performance, small size, low cost, and low power consumption [7]. The bibliography therefore contains many examples of microsensors onto which various sensitive layers have been deposited by different methods such as micropipetting [13–16], sputtering [17–24], precipitation–oxidation [25,26], stepwise-heating electrospinning [27], flame spray pyrolysis [28], spin coating [29], a carbo-thermal route [30], evaporation [31], metal-assisted chemical etching [32], and organic binder printing [33].

Radio-frequency (RF) sputtering is a method compatible with the industrial fabrication of miniaturized sensors by microelectronics and MEMS technologies. RF sputtering has many other advantages, such as the possibility of obtaining very thin films with nanometric scale grain sizes and very easily controlling the inter-granular porosity by varying the deposition parameters [34,35]. Films with a controlled nanostructure such as these are of great interest for their potential of acting as sensitive layers [36–38] and of being integrated into gas sensing devices.

In this work, we explore the use of fully compatible micromachining technologies to elaborate microheaters and deposit sensitive layers to obtain sensors at the micron scale. An elaboration of micro-hotplates was performed, and photolithographic steps and shadow masks for layer integration were investigated, with micromachining facilities at the CNRS-LAAS (Centre National de la Recherche Scientifique - Laboratoire d'Analyse et d'Architecture des Systèmes) laboratory. The sensor is based on semiconducting layers that were deposited via RF sputtering in the CIRIMAT laboratory. From many different semiconducting materials that could be deposited by this technique, an example of a very interesting one—the n-type ZnO:Ga—has been chosen in this study. Zinc oxide has received considerable attention from the scientific community for gas detection. Although ZnO is interesting because of its low cost, non-toxicity, and fast and strong response values, it can be greatly improved by doping [39]. Ga dopants have many advantages, such as the rather similar radius as compared to that of Zn, the easy substitution of Zn^{2+} by Ga^{3+} without lattice distortion, and the decrease in the resistivity of the sensor [40].

Acetaldehyde (C_2H_4O) is considered an air pollutant and its known to have a carcinogenic effect on humans—especially with respect to nose cancers [41]. Recent studies have highlighted the potential of pure [42–49] or doped [50–52] zinc oxide for the detection of this pollutant. While Ga doping has been shown to strongly improve CO sensing [53], there have hitherto been no results as to the detection of acetaldehyde using Ga-doped ZnO sensitive layers. In the present work, the objective was first to demonstrate the feasibility of the integration of a ZnO:Ga sensitive layer via a stencil mask technique and second to present its high level of sensing performances under acetaldehyde. The microsensors were tested with variable thermal sequences under a low-level concentration (0.5 ppm) of acetaldehyde.

2. Experimental

Thin sensitive films were deposited with an Alcatel SCM 400 (Alcatel, France) apparatus using a homemade sintered ceramic target of pure ZnO:Ga with a relative density around 70% (9 cm in diameter). The RF power was lowered at 50 W to avoid target reduction [54], and the pressure inside the chamber was lower than 2×10^{-5} Pa before deposition. During the deposition of the films, the target-to-substrate distance was fixed at 7 cm (Table 1). The thicknesses of deposited thin films have been set to 50 nm on microsensors and 100 nm on glass substrates for structural characterizations. A pressure of 2 Pa was set to promote the intergranular porosity [55].

Table 1. Deposition parameters of thin sensitive films.

Target Material	ZnO:Ga
Magnetron	Yes
Substrates	Glass and Micro-Hotplate
Power	50 W
Argon pressure	2 Pa
Target to substrate distance	7 cm
Deposition rate	2.3 nm/min

Film thicknesses were measured using a Dektak 3030ST stylus profilometer across a step obtained by the lift-off of a felt pen line in acetone after deposition. The structural properties were determined by X-ray diffraction (XRD) using a Siemens D4 diffractometer with the Cu K_α radiation ($K_\alpha = 1.5418$ Å). Microscopic studies were realized with a Veeco Dimension 3000 atomic force microscope (AFM) equipped with a super sharp TESP-SS Nanoworld tip (nominal resonance frequency 320 KHz, nominal radius curvature 2 nm).

For sensing measurements, sensors were placed into a chamber with an alternating flow of air and 0.5 ppm of acetaldehyde. The composition and humidity of the gas mixture were controlled via mass flow controllers (MFCs). The heating and the sensing resistors of each sensor were connected to a source measurement unit (SMU). The entire test bench was automatically controllable thanks to a suitable and compatible interface and dedicated software. After a period of stabilization of 2 h under synthetic air, 0.5 ppm of acetaldehyde was introduced 5 times for 15 min with return periods in air of 30 min between each exposure to the target gas. The global flow (200 sscm) and the relative humidity (50%) remained constant during both air and target gas sequences. The response of gas sensor was calculated according to the formula: $S = R_{gas}/R_{air}$ (where R_{air} and R_{gas} are the resistance in air and 0.5 ppm of C_2H_4O, respectively).

3. Preparation of Microheaters

The tested devices were developed on optimized micro-hotplates that can work at high temperatures and low power consumption (500 °C; ~55 mW) with a very good stability and reproducibility. These silicon structures were elaborated using standard photolithographic processes. In order to avoid edge effects and to improve thermo-mechanical behavior, a circular membrane and heater geometry (Figure 1) were chosen. The design was elaborated to optimize temperature homogeneity in the center of the heated area onto which the sensing electrodes were deposited.

The platform consists of a silicon bulk on which a thermally resistive bilayer SiO_2/SiN_x membrane was grown. Afterwards, Pt metallization was realized via lift-off to define a heating resistor and the sensing electrodes. Contacts were opened in a previously deposited passivation layer. Finally, the rear side of the bulk was etched to release the membrane in order to increase the thermal resistance and then to limit thermal dissipation. Figure 1b shows the top view of the final membrane. This technology can prepare multi-sensors on which more than one sensing chip can be obtained in the same device (a 4-chip sensor is presented as an example in Figure 1c). This type of multi-sensor is especially suitable for operation in complex atmospheres containing various interfering gases and obtaining a good selectivity.

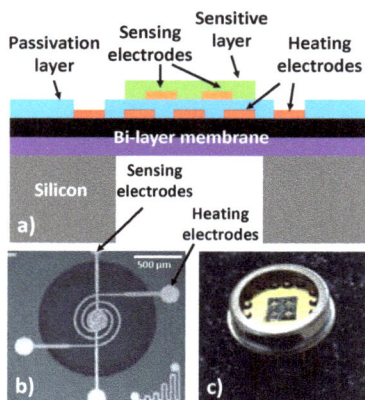

Figure 1. Micro-hotplate gas sensor: (**a**) a cross-sectional schematic view; (**b**) a chip top view; (**c**) the multi-sensor (4 chips) packaged on a TO-9 support.

Thermal measurement of the micro-hotplate surface with an IR camera allowed the calibration between the power applied and the resulting heating temperature of the membrane. The results given in Figure 2 show a good linear relation between the power applied and the temperature measured. The heating platform makes it possible to heat from room temperature to 550 °C in 25 ms, and the cooling time is of the same order of magnitude. This type of platform can thus generate very rapid temperature variations, which is suitable for operating the sensor in a pulsed mode. At the end of the process (before dicing the chips), it is possible to locally deposit a metal-oxide layer onto the electrodes to form the sensing thin film resistor. This will be described in the paragraph below, which is dedicated to the integration of ZnO:Ga by using a deposition through a shadow mask.

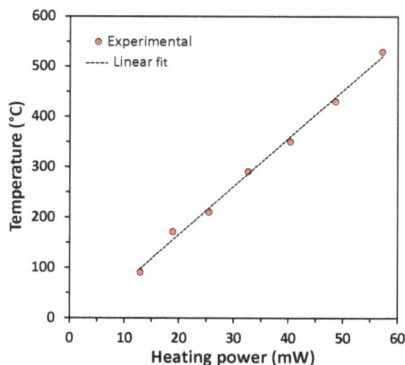

Figure 2. Temperature reached in the center of the microheater vs. the applied heating power.

4. Integration of N-type ZnO:Ga by the Shadow Mask Process

4.1. Characterization of the ZnO:Ga Layer

For structural characterization, 100-nm-thick films have been deposited on glass substrates in the condition defined in Table 1. The XRD patterns acquired at room temperature for as-deposited and annealed ZnO:Ga thin films have been reported in Figure 3. The results confirm that zinc oxide is crystallized (space group P6$_3$mc with lattice parameters a = 3.35(1) Å and c = 5.22(6) Å according to L. Weber [56]) with a single growth orientation along the (001) direction, which is largely reported in the bibliography for ZnO thin films deposited by physical vapor deposition techniques [57,58]. The lattice parameter c calculated from the (002) peak using a pseudo-voigt function is equal to 5.23(0) Å for the as-deposited thin film and becomes equal to 5.21(3) Å after a 400 °C annealing treatment under air for 4 h. The decrease of the c parameter after annealing could be due to a possible zinc substitution by silicon from the substrate [59], the removal of a lattice disorder, and the effect of the film/substrate interface strength due to the difference of the thermal expansion coefficients [60]. We noticed an increase in the (002) peak intensities after air annealing, which might be explained by the increase in the ZnO crystallized fraction caused by the annealing process.

The Scherrer relation [61] is defined by Equation (1):

$$d = \frac{K \times \lambda}{FWHM_{sample} \times \cos\theta} \tag{1}$$

where *d* is the size of the crystallites, *K* a shape correction factor, λ the X-ray wavelength, *FWHM*$_{sample}$ the width of the peak at its half maximum amplitude (corrected from instrumental contribution), and θ the peak position. The widths of the (002) peaks were calculated from a pseudo-voigt function after removing copper Kα_2 using EVA software. Considering isotropic shape crystallites, where *K* is equal to 0.9, and by neglecting the possible micro-strain component, the crystallites sizes were estimated. The average crystallites sizes of the 100 nm as-deposited ZnO:Ga thin film were about 33 nm. Thermal treatment increased the mean crystallite size by 5 nm, and the behavior is in agreement with the literature [62,63].

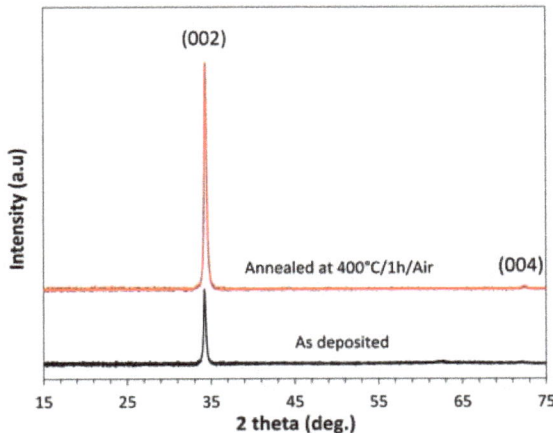

Figure 3. XRD pattern of a 100 nm ZnO:Ga film deposited on glass, before and after annealing under air at 400 °C.

In the sensor device, thinner zinc oxide layers were deposited to attain better sensitivity. As a result, the observation of the surface of the ZnO:Ga layer by AFM has been made on films deposited with a thickness of 50 nm (after annealing). The AFM image reported in Figure 4a shows a classical grain morphology, which consists of surface domes (top of the grown column). This characteristic is often mentioned in the literature [64,65]. The in-plane average grain size distribution (Figure 4b) was estimated by an immersion threshold thanks to the Gwyddion software [66].

The average grain size determined as the diameter of 50% of the total cumulative frequency (d_{50}) is 28 nm for the 50-nm-thick film annealed at 400 °C, and the maximum peak-to-valley amplitude was found to be equal to 11 nm. In the case of the 100-nm-thick films (used for the XRD analysis), the surface morphology (not exposed here) was similar to the 50-nm-thick films, except that the median grain size was larger (60 nm).

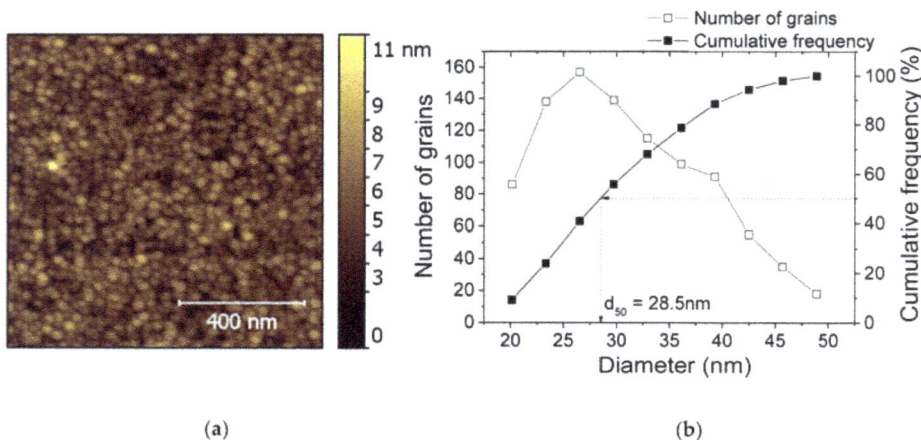

(a) (b)

Figure 4. (**a**) AFM image of a 50-nm-thick ZnO:Ga film annealed at 400 °C for 1 h under an air atmosphere. (**b**) Grain size distribution deduced from the image analysis.

4.2. Description of the Integration Process via Stencil Mask

The main disadvantage of the lift-off technique is the complexity of the various and necessary steps, which involve expensive equipment. Moreover, during the deposition of the photoresist, its development, and the removal of the remaining resist mask, the interaction of the solutions used with the sensitive layer can lead its dissolution and/or contamination. These are the reasons why the possibility of depositing the layer through a stencil mask has been evaluated. The entire process is shown in Figure 5. The mask has been opened in an adhesive film made of polyvinyl chloride with a thickness of 75 µm. The holes made with a simple cutting laser machine had a diameter close to 600 µm. The mask diameter was chosen to be lower than that of the membrane (1.2 mm) but higher than the active area (400 µm) where the interdigitated sensing electrodes are located. The mask with 100 cut out holes (Figure 5, Step 1) was aligned and stuck onto the surface of a quarter of a micro-machined silicon wafer (with 100 sensing chips). To achieve mask placement with high accuracy, this step was performed with an optical microscope and a manual pick-and-place machine (Figure 5, Step 2). After vacuum deposition of the 50-nm-thick ZnO:Ga sensitive layer via RF sputtering (Step 3), the stencil mask was simply peeled off after a 120 °C thermal treatment over a few tens of seconds (Step 4). Figure 6 shows the resulting ZnO:Ga layer, which is located on top of the membrane and, above all, well covering the sensing electrode area.

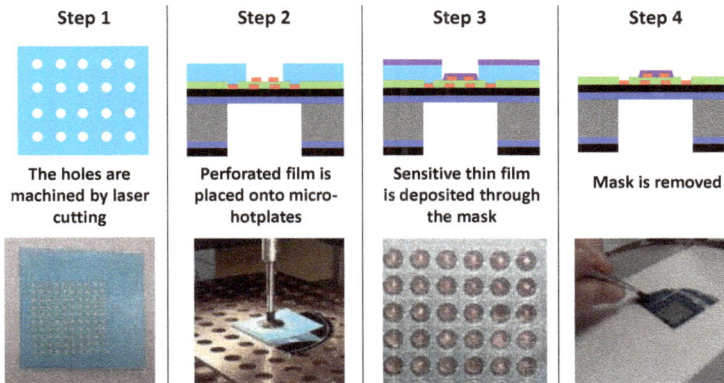

Figure 5. Main steps in the integration process of ZnO:Ga sensitive layers using a shadow mask.

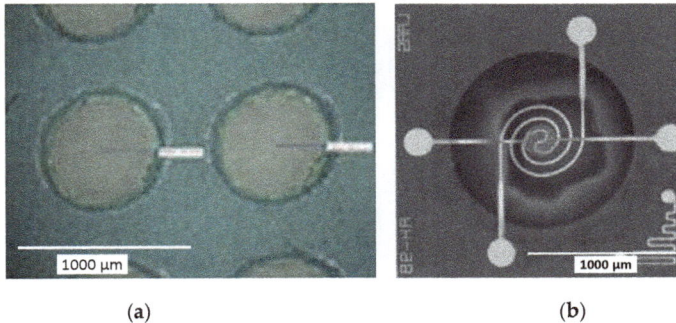

Figure 6. Optical microscopy image of (**a**) the stencil mask (hole diameter: 600 µm) and (**b**) the ZnO:Ga layer deposited onto the electrode area after the removal of the stencil mask.

5. Sensing Tests

In this first study, the microsensor based on ZnO:Ga semiconducting layers have only been tested under sub-ppm concentrations (500 ppb) of acetaldehyde (C_2H_4O). Five levels of power were applied to the heater to explore the sensing performances from 10 to 45 mW. The sensor was held at each heating step for 1 min, and every 5 min, this basic cycle was repeated. During the sensing test, a dissymmetrical procedure has been used: Three cycles were repeated under target gas (acetaldehyde), while 6 cycles were used in air to ensure total recovery of the signal. The target gas was introduced 5 times throughout the experiment to test the repeatability. Figure 7 shows the last 2 of the 6 previous cycles under air, the three cycles under 500 ppb of acetaldehyde, and the first 4 of the following 6 cycles under air.

We observed that, before the introduction of C_2H_4O, the evolution of the resistance during the power ramp was stable from one cycle to another. On the other hand, when the atmosphere switched from air to 500 ppb of acetaldehyde, the sensing resistance was substantially shifted to higher values. The stabilization of the resistance was typically obtained from the second cycle. When the atmosphere was switched again to pure air, recovery was achieved after the third power ramp cycle. The cycle obtained in the fourth cycle was similar to the last cycle under air before introducing acetaldehyde.

The estimation of the gain of resistance was difficult to obtain directly from the variation of the signal reported in Figure 7. This is the reason why the response R_{gas}/R_{air} was presented in Figure 8.

As the resistance of the sensitive layer was constantly changing with the value of the power and during each power step, it was therefore not possible to take a fixed value as a reference. The entire variation of the resistance during the last power ramp cycle before gas introduction was then taken as a reference (Figure 7). In Figure 8, for each cycle under acetaldehyde, the values of the resistance are divided by the values of the last cycle under air. From these curves, the very high response of the ZnO:Ga sensitive layer is highlighted. The best values were obtained from the lowest power heating, even though at 10 and 20 mW the stabilization could never be reached, while at 30, 40, and 45 mW the stabilization could be obtained after approximately 30 s. Regardless of the power applied, the signal was significant even at the highest power value of 45 mW for which the response R_{gas}/R_{air} remains around 20. It is difficult to compare the present results with those of the other authors because, to our knowledge, nobody has yet published a study concerning the sensitivity of ZnO:Ga toward acetaldehyde. However, the bibliography confirms that doping, for instance, with Co [50,51], Cu [50], or Ru [51], improves the sensing performance of ZnO toward acetaldehyde in comparison with undoped ZnO.

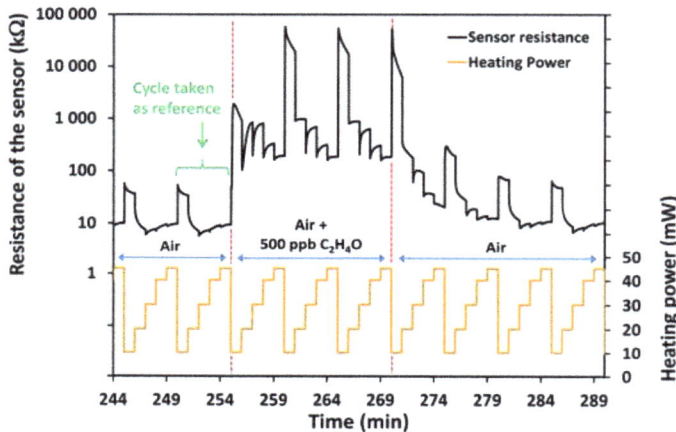

Figure 7. Variation of the resistance of the sensing layer with the heating power and the gas composition.

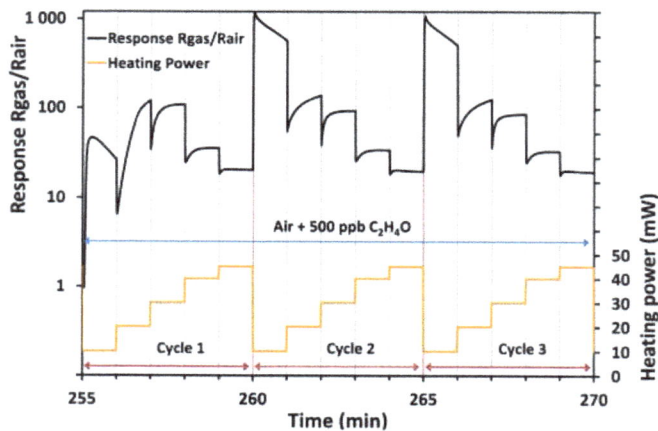

Figure 8. Response of the sensor (R_{gas}/R_{air}) under 500 ppb of acetaldehyde in a temperature-cycled mode.

For now, it is not yet known if heating cyclically to high power values (40 and 45 mW for instance) has a significant effect. This is the reason why more experiments and investigations are in progress. Actual power cycles will be compared to other measurement modes as a constant temperature mode and a cycled mode with a high-temperature baseline. Moreover, it is necessary to corroborate the current results with complementary experiments under various acetaldehyde concentrations.

6. Conclusions

Micro-hotplates were first prepared using silicon microtechnologies. Because the microheater was designed for use on a thin membrane, it was possible for us to generate very fast temperature variations (from room temperature to 550 °C in 25 ms), and a rapid temperature cycled mode could be applied. A method using a stencil mask was developed so that the sensitive layer can avoid contact with the products used during the photolithography steps. This process was successfully tested during the integration of the ZnO:Ga sensitive layer. The variation of the resistance of this 50-nm-thick sensitive layer was measured under 500 ppb of acetaldehyde. The very high response obtained was between 10 and 1000, depending on the working temperature. Using a rapid temperature cycled mode is a good opportunity to evaluate the selectivity of the sensor in other interfering gases, and such a study will be carried out in a next step.

Acknowledgments: This work was partly supported by LAAS-CNRS micro and nano technolgies platform member of the French RENATECH network and by University of Toulouse in the frame of NeoCampus program. Publishing costs are covered by the laboratories.

Author Contributions: L. Presmanes has coordinated the research work, processed and analyzed the gas sensing results and written the article. Y. Thimont has deposited ZnO:Ga thin film and characterized their microstructure. C. Bonningue has prepared the ZnO:Ga target devoted for the sputtering of the active material. A. Barnabé and P. Tailhades have characterized the structure of the ZnO thin film. P. Menini and A. Chapelle from LAAS-CNRS worked on the fabrication of microthotplate platforms and the stencil masks and on the sensors' characterizations under controlled atmospheres. F. Blanc and C. Talhi have developed the gas sensing setup.

Conflicts of Interest: The authors declare no conflict of interest.

References

1. Demarne, V.; Grisel, A. An integrated low-power thin-film CO gas sensor on silicon. *Sens. Actuators* **1988**, *13*, 301–313. [CrossRef]
2. Courbat, J.; Canonica, M.; Teyssieux, D.; Briand, D.; de Rooijet, N.F. Design and fabrication of micro-hotplates made on a polyimide foil: Electrothermal simulation and characterization to achieve power consumption in the low mW range. *J. Micromech. Microeng.* **2010**, *21*, 015014. [CrossRef]
3. Vaihinger, S.; Göpel, W. Chapter 6: Multi-Component Analysis in Chemical Sensing. In *Sensors Set: A Comprehensive Survey*; Göpel, W., Hesse, J., Zemel, J.N., Eds.; Wiley-VCH Verlag GmbH: Weinheim, Germany, 1995.
4. Sofian, M.K.; Oussama, M.E.; Imad, A.A.; Marsha, C.K. Semiconducting metal oxide based sensors for selective gas pollutant detection. *Sensors* **2009**, *9*, 8158–8196.
5. Drobek, M.; Kim, J.-H.; Bechelany, M.; Vallicari, C.; Julbe, A.; Kim, S.S. MOF-Based Membrane Encapsulated ZnO Nanowires for Enhanced Gas Sensor Selectivity. *ACS Appl. Mater. Interfaces* **2016**, *8*, 8323–8328. [CrossRef] [PubMed]
6. Liu, X.; Cheng, S.; Liu, H.; Hu, S.; Zhang, D.; Ning, H. A Survey on Gas Sensing Technology. *Sensors* **2012**, *12*, 9635–9665. [CrossRef] [PubMed]
7. Briand, D.; Courbat, J. Chapter 6: Micromachined semiconductor gas sensors. In *Semiconductor Gas Sensors*; Jaaniso, R., Tan, O.K., Eds.; Woodhead Publishing: Sawston, Cambridge, UK, 2013; pp. 220–260.
8. Sears, W.M.; Colbow, K.; Consadori, F. General characteristics of thermally cycled tin oxide gas Sensors. *Semicond. Sci. Technol.* **1989**, *4*, 351–359. [CrossRef]
9. Llobet, E.; Brezmes, J.; Ionescu, R.; Vilanova, X.; Al-Khalifa, S.; Gardner, J.W.; Bârsan, N.; Correig, X. Wavelet transform and fuzzy ARTMAP-based pattern recognition for fast gas identification using a micro-hotplate gas sensor. *Sens. Actuators B Chem.* **2002**, *83*, 238–244. [CrossRef]

10. Parret, F.; Ménini, Ph.; Martinez, A.; Soulantica, K.; Maisonnat, A.; Chaudret, B. Improvement of Micromachined SnO₂ Gas Sensors Selectivity By Optimised Dynamic Temperature Operating Mode. *Sens. Actuators B Chem.* **2006**, *118*, 276–282. [CrossRef]

11. Faglia, G.; Comini, E.; Cristalli, A.; Sberveglieri, G.; Dori, L. Very low power consumption micromachined CO sensors. *Sens. Actuators B Chem.* **1999**, *55*, 140–146. [CrossRef]

12. Dufour, N.; Chapelle, A.; Talhi, C.; Blanc, F.; Franc, B.; Menini, P.; Aguir, K. Tuning the Bias Sensing Layer: A New Way to Greatly Improve Metal-Oxide Gas Sensors Selectivity. In Proceedings of the International Conference on Sensing Technology (ICST), Wellington, New Zealand, 3–5 December 2013.

13. Fong, C.-F.; Dai, C.-L.; Wu, C.-C. Fabrication and Characterization of a Micro Methanol Sensor Using the CMOS-MEMS Technique. *Sensors* **2015**, *15*, 27047–27059. [CrossRef] [PubMed]

14. Martinez, C.J.; Hockey, B.; Montgomery, C.B.; Semancik, S. Porous tin oxide nanostructured microspheres for sensor applications. *Langmuir* **2005**, *21*, 7937–7944. [CrossRef] [PubMed]

15. Liao, W.-Z.; Dai, C.-L.; Yang, M.-Z. Micro Ethanol Sensors with a Heater Fabricated Using the Commercial 0.18 μm CMOS Process. *Sensors* **2013**, *13*, 12760–12770. [CrossRef] [PubMed]

16. Yang, M.-Z.; Dai, C.-L. Ethanol Microsensors with a Readout Circuit Manufactured Using the CMOS-MEMS Technique. *Sensors* **2015**, *15*, 1623–1634. [CrossRef] [PubMed]

17. Behera, B.; Chandra, S. An innovative gas sensor incorporating ZnO–CuO nanoflakes in planar MEMS technology. *Sens. Actuators B Chem.* **2016**, *229*, 414–424. [CrossRef]

18. Stankova, M.; Ivanov, P.; Llobet, E.; Brezmes, J.; Vilanova, X.; Gràcia, I.; Cané, C.; Hubalek, J.; Malysz, K.; Correig, X. Sputtered and screen-printed metal oxide-based integrated micro-sensor arrays for the quantitative analysis of gas mixtures. *Sens. Actuators B Chem.* **2004**, *103*, 23–30. [CrossRef]

19. Lee, C.-Y.; Chiang, C.-M.; Wang, Y.-H.; Ma, R.-H. A self-heating gas sensor with integrated NiO thin-film for formaldehyde detection. *Sens. Actuators B Chem.* **2007**, *122*, 503–510. [CrossRef]

20. Stankova, M.; Vilanova, X.; Calderer, J.; Llobet, E.; Ivanov, P.; Gràcia, I.; Cané, C.; Correig, X. Detection of SO₂ and H₂S in CO₂ stream by means of WO₃-based micro-hotplate sensors. *Sens. Actuators B Chem.* **2004**, *102*, 219–225. [CrossRef]

21. Tang, Z.; Fung, S.K.H.; Wong, D.T.W.; Chan, P.C.H.; Sin, J.K.O.; Cheung, P.W. An integrated gas sensor based on tin oxide thin-film and improved micro-hotplate. *Sens. Actuators B Chem.* **1998**, *46*, 174–179. [CrossRef]

22. Sheng, L.Y.; Tang, Z.; Wu, J.; Chan, P.C.H.; Sin, J.K.O. A low-power CMOS compatible integrated gas sensor using maskless tin oxide sputtering. *Sens. Actuators B Chem.* **1998**, *49*, 81–87. [CrossRef]

23. Takács, M.; Dücső, C.; Pap, A.E. Fine-tuning of gas sensitivity by modification of nano-crystalline WO₃ layer morphology. *Sens. Actuators B Chem.* **2015**, *221*, 281–289. [CrossRef]

24. Zappa, D.; Briand, D.; Comini, E.; Courbat, J.; de Rooij, N.F.; Sberveglieri, G. Zinc Oxide Nanowires Deposited on Polymeric Hotplates for Low-power Gas Sensors. *Procedia Eng.* **2012**, *47*, 1137–1140. [CrossRef]

25. Yang, M.Z.; Dai, C.L.; Shih, P.J.; Chen, Y.C. Cobalt oxide nanosheet humidity sensor integrated with circuit on chip. *Microelectron. Eng.* **2011**, *88*, 1742–1744. [CrossRef]

26. Dai, C.L.; Chen, Y.C.; Wu, C.C.; Kuo, C.F. Cobalt oxide nanosheet and CNT micro carbon monoxide sensor integrated with readout circuit on chip. *Sensors* **2010**, *10*, 1753–1764. [CrossRef] [PubMed]

27. Tang, W.; Wang, J. Methanol sensing micro-gas sensors of SnO2-ZnO nanofibers on Si/SiO₂/Ti/Pt substrate via stepwise-heating electrospinning. *J. Mater. Sci.* **2015**, *50*, 4209–4220. [CrossRef]

28. Kuhne, S.; Graf, M.; Tricoli, A.; Mayer, F.; Pratsinis, S.E.; Hierlemann, A. Wafer-level flame-spray-pyrolysis deposition of gas-sensitive layers on microsensors. *J. Micromech. Microeng.* **2008**, *18*, 035040. [CrossRef]

29. Wan, Q.; Li, Q.H.; Chen, Y.J.; Wang, T.H.; He, X.L.; Li, J.P.; Lin, C.L. Fabrication and ethanol sensing characteristics of ZnO nanowire gas sensors. *Appl. Phys. Lett.* **2004**, *84*, 3654–3656. [CrossRef]

30. Nguyen, H.; Quy, C.T.; Hoa, N.D.; Lam, N.T.; Duy, N.V.; Quang, V.V.; Hieu, N.V. Controllable growth of ZnO nanowire grown on discrete islands of Au catalyst for realization of planar type micro gas sensors. *Sens. Actuators B Chem.* **2014**, *193*, 888–894. [CrossRef]

31. Pandya, H.J.; Chandra, S.; Vyas, A.L. Integration of ZnO nanostructures with MEMS for ethanol sensor. *Sens. Actuators B Chem.* **2011**, *161*, 923–928. [CrossRef]

32. Peng, K.Q.; Wang, X.; Lee, S.T. Gas sensing properties of single crystalline porous silicon nanowires. *Appl. Phys. Lett.* **2009**, *95*, 243112. [CrossRef]

33. Dong, K.Y.; Choi, J.K.; Hwang, I.S.; Lee, J.W.; Kang, B.H.; Ham, D.J.; Lee, J.H.; Ju, B.K. Enhanced H_2S sensing characteristics of Pt doped SnO_2 nanofibers sensors with micro heater. *Sens. Actuators B Chem.* **2011**, *157*, 154–161. [CrossRef]

34. Oudrhiri-Hassani, F.; Presmanes, L.; Barnabé, A.; Tailhades, P. Microstructure, porosity and roughness of RF sputtered oxide thin films: characterization and modelization. *Appl. Surf. Sci.* **2008**, *254*, 5796–5802. [CrossRef]

35. Sandu, I.; Presmanes, L.; Alphonse, P.; Tailhades, P. Nanostructured cobalt manganese ferrite thin films for gas sensor application. *Thin Solid Films* **2006**, *495*, 130–133. [CrossRef]

36. Chapelle, A.; El Younsi, I.; Vitale, S.; Thimont, Y.; Nelis, T.; Presmanes, L.; Barnabé, A.; Tailhades, P. Improved semiconducting $CuO/CuFe_2O_4$ nanostructured thin films for CO_2 gas sensing. *Sens. Actuators B Chem.* **2014**, *204*, 407–413. [CrossRef]

37. Chapelle, A.; Yaacob, M.; Pasquet, I.; Presmanes, L.; Barnabe, A.; Tailhades, P.; Du Plessis, J.; Kalantar, K. Structural and gas-sensing properties of $CuO–Cu_xFe_{3-x}O_4$ nanostructured thin films. *Sens. Actuators B Chem.* **2011**, *153*, 117–124. [CrossRef]

38. Presmanes, L.; Chapelle, A.; Oudrhiri-Hassani, F.; Barnabe, A.; Tailhades, P. Synthesis and CO Gas-Sensing Properties of CuO and Spinel Ferrite Nanocomposite Thin Films. *Sens. Lett.* **2013**, *9*, 587–590. [CrossRef]

39. Afaah, A.N.; Khusaimi, Z.; Rusop, M. A Review on Zinc Oxide Nanostructures: Doping and Gas Sensing. *Adv. Mater. Res.* **2013**, *667*, 329–332.

40. Yang, J.; Jiang, Y.; Li, L.; Gao, M. Structural, morphological, optical and electrical properties of Ga-doped ZnO transparent conducting thin films. *Appl. Surf. Sci.* **2017**. Available online: http://doi.org/10.1016/j. apsusc.2016.10.079 (accessed on 17 October 2016).

41. Zhou, Y.; Li, C.; Huijbregts, M.A.J.; Mumtaz, M.M. Carcinogenic Air Toxics Exposure and Their Cancer-Related Health Impacts in the United States. *PLoS ONE* **2015**, *10*, e0140013. [CrossRef] [PubMed]

42. Sivalingam, K.; Shankar, P.; Mani, G.K.; Rayappan, J.B.B. Solvent volume driven ZnO nanopetals thin films: Spray pyrolysis. *Mater. Lett.* **2014**, *134*, 47–50. [CrossRef]

43. Rai, P.; Raj, S.; Ko, K.-J.; Park, K.-K.; Yu, Y.-T. Synthesis of flower-like ZnO microstructures for gas sensor applications. *Sens. Actuators B Chem.* **2013**, *178*, 107–112. [CrossRef]

44. Giberti, A.; Carotta, M.C.; Fabbri, B.; Gherardi, S.; Guidi, V.; Malagù, C. High-sensitivity detection of acetaldehyde. *Sens. Actuators B Chem.* **2012**, *174*, 402–405. [CrossRef]

45. Zhang, L.; Zhao, J.; Lu, H.; Gong, L.; Li, L.; Zheng, J.; Li, H.; Zhu, Z. High sensitive and selective formaldehyde sensors based on nanoparticle-assembled ZnO micro-octahedrons synthesized by homogeneous precipitation method. *Sens. Actuators B Chem.* **2011**, *160*, 364–370. [CrossRef]

46. Rai, P.; Song, H.-M.; Kim, Y.-S.; Song, M.-K.; Oh, P.-R.; Yoon, J.-M.; Yu, Y.-T. Microwave assisted hydrothermal synthesis of single crystalline ZnO nanorods for gas sensor application. *Mater. Lett.* **2012**, *68*, 90–93. [CrossRef]

47. Zhang, S.-L.; Lim, J.-O.; Huh, J.-S.; Noh, J.-S.; Lee, W. Two-step fabrication of ZnO nanosheets for high-performance VOCs gas sensor. *Curr. Appl. Phys.* **2013**, *13*, S156–S161. [CrossRef]

48. Du, J.; Yao, H.; Zhao, R.; Wang, H.; Xie, Y.; Li, J. Controllable synthesis of prism- and lamella-like ZnO and their gas sensing. *Mater. Lett.* **2014**, *136*, 427–430. [CrossRef]

49. Zhang, L.; Zhao, J.; Zheng, J.; Li, L.; Zhu, Z. Shuttle-like ZnO nano/microrods: Facile synthesis, optical characterization and high formaldehyde sensing properties. *Appl. Surf. Sci.* **2011**, *258*, 711–718. [CrossRef]

50. Mani, G.K.; Rayappan, J.B.B. ZnO nanoarchitectures: Ultrahigh sensitive room temperature acetaldehyde sensor. *Sens. Actuators B Chem.* **2016**, *223*, 343–351. [CrossRef]

51. Xu, J.; Han, J.; Zhang, Y.; Sun, Y.; Xie, B. Studies on alcohol sensing mechanism of ZnO based gas sensors. *Sens. Actuators B Chem.* **2008**, *132*, 334–339. [CrossRef]

52. Shalini, S.; Balamurugan, D. Ambient temperature operated acetaldehyde vapour detection of spray deposited cobalt doped zinc oxide thin film. *J. Colloid Interface Sci.* **2016**, *466*, 352–359. [CrossRef] [PubMed]

53. Hjiri, M.; Dhahri, R.; El Mir, L.; Bonavita, A.; Donato, N.; Leonardi, S.G.; Neri, G. CO sensing properties of Ga-doped ZnO prepared by sol–gel route. *J. Alloy. Compd.* **2015**, *634*, 187–192. [CrossRef]

54. Bui, M.A.; Le Trong, H.; Presmanes, L.; Barnabé, A.; Bonningue, C.; Tailhades, P. Thin films of $Co_{1.7}Fe_{1.3}O_4$ prepared by radio-frequency sputtering - First step towards their spinodal decomposition. *CrystEngComm* **2014**, *16*, 3359–3365. [CrossRef]

55. Shang, C.; Thimont, Y.; Barnabe, A.; Presmanes, L.; Pasquet, I.; Tailhades, Ph. Detailed microstructure analysis of as-deposited and etched porous ZnO films. *Appl. Surf. Sci.* **2015**, *344*, 242–248. [CrossRef]

56. Weber, L. XII. Ein einfacher Ausdruck für das Verhältnis der Netzdichten der Bravaisschen Baumgitter. *Zeitschrift für Kristallographie-Cryst. Mater.* **1923**, *58*, 398. [CrossRef]

57. Dang, W.L.; Fu, Y.Q.; Luo, J.K.; Flewitt, A.J.; Milne, W.I. Deposition and characterization of sputtered ZnO films. *Superlattices Microstruct.* **2007**, *42*, 89–93. [CrossRef]

58. Thimont, Y.; Clatot, J.; Nistor, M.; Labrugere, C.; Rougier, A. From ZnF$_2$ to ZnO thin films using pulsed laser deposition: Optical and electrical properties. *Sol. Energ. Mater. Sol. Cells* **2012**, *107*, 136–141.

59. Clatot, J.; Campet, G.; Zeinert, A.; Labrugere, C.; Nistor, M.; Rougier, A. Low temperature Si doped ZnO thin films for transparent conducting oxides. *Sol. Energ. Mater. Sol. Cells* **2011**, *95*, 2357–2362. [CrossRef]

60. Thermal Expansion Coefficients at 20 °C. Available online: http://hyperphysics.phy-astr.gsu.edu/hbase/tables/thexp.html (accessed on 5 May 2017).

61. Scherrer, P. Bestimmung der Grösse und der inneren Struktur von Kolloidteilchen mittels Röntgenstrahlen. *J. Nachr. Ges. Wiss. Göttingen* **1918**, *26*, 98–100.

62. Sugapriya, S.; Lakshmi, S.; Senthilkumaran, C.K. Effect on Annealing Temperature on ZnO Nanoparticles. *Int. J. ChemTech Res.* **2015**, *8*, 297–302.

63. Husna, J.; Mannir Aliyu, M.; Aminul Islam, M.; Chelvanathan, P.; Radhwa Hamzah, N.; Sharafat Hossain, M.; Karim, M.R.; Amin, N. Influence of Annealing Temperature on the Properties of ZnO Thin Films Grown by Sputtering. *Energy Procedia* **2012**, *25*, 55–61. [CrossRef]

64. Barnabe, A.; Lalanne, M.; Presmanes, L.; Soon, J.M.; Tailhades, Ph.; Dumas, C.; Grisolia, J.; Arbouet, A.; Paillard, V.; BenAssayag, G.; et al. Structured ZnO-based contacts deposited by non-reactive rf magnetron sputtering on ultra-thin SiO$_2$/Si through a stencil mask. *Thin Solid Films* **2009**, *518*, 1044–1047. [CrossRef]

65. Lalanne, M.; Soon, J.M.; Barnabe, A.; Presmanes, L.; Pasquet, I.; Tailhades, P. Preparation and characterization of the defect-conductivity relationship of Ga-doped ZnO thin films deposited by nonreactive radio-frequency-magnetron sputtering. *J. Mater. Res.* **2010**, *25*, 2407–2414. [CrossRef]

66. Necas, D.; Klapetek, P. Gwyddion software. Available online: http://www.gwyddion.net/ (accessed on 8 March 2016).

sensors

MDPI

Article

The Impact of Bending Stress on the Performance of Giant Magneto-Impedance (GMI) Magnetic Sensors

Julie Nabias, Aktham Asfour * and Jean-Paul Yonnet

Université Grenoble Alpes, CNRS, Grenoble INP (Institute of Engineering Université Grenoble Alpes),
G2Elab, F-38000 Grenoble, France; Julie.Nabias@g2elab.grenoble-inp.fr (J.N.);
Jean-Paul.Yonnet@g2elab.grenoble-inp.fr (J.-P.Y.)
* Correspondence: aktham.asfour@g2elab.grenoble-inp.fr; Tel.: +33-476-826-395

Academic Editors: Nicole Jaffrezic-Renault and Gaelle Lissorgues
Received: 9 December 2016; Accepted: 10 March 2017; Published: 20 March 2017

Abstract: The flexibility of amorphous Giant Magneto-Impedance (GMI) micro wires makes them easy to use in several magnetic field sensing applications, such as electrical current sensing, where they need to be deformed in order to be aligned with the measured field. The present paper deals with the bending impact, as a parameter of influence of the sensor, on the GMI effect in 100 μm Co-rich amorphous wires. Changes in the values of key parameters associated with the GMI effect have been investigated under bending stress. These parameters included the GMI ratio, the intrinsic sensitivity, and the offset at a given bias field. The experimental results have shown that bending the wire resulted in a reduction of GMI ratio and sensitivity. The bending also induced a net change in the offset for the considered bending curvature and the set of used excitation parameters (1 MHz, 1 mA). Furthermore, the field of the maximum impedance, which is generally related to the anisotropy field of the wire, was increased. The reversibility and the repeatability of the bending effect were also evaluated by applying repetitive bending stresses. The observations have actually shown that the behavior of the wire under the bending stress was roughly reversible and repetitive.

Keywords: Giant Magneto-Impedance (GMI); amorphous wire; bending stress; flexible

1. Introduction

The use of Giant Magneto-Impedance (GMI) wires as sensing elements for electrical current sensors in real industrial environments raises important issues related to the parameters that influence the sensor. These parameters can dramatically affect the accuracy and reliability of the measurement. They may include—but are not limited to—the temperature and the surrounding magnetic fields produced, for example, by other conductors in close proximity to the conductor of interest.

Mechanical effects can also be encountered in a number of structures of GMI-based current sensors. In fact, in some situations, the GMI wire has to be deformed since it needs to be aligned with the magnetic field produced by the conductor carrying the measured electrical current. Indeed, this is typically the case in some prototypes of current sensors where the sensor has a toroidal structure [1–7], unlike other prototypes that do not involve deformation [8,9].

In the case of toroidal configuration, some studies have already been conducted to evaluate the impact of the deformation of the amorphous wire on the GMI effect [1].

The effects of tensile [10–17] and torsional [18,19] stresses have been intensively investigated in amorphous microwires with low (to vanishing or slightly negative) magnetostriction. Stresses induced during certain fabrication processes, such as cold drawing, can also affect the domain structure of the wires and influence the GMI ratio and field sensitivity [20].

By far, the bending stress effect in GMI amorphous wires is actually less known. Nevertheless, the investigation of this effect is particularly important in applications such as current sensors. In fact, it is important to ensure that the potential variations of the intrinsic relevant quantities of the GMI curve due to bending stress will not affect the final response of the sensor. Or at least, the knowledge of the impact of the bending on these relevant quantities has to be developed so as to allow for adequate solutions to minimizing this impact.

In a sensor application, the relevant quantities to be considered include the offset resulting from the field biasing, the intrinsic sensitivity at the bias point, and the GMI ratio.

These quantities are illustrated in Figure 1, which presents typical nonlinear GMI characteristics (modulus of the impedance, $|Z(H)|$, as a function of the magnetic field H) for an amorphous wire.

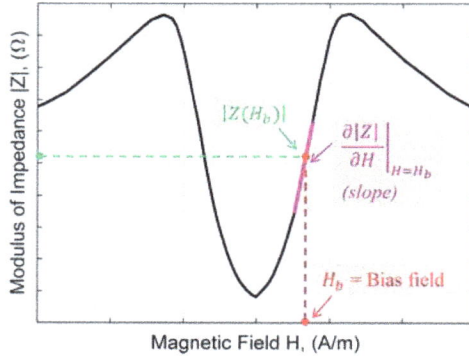

Figure 1. A typical Giant Magneto-Impedance (GMI) curve, $|Z(H)|$.

As is well-known, in the classical use of a GMI wire for developing a linear sensor, the wire must be biased by a magnetic field, H_b, in the region that exhibits near-linear behavior. This field biasing gives rise to an offset voltage at the final sensor output when the measured magnetic field is zero, since the output voltage is directly proportional to the value of the impedance at the bias field ($|Z(H_b)|$ in Figure 1). Electronic canceling of the offset voltage is usually employed. However, this canceling should be effective only if the value of $|Z(H_b)|$ does not change under the parameters of influence.

In addition to linearity considerations, the bias point is also chosen to obtain a maximum sensitivity. At the bias point, the intrinsic sensitivity of the sensor is defined by

$$S(H_b) = \left. \frac{\partial |Z|}{\partial H} \right|_{H=H_b}. \tag{1}$$

This quantity $\left. \frac{\partial |Z|}{\partial H} \right|_{H=H_b}$ will dramatically determine the final sensitivity of the sensor in open-loop operation and, if it is not high enough, it could also impact the sensitivity in closed-loop operation.

A third quantity to be considered is the GMI ratio, $\Delta Z/Z$, defined as

$$\frac{\Delta Z}{Z} (\%) = \frac{|Z(H) - Z(H_{max})|}{|Z(H_{max})|} \times 100 \tag{2}$$

where a commonly used value for H_{max} is the saturation field of the magnetic material. In practice, H_{max} is generally the maximum field available in a given setup. The GMI ratio obviously contributes to fixing the total dynamic range of the open-loop sensor. In fact, a higher GMI ratio allows more excursion of the measured field around the bias point.

Under the parameters of influence, the offset, the intrinsic sensitivity, S, and the GMI ratio are potentially subject to changes that can largely degrade the sensor's performance and reliability.

According to the requirements of each application, one or several of these quantities have to be optimized.

In this paper, an experimental study is conducted to evaluate the impact of the bending stress, as an influence parameter, on the offset, the intrinsic sensitivity, and the GMI ratio of an amorphous wire. A full description of the measurement setup and experimental conditions are given. The first obtained results are illustrated and discussed. Tests of the repeatability and reversibility of the bending stress effect are also included.

2. Materials and Methods

Figure 2 shows the developed experimental setup. The used amorphous Co-rich wires (Co-Fe-Si-B) with a 100 μm diameter were from Unitika Ltd., obtained with the in-rotating-water spinning process. No additional treatments were performed on these wires.

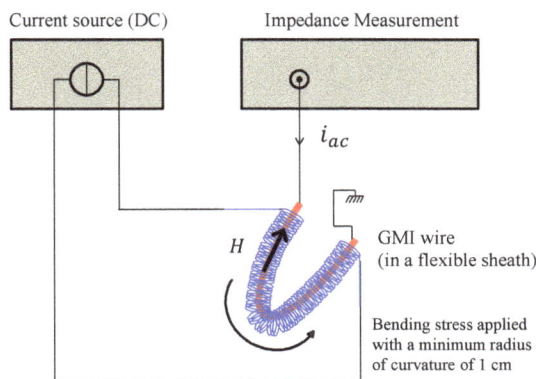

Figure 2. The experimental setup. The GMI wire was 90 mm in length and excited by an AC current with a 1 mA amplitude and a 1 MHz frequency. In this figure, the wire is in a bent position.

For all experiments, a wire 90 mm long was cut and placed inside a flexible envelope (sheath). The extremities of the wire were soldered to copper wires for electrical connections and measurement purposes. The external magnetic field, H, was applied to the wire using a 270-turn coil placed around the flexible sheath. The maximum available field H_{max} was 1 kA/m.

In all the experiments, the impedance was measured when the GMI wire was supplied by a high frequency (HF) current, i_{ac}, with a 1 MHz frequency and a 1 mA amplitude. These values were only justified by the requirements of our application, where the sensor's energy consumption and the simplicity of implementation of the conditioning electronics are issues. Additionally, with relatively low AC currents, the nonlinear effects of the GMI (nonlinearity between the current and the voltage across the wire) were avoided [21].

The bending stress was simultaneously applied to the GMI wire, sheath, and coil. The minimum measured a radius of curvature, which is representative of the applied strain, was about 10 mm. This yields a calculated stress of approximately 800 MPa in tension and compression along the wire.

3. Results and Discussion

Figure 3 shows the change of $|Z(H)|$ and $\Delta Z/Z$ with the applied bending stress.

One could note that the $|Z(H)|$ curves of Figure 3 are slightly asymmetric, in both straight and bent positions. This intrinsic asymmetry, which has been intensively investigated in the literature, could have several causes [22]. While the study of the asymmetry is well beyond the scope of this paper, it seems that its origin is related to the hysteresis of the GMI material. The curves of Figure 3 have actually been plotted in the case of an increasing magnetic field (the asymmetry is reversed for a decreasing field). The asymmetry does not, however, withdraw the validity and the generality of the obtained results.

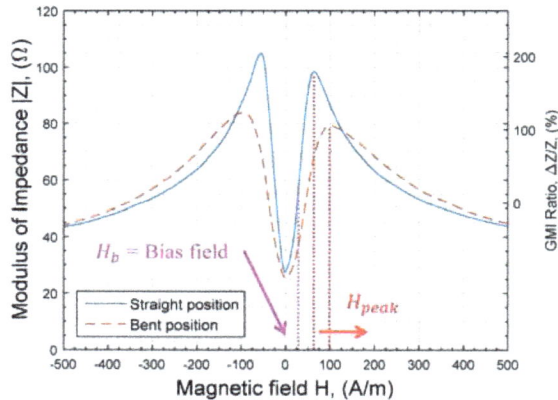

Figure 3. Change of $|Z(H)|$ and $\Delta Z/Z$ with the bending stress.

3.1. GMI Ratio and Anisotropy Field

For simplicity of illustration and explanation, we are interested, without a loss of generality, in the positive fields region only.

When the used sample was not bent (straight position), a typical maximum GMI ratio, $(\Delta Z/Z)_{max}$, of more than 220% was obtained. The field of the maximum impedance and the GMI ratio, H_{peak}, which is related to the anisotropy field H_k, was nearly equal to 64 A/m.

In the bent position, a decrease in the ratio $(\Delta Z/Z)_{max}$ to less than 160% was observed with H_{peak}, increasing to more than 100 A/m.

This result seems to be in agreement with the one that could be obtained with a tensile stress [10–17]. The used amorphous wires have a nearly zero magnetostriction [23]. Nevertheless, the domain structure of these wires is often considered to be similar to that of negative magnetostrictive wires [22]. In such a case, the quenched-in stresses due to fabrication by rapid solidification may cause the surface anisotropy to be circular and the inner anisotropy to be perpendicular to the wire axis, thereby leading to the formation of a specific domain structure, which consists of outer shell circular domains and inner core roughly axial domains [22]. This means that the anisotropy is circumferential in the outer shell of the wire.

In a first approximation, a bending stress could be assumed to be a combination of tensile stress along the outer fiber and compressive stress along the inner fiber of the wire, as illustrated by Figure 4. There is no deformation of the fiber along the neutral axis.

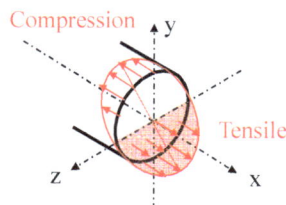

Figure 4. Representation of a bending stress.

In fact, the amplitude of the strain, σ, applied on the section of the wire can be calculated using the following formula:

$$\sigma = -E \times \frac{y}{\rho} \tag{3}$$

where E is the Young's modulus of the wire (16,000 kg/mm^2 in our case), y is the distance of a given fiber to the neutral axis, and ρ is the radius of curvature.

According to this formula, σ is positive when the distance y is counted negatively (outer fibers of the wire), and σ is negative when the distance y is counted positively (inner fibers of the wire). Therefore, the strain should theoretically be compensated in the total volume of the wire, as there is the exact same amount of tensile and compressive stress along the wire. However, this simplified model should be considered cautiously, as it appears from our experiments that there is a change in $|Z(H)|$ with the applied bending stress. This would actually suggest that the effect of the compressive stress is not exactly opposed to the effect of the tensile stress. It seems that the two effects are mixed in some manner, resulting in a dominating tensile stress effect.

The effect of the tensile stress has been intensively studied [10–17]. For a negative magnetostrictive wire, a tensile stress applied along the longitudinal axis will inhibit the orientation of the domains along this same axis, giving rise to a favorable domain orientation along the radial and azimuthal axes. This could cause an enhancement of the GMI ratio at low frequencies. However, at high frequencies, the wall mobility is extremely reduced, resulting in a very strong decrease in magnetic permeability [12]. The transverse anisotropy caused by the tensile stress in amorphous wires with vanishing (or slightly negative) magnetostriction reduces the GMI change when the value of the applied stress is large enough. For smaller stresses, the circumferential domain structure of the wires is refined, and the magnitude of the GMI effect can increase [15]. Many works have demonstrated that a commonly observed aspect, resulting from tensile stress, is that the field of the maximum impedance, H_{peak} (which is related to the anisotropy field H_k), is shifted towards higher values [10–17]. This shift is generally accompanied by a decrease in the value of maximum impedance and consequently of the GMI ratio. In our bending stress experiments, all of these observations were verified.

While the current study deals only with amorphous wires, it could be of great value to perform a similar investigation for the case of nanocrystalline materials in the future. These materials are usually processed with a controlled annealing of an amorphous precursor. Due to their nanometric grain sizes, these alloys exhibit outstanding soft magnetic proprieties, which make them good candidates for the occurrence of the GMI effect. In Fe-rich alloys, nanocrystallization produced by annealing largely enhances the magnitude of the GMI effect [24]. The same behavior as in amorphous alloys could appear in nanocrystalline alloys with an applied tensile stress, provided that they have the same sign of magnetostriction constant [25]. The tensile stress would enhance the transverse anisotropy of the material. The effect could be larger in amplitude, as the transverse anisotropy induced by tensile stress is larger in nanocrystalline alloys than in amorphous alloys [26]. Nevertheless, nanocrystalline materials are less adapted than their amorphous counterparts for our sensing application, as they exhibit poor mechanical properties.

The behavior of the GMI amorphous wires under compressive stress is, to our knowledge, not frequently investigated, despite a very recent experimental study in ribbons [27]. One might, however, argue that, as the magnetostriction constant is negative, a compressive stress applied along the longitudinal axis of the wire will induce a favorable orientation of the domains along this axis, reducing consequently the GMI effect, as the anisotropy is no more circumferential. More quantitative analysis of the wire's behavior under compressive stress is still, however, necessary.

The effect of bending or flexion stress on the GMI is even less known than the tensile and compressive stresses. In our results, and for the used length and bending curvature of the wire, it seems that this effect gives similar results to those of tensile stress. This observation should be considered cautiously, since the observed effect could be actually dependent on the bending curvature and on the parameters (frequency and amplitude) of the HF excitation current i_{ac}. According to this curvature, some competition could exist between tensile and compressive stress.

3.2. Sensitivity, Offset, and Reversibility

The change of the GMI ratio and the anisotropy under a bending stress is important to allow for a fundamental and rigorous understanding of the involved physical phenomena. However, the change in the local sensitivity at a given bias field, the change of the offset, and the reversibility are, from a practical point of view, crucial criteria in a real implementation of a GMI sensor involving bending. The quantification of these parameters is therefore essential.

In the absence of a bending stress, the maximum sensitivity for the used sample was achieved at a field of about 32 A/m (remember that we consider, and without loss of generality, the positive fields region only). In a sensor realization, this is generally chosen as a bias field (H_b in Figures 3 and 5). The associated sensitivity is noted $S(H_b)$.

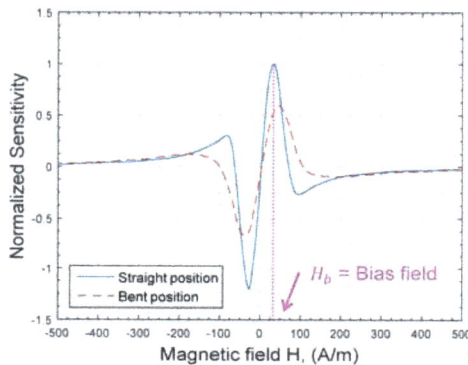

Figure 5. Change of the normalized sensitivity $\frac{1}{S(H_b)} \frac{\partial |Z(H)|}{\partial H}$, under bending stress. The sensitivity was normalized with respect to the maximum sensitivity, $S(H_b)$, obtained in a straight position of the wire.

In our experiments, the sensitivity $S(H) = \frac{\partial |Z(H)|}{\partial H}$ was obtained by differentiation of the $|Z(H)|$ curve of Figure 3. For simplicity reasons, we consider a normalized sensitivity with respect to the maximum sensitivity obtained in a straight position of the wire at H_b (i.e., $S(H_b)$).

This normalized sensitivity is then defined by $\frac{S(H)}{S(H_b)} = \frac{1}{S(H_b)} \frac{\partial |Z(H)|}{\partial H}$ and is plotted in Figure 5 for both straight and bent positions of the wire.

Then, when the wire is bent, the maximum sensitivity (in the region of positive fields) decreases to about 58% of $S(H_b)$. Moreover, this new maximum is obtained at a higher field of about 45 A/m (Figure 5). This obviously means that, for the design of GMI-based sensors, the value of the initially chosen bias field H_b (in the straight position of the wire) will not allow a maximum of sensitivity in bent positions. In fact, at this bias field H_b, the sensitivity in our sample was reduced to about 52% of its initial value.

The reduction of the maximum sensitivity was roughly consistent with the change of the field of maximum impedance, H_{peak}, and the value of the impedance at this same field $\left|Z\left(H_{peak}\right)\right|$. In fact, a rough estimation of the maximum sensitivity could be given by $\left|Z\left(H_{peak}\right)\right|/H_{peak}$ [28]. In our experiments, this ratio was decreased in the bent position to about 51% of its value in a straight position. The general trend between the changes of this ratio and the sensitivity is at least consistent.

In the applications of the GMI sensor that involve repetitive bendings, the quantification of the loss of sensitivity is then required. When the sensor is operating in a closed loop, it is important to ensure that the lower sensitivity, obtained in the bent position, will still be "high" so that the final sensor response will be dependent only on the parameters of the feedback (which could be developer-defined) [6].

Another practical issue to be considered is the change of the offset. At the bias field, H_b, the impedance $|Z(H_b)|$ was roughly equal to 64 Ω. Under bending, the change of $|Z(H)|$ at this same field was as high as 38% with $|Z(H_b)| = 40$ Ω. The instability of the offset is actually a major issue in a practical realization. An efficient offset canceling device has to be incorporated. Solutions could include the use of a difference amplifier [5] or a symmetrical AC bias field [4]. Each of these techniques has limitations that will not be discussed in this paper.

All the experimental results are summarized in Table 1.

Table 1. Summary of the results obtained with an amorphous Co-rich wire, with a bending stress applied.

Physical Quantities	With No Bending Stress Applied	With a Bending Stress Applied (Maximum Strain Applied along the Wire: 800 MPa)	Relative Change (%)
Maximum GMI ratio, $(\Delta Z/Z)_{max}$	220%	160%	−27%
Peak field, H_{peak} ($H_{peak} \approx H_k$)	64 A/m	100 A/m	+56%
Maximum normalized sensitivity, $\frac{1}{S(H_b)}\frac{\partial |Z(H)|}{\partial H}_{max}$	1	0.58	−42%
Normalized sensitivity at H_b (bias field), $\frac{1}{S(H_b)}\frac{\partial |Z(H)|}{\partial H}(H_b)$	1	0.52	−48%
Offset, $|Z(H_b)|$	64 Ω	40 Ω	−38%

While some solutions could be used to reduce the influence of the variations of the offset and to compensate for the loss in sensitivity in the practical utilization of the sensor (utilization that continuously involves bending stress of the sensitive element), one question must be carefully addressed. The reversibility and repeatability of the bending effect is actually a major concern. In the case of poor reversibility and repeatability, the efficiency of solutions that cancel the offset and compensate for the loss of sensitivity will actually be limited. This is why experimental tests of reversibility and repeatability were conducted on the same sample.

The wire was bent under the same conditions described in Section 2 and then relaxed to return to its initial straight position. For example, Figure 6 presents the change in the impedance $|Z(H)|$. with a first applied stress followed by a relaxation step (a return to the straight position) and then a second applied bending followed by relaxation. Figure 7 shows also the related curves of normalized sensitivities.

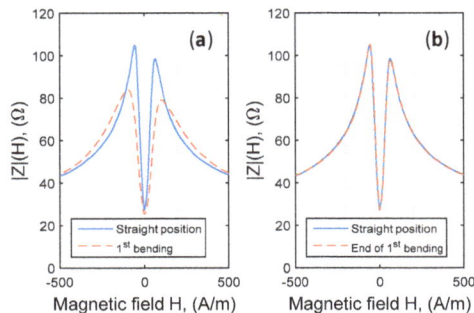

Figure 6. *Cont.*

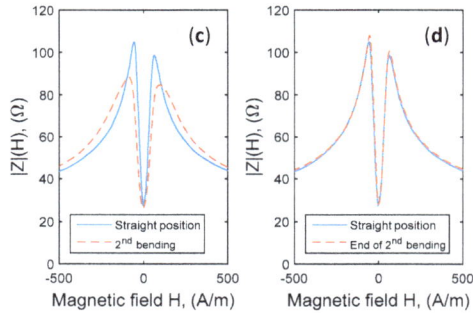

Figure 6. Reversibility and repeatability of the impedance curve, $|Z(H)|$, following two consecutive bending stresses. (**a**) Change in the modulus of the impedance, $|Z(H)|$, with a 1st bending stress applied; (**b**) Comparison between the modulus of the impedance after the relaxation of the bending stress (1st bending) and the initial straight position; (**c**) Change in the modulus of the impedance, $|Z(H)|$, with a 2nd bending stress; (**d**) Comparison between the modulus of the impedance after the relaxation of the bending stress (2nd bending) and the initial straight position.

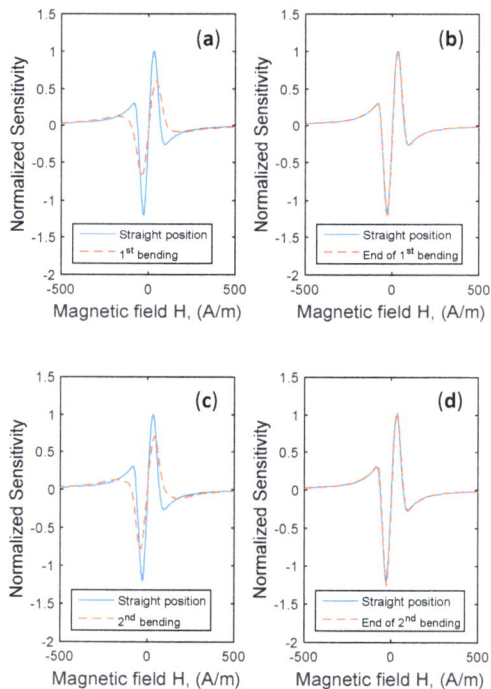

Figure 7. Reversibility and repeatability of the sensitivity following two consecutive bending stresses. (**a**) Change in the normalized sensitivity with a 1st bending stress; (**b**) Comparison between the normalized sensitivity after the relaxation of the bending stress (1st bending) and the initial straight position; (**c**) Change in the normalized sensitivity with a 2nd bending stress; (**d**) Comparison between the normalized sensitivity after the relaxation of the bending stress (2nd bending) and the initial straight position.

We have shown, for illustration purposes only, the effect of two consecutive bendings. About 10 consecutive bendings were conducted and showed that the change in the $|Z(H)|$ curve did not exceed a few percent of the absolute value of $|Z(H)|$. This value can be included in the error margin of two consecutive measurements (within our current experimental setup, two consecutive applied stresses can slightly differ).

Nevertheless, we can conclude that the first observations showed that the effect of the bending stress was roughly reversible and repetitive (obviously within the limit of elasticity of the wire). This result is actually of great value for sensor applications.

While this study has addressed an initial overall behavior of amorphous GMI Co-rich wires under bending stress, it is still obviously required that these experimental results be confirmed by theoretical analysis of model and completed by further experimental investigations. First of all, and unlike the effect of tensile stress, which has been largely studied in the literature, the effect of compressive stress should be carefully addressed from theoretical and experimental points of view. The general approximation made in considering that the bending effect is a combination of tensile and compressive effects must be refined. One of the two effects could dominate according to the diameter of the curvature of the bent wire. The influence of the frequency and of the amplitude of high frequency excitation current, i_{ac}, should also be considered.

All of these issues comprise the subject of our ongoing and future works.

4. Conclusions

We addressed an experimental study to evaluate the impact of the bending stress on the GMI effect in amorphous Co-rich wires. This bending stress was considered an influential parameter in our application of electrical current GMI sensors. Its effect on the relevant quantities for GMI sensor implementation was investigated. The results show that the GMI ratio, the intrinsic sensitivity, and the offset in the studied sample were affected by the bending. For the considered experimental conditions of bending and the used parameters of excitation current of the wire, both GMI ratio and maximum sensitivity were reduced (by roughly 50%). A net change in the offset was also observed. It has also been shown that the effect of bending seemed to be reversible and repetitive. However, more intensive work is still necessary to allow for a better understanding of the physical phenomena related to bending and compression. This understanding must also be confirmed by more experimental investigations.

Author Contributions: Julie Nabias contributed in the ideas of this work. She designed the setup and realized the measurements. She also performed the data processing, contributed to the interpretation of the results, and wrote the paper. Aktham Asfour and Jean-Paul Yonnet contributed ideas and participated in the setup design, results interpretation, and the paper writing.

Conflicts of Interest: The authors declare no conflict of interest.

References and Note

1. Rheem, Y.W.; Kim, C.G.; Kim, C.O.; Yoon, S.S. Current sensor application of asymmetric giant magnetoimpedance in amorphous materials. *Sens. Actuators A Phys.* **2003**, *106*, 19–21. [CrossRef]
2. Mapps, D.J.; Panina, L.V. Magnetic Field Detector and a Current Monitoring Device Including Such a Detector. U.S. Patent No. 7,564,239, 21 July 2009.
3. Ripka, P. Current sensors using magnetic materials. *J. Optoelectron. Adv. Mater.* **2004**, *6*, 587–592.
4. Malátek, M.; Ripka, P.; Kraus, L. Double-core GMI current sensor. *IEEE Trans. Magn.* **2005**, *41*, 3703–3705. [CrossRef]
5. Zhan, Z.; Yaoming, L.; Jin, C.; Yunfeng, X. Current sensor utilizing giant magneto-impedance effect in amorphous ribbon toroidal core and CMOS inverter multivibrator. *Sens. Actuators A Phys.* **2007**, *137*, 64–67. [CrossRef]
6. Asfour, A.; Yonnet, J.-P.; Zidi, M. A High Dynamic Range GMI Current Sensor. *J. Sens. Technol.* **2012**, *2*, 165–171. [CrossRef]

7. Fisher, B.; Panina, L.V.; Fry, N.; Mapps, D.J. High Performance Current Sensor Utilizing Pulse Magneto-Impedance in Co-Based Amorphous Wires. *IEEE Trans. Magn.* **2013**, *49*, 89–92. [CrossRef]

8. Han, B.; Zhang, T.; Zhang, K.; Yao, B.; Yue, X.; Huang, D.; Huan, R.; Tang, X. Giant magnetoimpedance current sensor with array-structure double probes. *IEEE Trans. Magn.* **2008**, *44*, 605–608.

9. Kudo, T.; Tsuji, N.; Asada, T.; Sugiyama, S.; Wakui, S. Development of a small and wide-range three-phase current sensor using an MI element. *IEEE Trans. Magn.* **2006**, *42*, 3362–3364. [CrossRef]

10. García, C.; Zhukov, A.; Blanco, J.M.; Zhukova, V.; Ipatov, M.; Gonzalez, J. Effect of Tensile Stresses on GMI of Co-Rich Amorphous Microwires. In Proceedings of the INTERMAG Asia 2005, Digests of the IEEE International Magnetics Conference, Nagoya, Japan, 4–8 April 2005; pp. 1273–1274.

11. Ciureanu, P.; Khalil, I.; Melo, L.G.C.; Rudkowski, P.; Yelon, A. Stress-induced asymmetric magneto-impedance in melt-extracted Co-rich amorphous wires. *J. Magn. Magn. Mater.* **2002**, *249*, 305–309. [CrossRef]

12. Knobel, M.; Sanchez, M.L.; Velazquez, J.; Vázquez, M. Stress dependence of the giant magneto-impedance effect in amorphous wires. *J. Phys. Condens. Matter* **1995**, *7*, L115. [CrossRef]

13. Knobel, M.; Pirota, K.R. Giant magnetoimpedance: Concepts and recent progress. *J. Magn. Magn. Mater.* **2002**, *242*, 33–40. [CrossRef]

14. Knobel, M.; Vázquez, M.; Kraus, L. Giant magnetoimpedance. *Handb. Magn. Mater.* **2003**, *15*, 1–92.

15. Atkinson, D.; Squire, P.T. Experimental and phenomenological investigation of the effect of stress on magneto-impedance in amorphous alloys. *IEEE Trans. Magn.* **1997**, *33*, 3364–3366. [CrossRef]

16. Knobel, M.; Vazquez, M.; Hernando, A.; Sánchez, M.L. Effect of tensile stress on the field response of impedance in low magnetostriction amorphous wires. *J. Magn. Magn. Mater.* **1997**, *169*, 89–97.

17. Mandal, K.; Puerta, S.; Vazquez, M.; Hernando, A. The frequency and stress dependence of giant magnetoimpedance in amorphous microwires. *IEEE Trans. Magn.* **2000**, *36*, 3257–3259. [CrossRef]

18. Blanco, J.M.; Zhukov, A.; Gonzalez, J. Effect of tensile and torsion on GMI in amorphous wire. *J. Magn. Magn. Mater.* **1999**, *196*, 377–379. [CrossRef]

19. Kim, C.G.; Yoon, S.S.; Vazquez, M. Evaluation of helical magnetoelastic anisotropy in Fe-based amorphous wire from the decomposed susceptibility spectra. *J. Magn. Magn. Mater.* **2001**, *223*, 199–202. [CrossRef]

20. Zhao, Y.; Hao, H.; Zhang, Y. Preparation and giant magneto-impedance behavior of Co-based amorphous wires. *Intermetallics* **2013**, *42*, 62–67. [CrossRef]

21. Seddaoui, D.; Ménard, D.; Movaghar, B.; Yelon, A. Nonlinear electromagnetic response of ferromagnetic metals: Magnetoimpedance in microwires. *J. Appl. Phys.* **2009**, *105*, 083916. [CrossRef]

22. Phan, M.; Peng, H. Giant magnetoimpedance materials: Fundamentals and applications. *Prog. Mater. Sci.* **2008**, *53*, 323–420. [CrossRef]

23. Co-Based "SENCY"TM, Unitika Ltd. Datasheet

24. Hernando, B.; Sanchez, M.L.; Prida, V.M.; Tejedor, M.; Vázquez, M. Magnetoimpedance effect in amorphous and nanocrystalline ribbons. *J. Appl. Phys.* **2001**, *90*, 4783–4790. [CrossRef]

25. Li, Y.F.; Vazquez, M.; Chen, D.X. Giant magnetoimpedance effect and magnetoelastic properties in stress-annealed FeCuNbSiB nanocrystalline wire. *IEEE Trans. Magn.* **2002**, *38*, 3096–3098. [CrossRef]

26. Yang, X.L.; Yang, J.X.; Chen, G.; Shen, G.T.; Hu, B.Y.; Jiang, K.Y. Magneto-impedance effect in field-and stress-annealed Fe-based nanocrystalline alloys. *J. Magn. Magn. Mater.* **1997**, *175*, 285–289. [CrossRef]

27. Beato, J.J.; Algueta-Miguel, J.M.; de la Cruz Blas, C.; Santesteban, L.G.; Perez-Landazabal, J.I.; Gomez-Polo, C. GMI Magnetoelastic Sensor for Measuring Trunk Diameter Variations in Plants. *IEEE Trans. Magn.* **2016**. [CrossRef]

28. Ménard, D.; Seddaoui, D.; Melo, L.G.C.; Yelon, A.; Dufay, B.; Saez, S.; Dolabdjian, C. Perspectives in Giant Magnetoimpedance Magnetometry. *Sens. Lett.* **2009**, *7*, 339–342. [CrossRef]

sensors

MDPI

Review

Recent Advances in Electrospun Nanofiber Interfaces for Biosensing Devices

Eleni Sapountzi [1], Mohamed Braiek [1,2], Jean-François Chateaux [3], Nicole Jaffrezic-Renault [1] and Florence Lagarde [1,*]

[1] Université Lyon, CNRS, Université Claude Bernard Lyon 1, ENS de Lyon, Institute of Analytical Sciences, UMR 5280, 5 Rue la Doua, F-69100 Villeurbanne, France; elena.sapountzi@gmail.com (E.S.); mohamed_braiek@yahoo.fr (M.B.); nicole.jaffrezic@isa-lyon.fr (N.J.-R.)

[2] Laboratoire des Interfaces et des Matériaux Avancés, Faculté des Sciences de Monastir, Avenue de l'Environnement, University of Monastir, Monastir 5019, Tunisia

[3] Université Lyon, Université Claude Bernard Lyon 1, CNRS, Institut des Nanotechnologies de Lyon, UMR5270, Bâtiment Léon Brillouin, 6, rue Ada Byron, F-69622 Villeurbanne CEDEX, France; jean-francois.chateaux@univ-lyon1.fr

* Correspondence: florence.lagarde@isa-lyon.fr; Tel.: +33-437-423-556

Received: 4 July 2017; Accepted: 13 August 2017; Published: 16 August 2017

Abstract: Electrospinning has emerged as a very powerful method combining efficiency, versatility and low cost to elaborate scalable ordered and complex nanofibrous assemblies from a rich variety of polymers. Electrospun nanofibers have demonstrated high potential for a wide spectrum of applications, including drug delivery, tissue engineering, energy conversion and storage, or physical and chemical sensors. The number of works related to biosensing devices integrating electrospun nanofibers has also increased substantially over the last decade. This review provides an overview of the current research activities and new trends in the field. Retaining the bioreceptor functionality is one of the main challenges associated with the production of nanofiber-based biosensing interfaces. The bioreceptors can be immobilized using various strategies, depending on the physical and chemical characteristics of both bioreceptors and nanofiber scaffolds, and on their interfacial interactions. The production of nanobiocomposites constituted by carbon, metal oxide or polymer electrospun nanofibers integrating bioreceptors and conductive nanomaterials (e.g., carbon nanotubes, metal nanoparticles) has been one of the major trends in the last few years. The use of electrospun nanofibers in ELISA-type bioassays, lab-on-a-chip and paper-based point-of-care devices is also highly promising. After a short and general description of electrospinning process, the different strategies to produce electrospun nanofiber biosensing interfaces are discussed.

Keywords: electrospinning; biosensing devices; bioreceptor immobilization; carbon nanofibers; metal oxide nanofibers; polymer nanofibers; metal nanoparticles; carbon nanotubes

1. Introduction

Recent advances in nanoscience and nanotechnology have opened up new horizons for the development of biosensors of enhanced sensitivity, specificity, detection time, and low cost. Sensors miniaturization provides great versatility for incorporation into multiplexed, portable, wearable, and even implantable medical devices [1–4]. Engineering of the transducer surface using nanomaterials (i.e., nano-sized objects or nanoengineered/nanostructured materials) brings novel and sometimes unique properties that have been extensively harnessed in the past few years and addressed in several recent reviews [5–9]. The nanostructures used in the construction of biosensor devices vary in size (1 to 100 nm in at least one of their dimensions), shape (nanoparticles, nanotubes, nanorods, nanowires, nanofibers, nanosheets . . .), chemical nature (carbon-based materials, metals, metal oxides, polymers

. . .) and physicochemical properties (electronic, optical, magnetic, mechanical, thermal . . .). The attractiveness of such nanomaterials relies not only on their ability to act as efficient and stabilizing platforms for the biosensing elements, but also on their small size, large surface area, high reactivity, controlled morphology and structure, biocompatibility, and in some cases electrocatalytic properties. Structuring the transducer at the nanoscale contributes to enlarge the overall surface available for bioreceptor immobilization. Incorporation of nanomaterials into the sensing layers is also often associated with higher mass transfer rates, acceleration and magnification of the transduction process, contributing to signal amplification and faster biosensor response.

Among the variety of nanostructuring materials, nanofibers (NFs) produced by electrospinning have been the object of growing interest during the past decade and the technique has been extensively reviewed with respect to its setup, mechanism, applications, advantages, technical issues, and prospective developments [10–27]. Electrospinning is a convenient and powerful technique to generate uniform sub-micron fibers in a continuous process and at large scale from a rich variety of polymers. Fiber mats, exhibiting large surface areas, may be produced on a support or used as self-standing substrates [10–12]. Compared to the techniques commonly used for fibers production (i.e., drawing, phase separation, template assisted or self-assembly techniques), electrospinning combines simplicity, versatility and low cost with superior capabilities to elaborate scalable ordered and complex nanofibrous assemblies [13]. Electrospun NFs have demonstrated high potential for a wide spectrum of applications such as drug delivery [13–16], tissue engineering [13–15,17], water treatment [18], energy conversion and storage [19,20], or electronics [21]. Due to their large surface areas, high porosity and their ability to be easily functionalized, nanofiber mats produced by electrospinning have also been increasingly exploited to enhance performances of analytical devices. Many papers have been recently dedicated to the various and exciting potentialities existing in the field. NFs can be used as sorbents in microextraction techniques [22], as chromatographic phases [23], as separators, concentrators or mixers in microfluidic systems [24–26]. A large number of physical and chemical sensors based on electrospun NFs have been reported, mostly for gas sensing [27,28]. The number of works related to the fabrication of electrospun NFs-based biosensing devices is also growing fast. One of the critical points to achieve efficient biosensing devices, regardless of their type (biosensors or bioassays, static of flow systems) and of the nature of biosensing elements (enzymes, antibodies . . .), is to elaborate functional NFs biointerfaces. This means biointerfaces where the bioreceptors' activity and accessibility, binding events, and signal transductions are optimal. The optimization should be performed considering the analytical performances targeted, some of them (e.g., accuracy, sensitivity, selectivity) being often prioritized compared to others (e.g., response time, long-term stability).

Herein, we intend to provide the reader with a comprehensive overview of the application of electrospun NFs to the biosensing area, reporting the more recent contributions and advances in the field. Very few reviews have been dedicated to the topic [29–32] and none of them provide a complete survey, focusing more on some specific materials used to produce NFs scaffolds (e.g., metal oxides [29], nanomaterial/polymer composites [30]), the formation of biomolecules/NFs biocomposites [31], or the detection of specific target analytes (e.g., glucose [32]). After a short and general description of electrospinning process, the different strategies used in the elaboration of electrospun NFs biointerfaces for biosensor applications will be discussed. Then, recent developments in the area of ELISA-type bioassays and point-of-care devices, with a special attention to lab-on-a chip and paper-based systems, will be addressed. All kinds of bioreceptor and transduction mode, immobilization strategy, NFs material and field of application will be considered.

2. General Overview of Electrospinning Process

Electrospinning is an electrostatically-driven process ideally conducted under controlled temperature and humidity conditions [10]. Two standard set-ups, i.e., vertical or horizontal, are currently available to produce single fibers or NF mats. Both consist of four major components: a high-voltage power supply, a spinneret with a metallic needle, a syringe pump and a grounded collector (Figure 1). A polymer melt or solution is dispensed at a constant and controlled rate through the spinneret into a high voltage electrical field generated between the end of the needle and the collector. The droplet formed at the end of the needle tip, subjected to the electrical field, is first elongated into a conical shape termed as the Taylor cone. When the repulsive electrical forces overcome the surface tension forces, an electrified polymer jet ejects from the apex of the cone. The jet is then elongated and whipped continuously by electrostatic repulsion forces. Thinning of the jet results in solvent evaporation or solidification of the melt, and the deposition of a nonwoven web of solid NFs on the collector. The electrospinning process is therefore governed by a variety of forces, e.g., the Coulomb force between charges on the jet surface, the electrostatic force due to the external electric field, the viscoelastic force of the solution, the surface tension, the gravitational force (Figure 1b). The technique is highly versatile since, besides the conventional non-oriented fibers, a variety of morphologies and structures (e.g., aligned or crossed fiber arrays, ribbon, necklace-like, nanowebs, hollow, helical or coil fibers ...) can be obtained by modifying electrospinning conditions or set-up (e.g., collector configuration) [33].

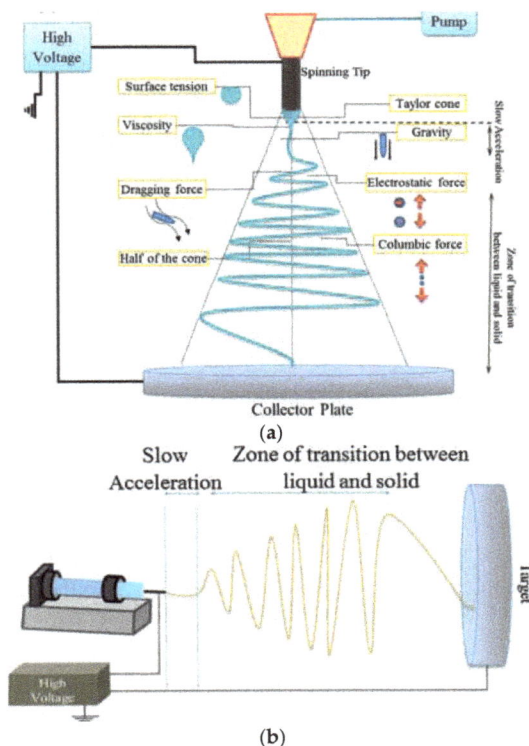

Figure 1. Typical horizontal (**a**) and vertical (**b**) electrospinning set-ups. They are represented with a static collector plate but other configurations exist. Reprinted with permission from [16]. Copyright 2016 Elsevier.

For specific applications, e.g., drug delivery or biosensing, where bioactive molecules should be entrapped in the fibers but should not be in direct contact with organic solvents, core-shell type NFs can be produced using modified spinneret/pump configurations, as shown in Figure 2 [16]. The thickness of the sheath and the size of the inner part of the fiber can be easily controlled by tuning the ratio of inner-outer injection speed during electrospinning.

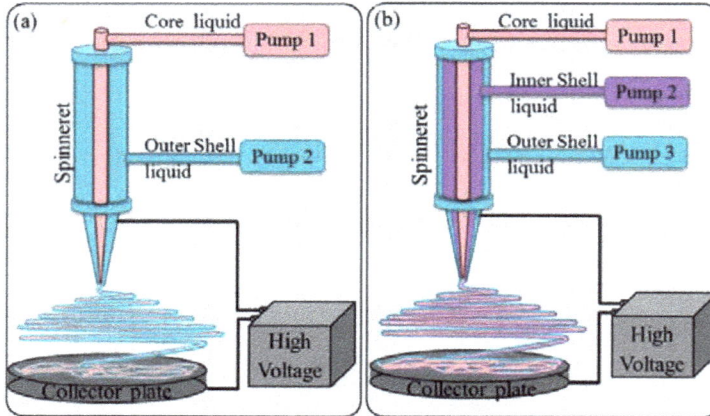

Figure 2. Coaxial (**a**) and triaxial (**b**) electrospinning set-ups. Reprinted with permission from [16]. Copyright 2016 Elsevier.

NFs' diameter and morphology can be controlled by a proper selection of a variety of parameters related to spinning process (i.e., applied voltage, flow-rate, collector type, needle diameter, tip-to-collector distance), polymer solution (i.e., concentration and molecular weight of the polymer, solution viscosity, surface tension, conductivity), and ambient conditions (humidity, temperature, pressure, type of atmosphere). Generally speaking, NFs' diameter increases with increasing viscosity and polymer concentration and with decreasing conductivity, while fiber beading can be avoided in rising viscosity or using polymers with higher molecular weight. High surface tension favors jet instability. As regards as processing parameters, high voltage and low feed rate tend to decrease NFs diameter, while fiber beading is observed when the tip-to-collector distance and the polymer solution feed rate are not sufficient. Humidity and temperature conditions are also of uppermost importance, since excess of humidity results in pores formation and too high temperature leads to decreased fibers diameter [10,13].

Electrospun NFs may be fabricated from a remarkably wide range of polymers, either natural (e.g., silk fibroin, collagen, chitosan, gelatin ...) or synthetic (e.g., polyacrylonitrile (PAN), polystyrene (PS), polyvinylalcohol (PVA), polyvinylpyrrolidone (PVP), polyethylenimine (PEI) ...) incorporating or not fillers such as metal salts, metal nanoparticles (NPs), carbon nanotubes (CNTs), graphene, fluorescent or photoluminescent markers to provide additional advanced functionalities [28,30,34].

3. Electrospun NFs in Biosensors

Electrospun NF-based biosensors reported in the literature mostly rely on electrochemical transduction. Sensing elements are principally enzymes, antibodies, more scarcely DNA strands or aptamers.

The bioreceptors can be immobilized using various strategies, depending on the physical and chemical characteristics of both the recognition elements and the NF scaffolds, and on their interfacial interactions. The most common methods proposed to generate bioreceptor-NF hybrid assemblies consist in the attachment of the biomolecules onto the fiber surface by physical or chemical sorption, covalent binding, cross-linking or entrapment in a membrane. This approach has been extensively used to immobilize enzymes [35–61], antibodies [62–70], DNA strands [71–73] and aptamers [74,75].

Another way to proceed, more specifically developed for enzyme biosensors, consists in entrapping the bioactive molecules inside the NFs by electrospinning a blend of enzymes and polymer [76–85].

Retaining the bioreceptor functionality is one of the main challenges associated with the production of NF-based biosensing interfaces. An important requirement for the immobilization step is that the matrix provide a favorable and inert environment for the biomolecules, i.e., it should not induce severe modifications in their native structure, which would compromise their biological activity (recognition capacities, reactivity, and/or selectivity). From that point-of-view, hydrophilic polymers such as PVA or PEI are particularly well-suited and have been extensively used in the entrapment immobilization strategy. However, the resulting water-soluble NFs must be treated to guarantee further operational stability of the biosensor in aqueous media. This is generally done by cross-linking using liquid or gaseous glutaraldehyde (GA). The operational stability issue can also be solved by generating NFs of low water solubility (hydrophic polymers, carbon, or metal oxide NFs) further used to attach the bioreceptors. In this case, a post-treatment of the fibers is generally performed to improve the biocompatibility of the surface and introduce functional groups if needed.

To obtain highly sensitive biosensors, the fiber mats produced by electrospinning should also provide a large active surface area. This means that high bioreceptor loadings are requested and that the biomolecules should, not only keep their biological functionality, but also remain accessible to the molecules to be detected.

To achieve highly effective biosensors, it is finally very important that the biorecognition processes occurring at biointerfaces should be efficiently transduced into measurable signals. Different strategies have been proposed, e.g., the production of NFs doped with conducting NMs for electrochemical biosensors. Electrospinning process enables for the homogeneous distribution of nanoobjects (NPs, CNTs . . .) inside the fibers, resulting in enhanced electron transfers from the recognition sites to the electrodes.

3.1. Attachment of Sensing Biomolecules onto Electrospun NFs

NFs fabricated from a variety of materials, including metal oxides, carbon, and polymers doped or not with conductive nanomaterials (e.g., metal NPs, CNTs) or covered with a conductive polymer layer, have been explored as functional platforms for immobilizing the biosensing molecules. A summary of the enzymatic biosensors prepared this way is given in Table 1. As shown in the table, the different immobilization strategies, i.e., covalent binding, cross-linking, sorption, and entrapment in a membrane, have been used. Due to its simplicity and versatility, adsorption approach is the most commonly employed, but entrapment and covalent strategies may be preferred to avoid leaching issues. It has been shown, in many cases, that appropriate grafting of the biomolecules onto surfaces can prevent molecular movements that typically lead to conformational changes and enzymes inactivation [86].

Table 1. Electrospun NFs-based enzyme biosensors fabricated by attachment of the enzymes onto the fibers. Unless otherwise stated, the detection is performed by amperometry.

NFs	Treatment after Electrospinning	Bioreceptor (Analyte)	Fixation of the Bioreceptor	References
Metal oxide NFs				
ZnO (from PVP and ZnAc)	Thermal (700 °C). Addition of a PVA film	Gox (Glucose)	Adsorption	[35]
TiO$_2$ (from PVP and Ti(BuO)$_4$)	Thermal (500 °C, 5 min)	Gox (Glucose)	Adsorption followed by coverage with Chit membrane	[36]
TiO$_2$ (from PVP and Ti(PrO)$_4$) [1]	Thermal (470 °C, 4 h). Oxygen plasma (COOH)	ChEt and ChOx (CholOl)	Covalent binding (EDC/NHS)	[37]
Mn$_2$O$_3$–Ag (from PVP, Mn(NO$_3$)$_2$, AgNO$_3$)	Thermal (500 °C, 3 h). Dispersion of NFs in Nafion/GOx and casting	Gox (Glucose)	Entrapment into Nafion + cross-linking with GA	[38]
Carbon NFs				
Carbon-Cu (from PAN/PVP and CuAc$_2$)	Thermal (280 °C, 2 h in air; 900 °C, 1 h in N$_2$). Dispersion of NFs in Nafion/Lac and casting	Lac (catechol)	Entrapment into Nafion	[42]
Carbon-Ni (from PAN and NiAc$_2$)	Thermal (280 °C, 2 h and 900 °C, 1 h in N$_2$). Dispersion of NFs in Lac/DA and casting	Lac (catechol)	Entrapment into poly(DA)	[43]
Carbon-NCNS	Thermal (250 °C, 2 h in air; 900 °C, 30 min in N$_2$)	Gox (Glucose)	Adsorption followed by coverage with Nafion membrane	[44]
Mesoporous carbon (from PAN and SiO$_2$ NPs)	HF, 24 h Thermal	Gox (Glucose)	Adsorption	[45]
Carbon (from PAN)	Thermal (230 °C, 3 h in air; 300 °C, 2 h in H$_2$/Ar; 1200 °C, 0.5 h in Ar); HNO$_3$ 12 h (COOH); Growth of HA on NFs	CytC (H$_2$O$_2$)	Adsorption	[46]
Carbon (from PAN)	Thermal (230 °C, 3 h in air; 300 °C, 2 h in H$_2$/Ar; 1200 °C, 0.5 h in Ar); HNO$_3$ 12 h (COOH); Growth of PBNs on NFs	Gox (Glucose)	Entrapment in Chit membrane	[47]

Table 1. *Cont.*

NFs	Treatment after Electrospinning	Bioreceptor (Analyte)	Fixation of the Bioreceptor	References
Polymer NFs				
PMMA-MWCNTs(PDDA)		Gox (Glucose)	Adsorption	[48]
PANCAA-MWCNTs		Gox (Glucose)	Covalent binding (EDC/NHS)	[49]
Chit-MWCNTs (from Chit/PVA/MWCNTs)	0.5 M NaOH, 4 h (PVA removal)	Uricase (uric acid)	Cross-linking with GA	[50]
PAN/MWCNTs [1]	Chemical reduction (LiAlH₄)	PPO (catechol)	Cross-linking with GA	[51]
PAA/Nafion/Au NPs [1]		HRP (H₂O₂)	Electrostatic interactions	[52]
PVA/AgNPs [1] (from PVA-AgNO₃)	Reduction of AgNO₃ with EGCG	HRP (H₂O₂)	Adsorption	[53]
PVA/PEI/AgNPs (from PVA/PEI)	Immersion in AgNO₃ Reduction with EGCG	HRP (H₂O₂)	Adsorption	[53]
PVA/PEI-PdNPs (from PVA/PEI) [1]	Immersion in PdCl₂ and reduction with NaBH₄	HRP (H₂O₂)	Adsorption	[54]
PAN-Au NPs-MWCNTs (from PAN)	Electrodeposition of AuNPs and electrophoretic deposition of MWCNTs	Gox (Glucose)	Covalent binding (EDC/NHS)	[55]
PAN/PANI [1]	-	GDH (Glucose)	Covalent binding (EDC/NHS)	[56]
CA-CMC-PANI (from CA)	Immersion in CMC Immersion in ANI and polymerization	Lac (catechol)	Entrapment in Nafion	[57]
PLLA-PEDOT/PSS (from PLLA)	Immersion in EDOT-GOx and electropolymerization	Gox (Glucose)	Entrapment in PEDOT/PSS	[58]
Nylon 6,6-MWCNTs-poly(BIBA) (from Nylon 6,6-MWCNTs)	Immersion in BIBA and polymerization	Gox (Glucose)	Cross-linking with GA	[59]
PAN-MWCNTs/Ppy (from PAN_MWCNTs)	Immersion in FeTos and Py vapour polymerization	Gox (Glucose)	Adsorption	[60]
PS/Ir complex [2]		Gox (Glucose)	Adsorption	[61]

[1] Cyclic voltammetry detection; [2] luminescence detection; BIBA: 4-(4,7-di(thiophen-2-yl)-1H-benzo[d]imidazol-2-yl)benzaldehyde, CA: cellulose acetate, ChEt: cholinesterase, Chit: chitosan, CholOl: cholesterol oleate, CMC: carboxymethylcellulose, ChOx: choline oxidase, CA: dopamine, EGCG: epigallocatechin gallate, FeTos : Fe(III) p-toluenesulfonate, GDH: glucose dehydrogenase, GOx: glucose oxidase, HA: hydroxyapatite, HRP: horseradish peroxidase, Lac: laccase, MWCNTs: Multiwall carbon nanotubes, NCNS: nitrogen doped carbon nanosphere, PAN: polyacrylonitrile, PBNs: Prussian blue nanostructures, PDDA: poly(diallyldimethylammonium chloride), PEDOT: poly(3,4-ethylenedioxythiophene), PEI: polyethylenimine, PLLA: poly(L-lactide), PMMA: poly(methylmethacrylate), PPy: polypyrrole, PS: polystyrene, PSS: poly(sodium-p-styrene sulfonate), PVA: poly(vinylalcohol), PVP: polyvinylpyrrolidone.

3.1.1. Metal-Oxide NFs

New developments in metal-oxide NFs-based electrochemical biosensors have been discussed in a recent review [29]. Metal-oxide NFs can be produced by electrospinning a solution containing an inorganic precursor (metal alkoxide or metal salt) and a sacrificial polymer carrier. PVP is one of the most current polymers used for this purpose. The as-spun inorganic/organic composite is further calcinated at high temperature to remove the polymer and oxidize the precursor to produce the metal-oxide phase by nucleation and growth. The fabrication of metal-oxide NFs is more complicated than the production of polymer NFs. Apart from solvent evaporation, reactions such as hydrolysis, condensation and gelation of the inorganic precursors occur once the liquid jet is ejected from the needle. If gelation process is too rapid, clogging effects are observed, and the jet becomes less elastic, preventing the electrospinning process from running properly. Controlling all the solution and process parameters precisely (i.e., type and concentration of the precursor, concentration of the polymer host, incorporation of suitable additives) is therefore essential [87]. During calcination, the polymers decompose while the inorganic precursors oxidize and crystallize forming nanocrystals aligned along what used to be the as-spun fibers. The resulting polycrystalline metal-oxide NFs exhibit unique morphologies with large surface areas, nanopores coexisting with larger pores in the structure. As such, they possess enhanced mass transfer capacities and improved electronic and optical properties compared to thin films of the same nature. Moreover, the metal-oxide NFs can be collected using a metal-frame collector to obtain aligned configurations, which facilitates the production of transistor based biosensors.

Due to the removal of the polymer and sintering of the metal oxide phase during calcination, shrinkage in NF diameter is observed, accompanied by thermal and internal mechanical stress, resulting in brittle NFs. The poor mechanical strength of metal-oxide NFs might impede the long-term stability of derived biosensors, but this issue can be solved by incorporating suitable additives in the electrospinning solution or to the NFs before heating [29].

ZnO NFs-based biosensors have been produced for the detection of various target analytes of biomedical interest. For example, Ahmad et al. [35] reported the successful fabrication of ZnO NFs from PVP and Zn acetate for amperometric glucose determination. After calcination, the NFs were transferred onto a gold electrode and covered with a PVA film to ensure their firm attachment to the electrode. Glucose oxidase (GOx) was further immobilized on the surface through physical adsorption. The biosensor response to glucose injection was very rapid (4 s) and a linear range from 0.25 to 19 mM with a low limit of detection (LOD) of 1 μM could be achieved. Immunosensors based on ZnO NFs prepared from PAN/Zn acetate blends have been also developed for the electrochemical detection of histidine-rich protein-2, synthesized and released into the blood stream by the most lethal and prominent malarial parasite *Plasmodium falciparum* [62], epidermal growth factor receptor 2ErbB-2, a breast cancer biomarker [63] and carcinoma antigen-125, an ovarian cancer biomarker [64]. To decrease the intrinsic resistivity of ZnO, and therefore improve the biosensor sensitivity, it was proposed to dope the NFs with Cu [62] or multiwall carbon nanotubes (MWCNTs) [64] by incorporating copper nitrate or MWCNTs in the electrospinning solution. Reactive carboxyl groups were generated on the NF surface by oxygen plasma treatment [64], thermal oxidation of MWCNTs [62], or by assembling mercaptopropionic acid on ZnO via interaction of hydrosulphide groups (HS^-) with Zn^{2-} [63]. These groups were subsequently used to conjugate antibodies with the NF platform via the amino groups of the proteins using the well-established EDC/NHS coupling reaction.

In a similar way, TiO_2, Mn_2O_3 and IrO_x NFs produced by electrospinning have been reported as very promising interfaces for biomolecules immobilization and successfully used for the elaboration of ultrasensitive biosensors. The TiO_2 NFs were prepared from PVP and Ti butoxide or propoxide [36,37,66]. Tang et al. [36] modified Pt electrodes with TiO_2 electrospun fiber mats and adsorbed GOx on them by drop coating. They proposed to improve immobilization by adding a chitosan film (Chit) on the GOx/TiO_2 NFs/Pt electrodes. The addition of the TiO_2 NFs helped increasing the biosensor response towards glucose by a factor 2.7. The Chit/GOx/TiO_2 NFs/Pt electrode response to 100 µM glucose was respectively 4.6 and 74 times higher than those of Chit/GOx/Pt and Chit/GOx/TiO$_2$ film/Pt, demonstrating the positive effect of NFs structuring. The same authors investigated the influence of TiO$_2$ NFs density on the biosensor signal and demonstrated that there was an optimal value. The quantity of TiO$_2$ NFs had to be sufficient to guarantee the immobilization of a large amount of GOx molecules, but too high densities resulted in a strong electron transfer resistance at the interface. More recently, Mondal et al. [37] produced partially aligned mesoporous TiO$_2$ NFs at the surface of an indium tin oxide (ITO) electrode. Two enzymes, cholesterol oxidase and cholesterol esterase, were immobilized covalently to the NFs previously treated by oxygen plasma. Esterified cholesterol could be successfully detected by cyclic voltammetry with a low limit of detection (0.49 mM) and response time (20 s). In another study, a cell capture immunoassay based on electrospun TiO$_2$ NF mats was proposed and applied to the detection of circulating tumor cells from colorectal and gastric cancer patients [65].

Mn$_2$O$_3$ NFs have been also reported as efficient platforms for the immobilization of GOx or DNA probes in view of the electrochemical detection of glucose [38] or dengue consensus primer [71]. Mn salts and PVP [38] or PAN [71] were used as inorganic precursors and sacrificial polymers, respectively. In the first work, AgNO$_3$ was also added to the electrospinning solution. The morphology of Mn$_2$O$_3$-Ag NFs produced after calcination of electrospun PVP/Mn(NO$_3$)$_2$/AgNO$_3$ was different from that of the Mn$_2$O$_3$ NFs obtained from PVP/Mn(NO$_3$)$_2$. Ag NPs coalesced together, generating highly porous NFs with enhanced enzyme loading capacity and improved electrochemical features, could be observed by TEM. The NFs, dispersed into a 1 wt % Nafion solution and mixed with GOx, were further dropped onto the electrode and GOx was finally cross-linked using glutaraldehyde (GA) vapour. Glucose could be detected by amperometry with a low limit of detection (1.73 µM) and selectivity towards uric and ascorbic acids was demonstrated. In the second work, label-free zeptomolar detection of DNA hybridization was achieved and a dengue consensus primer was successfully quantified both in control and spiked serum samples (limit of detection: 120×10^{-21} M).

Li et al. [66] recently proposed a label-free immunosensor based on a glassy carbon electrode (GCE) modified by IrOx NFs/CS (Figure 3). IrOx composition ($0 \leq x \leq 2$) could be controlled by changing the annealing temperature. At 500 °C, wire-in-tube nanostructures of high surface area and rapid electron transfer kinetics were produced. TEM analysis evidenced that independent nanowires are embedded in the NFs and that the wire-intube nanostructures possess separated walls along nearly their entire length (Figure 4). The average diameter of the inside wire and whole NF was around 70 and 110 nm, respectively, and the detailed nanostructure can be observed more clearly in the inset of Figure 4a. Antibodies were adsorbed on the modified electrode and the biosensor was successfully used for amperometric detection of the cancer biomarker α-fetoprotein in human serum, with a limit of detection of 20 pg mL^{-1}.

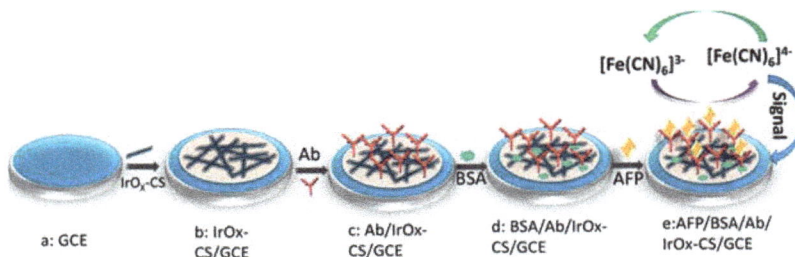

Figure 3. Principle of the electrospun NFs-based immunosensor for amperometric detection α-Fetoprotein proposed by Li et al. Reprinted with permission from [66]. Copyright 2015 American Chemical Society.

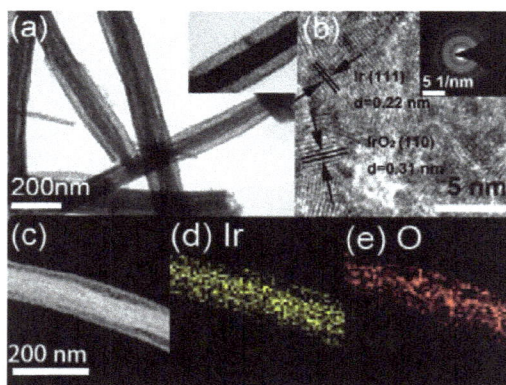

Figure 4. (**a**) TEM, (**b**) HRTEM, and (**c**) STEM images of IrO*x* nanofibers after annealing at 500 °C. The inset of panel (**a**) is the enlarged TEM image and inset of panel (**b**) is the SAED pattern. Panels (**d**,**e**) are the corresponding EDX mapping of Ir and O elements. Reprint with permission from [66]. Copyright 2015 American Chemical Society.

3.1.2. Carbon NFs

Owing to its excellent mechanical and chemical resistance as well as its superior electronic properties, carbon is one of the most commonly used electrode materials in the elaboration of electrochemical sensors and biosensors. Carbon-based NMs, such as CNTs, have been extensively integrated within electrochemical biosensors to improve their sensitivity. Compared to CNTs, carbon electrospun NFs (CENFs) can provide higher surface areas for biomolecule immobilization and are more easily produced. They offer similar conductivities and can be modified to introduce suitable functional groups or combined with conducting NMs, e.g., metal NPs, to generate hydrophilic surfaces of enhanced electronic properties and larger biomolecules loading capacities.

Among the variety of polymers usable to produce CENFs, PAN is the most commonly employed, as it offers both a high carbon yield and the capacity to generate NFs of superior mechanical properties [39–41]. Carbonaceous NFs are fabricated from electrospun polymer NFs following a two-step process. First, NFs are stabilized at relatively low temperatures (200–300 °C) in an oxidative atmosphere to convert thermoplastic polymer NFs to condensed thermosetting NFs. Second, the NFs are carbonized at higher temperatures (typically 800–1300 °C) in an inert atmosphere. The microstructural and electronic properties of the fibers as well as their surface chemistry can be modulated by changing the composition of the electrospun solution, adapting the thermal treatment conditions or by adding a suitable post-modification process.

Doping CENFs with metallic nanoparticles (MNPs), for example, has been shown to be an efficient method to enhance the sensitivity and stability of CENF-based biosensors. CENF-MNP composites are generally produced by electrospinning a precursor containing the polymer and a metal salt, the latter being converted into MNPs during the thermal treatment process [88]. Fu et al. [42] reported the fabrication of CENFs doped with Cu from a solution containing PAN, PVP and copper acetate. The electrospun NFs were further calcinated and dispersed in 0.1 M acetate buffer (pH 4), mixed with a laccase (Lac)/Nafion solution and casted on a glassy carbon electrode for amperometric detection of catechol. A CENF/Lac/Nafion/GCE without Cu was also produced for comparison. The CENFs, as observed by SEM (Figure 5a), exhibited a fibrous structure with an average diameter of 170 nm and fiber to fiber interconnections due to PVP carbonization.

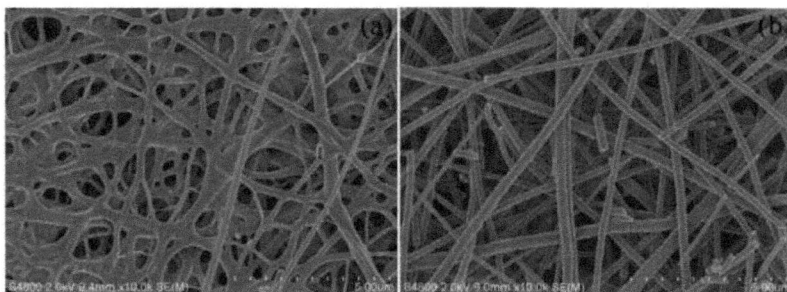

Figure 5. SEM images of (**a**) CENFs and (**b**) Cu/CENFs. Reprint with permission from [42].

In comparison, Cu NP/CNFs were more uniform, did not exhibit any interconnections, and the average diameter was significantly higher (300 nm, Figure 5b). The Cu NPs/CENFs/Lac/Nafion/GCE response to catechol was linear over a wide range of concentrations (9.95 µM–9.76 mM) and a low LOD (1.18 µM) was achieved, this value being 2.8 times lower than the LOD obtained in absence of Cu NPs. The stability was also slightly improved by loading the NFs with Cu NPs. Cu NPs/CENFs/Lac/Nafion biosensor retained 95.9% of the initial response after 22 days, while the CENFs/Lac/Nafion biosensor ended up with only 89.1%. The same group proposed another strategy to improve the performance of the biosensor [43]. The NFs were doped with Ni instead of Cu. A PAN-Ni acetate solution was first electrospun, then carbonized into a nitrogen atmosphere, and the resulting Ni NPs/ECNFs were mixed, after grinding, with Lac and dopamine (DA). The Lac-catalyzed oxidation of DA enabled the production of polydopamine (PDA) which efficiently embedded the enzymes into the Ni NPs/ECFs composite. PDA/Lac/Ni NPs/ECNFs aqueous suspension was finally casted onto a bare magnetic GCE and the resulting biosensor was assessed for catechol detection. Again, a control biosensor without Ni was prepared. LOD was improved by a factor of 1.7, compared to that of the Cu-doped CENFs mentioned before, and the linear range was also slightly wider (1 µM–9.1 mM). A good selectivity and stability was observed and the biosensor was successfully applied to the detection of catechol in spiked real water samples.

In another study, nitrogen-doped carbon nanosphere/ECNF composites (NCNS@ECNFs) were prepared by electrospinning polypyrrole nanosphere/PAN suspensions with subsequent controlled thermal treatment in a N_2 atmosphere and GOx immobilization by absorption [44]. Then, the NCNSs@ECNFs composite was fixed on a GCE and recovered with Nafion. Doping CNSs with nitrogen was an effective way to improve their hydrophilicity (and therefore biocompatibility), as well as their electron-donor ability and electrical conductivity. Combining the large surface area of ECNFs and the enhanced electrocatalytic activity of NCNSs allowed the development of a sensitive, stable and selective glucose biosensor based on the direct electron transfer of GOx, with a low LOD (2 µM) and wide linear range (12–1000 µM). SEM images revealed that the NCNS@ECNFs exhibit a three-dimensional incompact porous structure, providing a large effective area for GOx immobilization. NCNSs of 53 ± 9 nm average diameter, well-dispersed on the surface of ECNFs or embedded within the matrix, could be also observed. The hydrophilicity of the ECNF surface was considerably increased by incorporating NCNSs, creating a favorable environment for GOx and for the direct electron transfer between the enzyme and the carbon electrode. The electron-transfer was higher with NCNS@CNFs/GCE than with the bare GCE or the CNFs/GCE, demonstrating the positive effect of doping.

Another method to produce ECNF-based enzyme biosensors with enhanced analytical performances was recently reported by Bae et al. [45]. Higher and controlled porosity was generated by incorporating silica NPs (average size: 16 ± 2 nm) into the precursor PAN solution. After electrospinning, the NPs were removed by HF treatment and the resulting PAN NFs were carbonized. Mesosized pore carbon structures fabricated this way were very beneficial for an efficient immobilization of GOx and the embedded enzyme remained very accessible to glucose substrate. Characterization of the ECNFs by Raman spectroscopy evidenced the positive effect of heat treatment on the crystallinity and orientation of carbon, and a significant increase of the conductivity was observed after thermal treatment.

Another strategy to improve bioreceptor loading or attachment to the ECNFs, and therefore biosensors' analytical features, is to introduce post-treatments of the fibers using physical or wet chemical processes.

Cui et al. [46] functionalized carbonized PAN NFs with carboxylic groups using a wet chemical acidic treatment and used the carboxylated CENFs to synthetize hydroxyapatite (HA)-CENFs. Cytochrome *c*, further cast onto the microporous HA-CENFs composite, exhibited good electrocatalytic activity and fast response to H_2O_2.

Wang et al. [47] used the same protocol to produce ECNFs. Prussian Blue (PB) nanostructures could be subsequently grown in a controllable manner onto the carboxylic group-functionalized ECNFs. The PB-ECEF composite, coated on the surface of a GCE, was covered with a GOx/Chit film and an amperometric biosensor offering a wide linear range (0.02–12 mM) and low limit of detection (0.5 μM) was proposed for glucose detection.

Mondal et al. [67] first electrospun PMMA onto Si substrate, then the NFs were covered with a PAN/chloroplatinic acid film. High-temperature treatment of the composite film decomposed the PMMA NFs and generated embedded microchannels in the PAN-derived amorphous monolithic carbon electrode. The channels were decorated with Pt NPs by in situ thermal decomposition of the precursor metal salt. Carboxylic groups were generated at the composite surface by plasma treatment and anti-aflatoxinB1 (AFB1) antibodies were grafted using the EDC/NHS conjugation chemistry. The nanochannels aligned in the porous carbon film acted as a reaction chamber for antigen–antibody interactions and was beneficial for fast electron transport toward the electrode. AFB1 could be detected as low as 1 pg/mL with a linear range of 10^{-12}–10^{-7} g/mL.

Kim et al. [74] fabricated a potentiometric aptasensor based on ECNFs for bisphenol A (BPA) detection. Electrospinning of a PAN/PPMA phase-separated blend, followed by thermal treatment, generated multi-channel ECNFs of large surface area, clearly evidenced by SEM. The fibers were subsequently functionalized with carboxylic groups using an acidic oxidative treatment and BPA-binding aptamers were attached to the surface via covalent binding. The biosensor was highly sensitive (LOD: 1 fM) and could be reused over a period of 4 weeks.

3.1.3. Electrospun Polymeric and Composite NFs

As already mentioned, among the NF-based biosensors reported in the literature, only a few of them rely on optical [61,69] or mechanical [70] detection modes, and most of them are associated with electrochemical transduction. In this case, NFs must be conductive. Apart from the metal-oxide and carbon NFs described in the previous sections, composite NFs based on polymers incorporating conductive materials have been reported. The conductive materials may be metallic NPs (MNPs), carbon nanotubes (CNTs), conducting polymers or combinations [28].

Electrospun Polymer NFs Doped with CNTs

The straightforward strategy to produce CNT-polymer NF composites is to disperse CNTs into the polymer solution before electrospinning. Following this approach, Manesh et al. [48] fabricated a glucose biosensor by immobilizing GOx onto a polymer-CNTs composite. Multiwall carbon nanotubes (MWCNTs), wrapped by a cationic polymer [poly(diallyldimethylammonium chloride)] (PDDA), were dispersed into PMMA, and a nanofibrous membrane was produced by electrospinning the blend onto an ITO electrode. Wrapping PDDA over the surface of the MWCNTs helped prevent the aggregation of the MWCNTs and was used to attach the negatively charged enzyme onto the modified electrode surface. A thin layer of Nafion was added over the electrode surface to decrease the interferences caused by anions present in biological media. The fabricated Nafion/GOx/MWCNT(PDDA)-PMMA

NFs/ITO electrode exhibited excellent electrocatalytic activity towards hydrogen peroxide (H_2O_2) with a pronounced oxidation current at +100 mV. Glucose was amperometrically detected at +100 mV in 0.1 M phosphate buffer solution (PBS, pH 7), with a fast response time (4 s). The response to glucose was linear in the 20 μM–15 mM range and LOD was 1 μM.

Wang et al. [49] prepared CNT-doped poly(acrylonitrile-co-acrylic acid) (PANCAA) NFs by electrospinning and further modified the NFs with GOx by covalently immobilizing GOx on the membranes through the activation of carboxylic groups on the PANCAA NF surface. The electrochemical properties of enzyme electrodes were characterized by chronoamperometric measurements, which showed that MWCNT filling enhances the electrode current and sensitivity. Combined with the results of kinetic studies, it was concluded that the interactions between MWCNT and FAD play a significant role in enhancing the electroactivity of the immobilized GOx, even though the secondary structure of the immobilized GOx is disturbed in the presence of MWCNT.

Numnuan et al. [50] proposed an amperometric biosensor based on electrospun Chit-CNTs NFs for uric acid detection. A Ag NPs layer was first electrodeposited on a gold electrode. A Chit/PVA/MWCNTs mixture was then electrospun and, after removal of PVA by NaOH treatment, uricase was immobilized on the Chit-CNTs NF film through cross-linking between its amine groups and the Chit amino groups. The fabricated uric acid biosensor had a wide linear range (1.0–400 μM), with a LOD of 1.0 μM and a storage life of more than six weeks. The values measured for blood plasma samples using the proposed biosensor were in good agreement with those obtained by a standard enzymatic colorimetric method.

More recently, Bourourou et al. [51] dispersed MWCNTs into a PAN solution to produce electrospun NF mats that were used directly as electrodes. The nitrile groups of PAN polymer were subsequently reduced into amino groups, and polyphenol oxidase (PPO) was immobilized onto the PAN-MWCNTs NFs through covalent bindings using GA. The PAN-MWCNTs-PPO electrode was successfully used for the sensitive amperometric detection of catechol, with a wide linear range (1 μM–0.4 mM) and a LOD of 0.9 μM.

Electrospun Polymer NFs Doped with MNPs

The electrochemical properties of MNPs are extremely sensitive to their sizes, shape, and dispersion. A high dispersion of MNPs in functional materials is important to provide high electrochemical activity, while the associated tendency of MNPs to aggregate would lower their catalytic activity and reuse lifetime. Therefore, how to design and prepare MNP-based materials with long-term dispersion stability and high catalytic efficiency is a primary challenge for their wide applications. MNP-doped NFs can be prepared, as described for CNTs, by dispersing the NPs into a polymer solution and spinning subsequently. Devadoss et al. [52] synthetized Au NPs-Nafion-polyacrylic acid (PAA) NFs by electrospinning a blend of the three components. *N,N'*-(4-dimethylamino) pyridine (DMAP)-protected Au NPs (5.0 ± 0.5 nm) were used. TEM images evidenced the uniform inclusion of Au NPs in the composite NFs, which was attributed to a strong electrostatic interaction between the positively charged DMAP-protected Au NPs and the negatively charged sulfonate groups in Nafion. Au NP-composite NFs exhibited higher conductivity than the Nafion-PAA NFs. Horseradish peroxidase (HRP) was further immobilized on the nanofibrous electrode via electrostatic interactions with the negatively charges of PAA. It was demonstrated that the incorporation of Au NPs in the NFs improved the amperometric detection of H_2O_2. The biosensor LOD was decreased by a factor of 2.6.

MNPs-polymer hybrid NFs can be also synthetized through in situ reduction of metallic precursor ions, either introduced in the electrospun solution, or dropped on the polymer NFs after electrospinning. Zhu et al. [53] reported a facile and green approach to prepare Ag NPs-PVA NFs and Ag NPs-PVA/PEI NFs using the first and second strategy, respectively (Figure 6). The freshly prepared NFs were cross-linked using GA vapors to improve their water stability. Ag NPs were generated within the PVA NFs or on the surface of the PVA/PEI NFs by in situ reduction of AgNO$_3$

precursor with the green reductant, epigallocatechin gallate. Then, HRP was dropped onto the fiber mats. The HRP/AgNPs/PVA/GCE and HRP/AgNPs/(PVA/PEI)/GCE biosensors exhibited high amperometric sensitivity to H_2O_2 and glucose, the best analytical performances being obtained for the HRP/AgNPs/(PVA/PEI)/GCE biosensor. In another work of the group, PVA/PEI NFs were elaborated and immersed into a $PdCl_2$ solution for subsequent reduction of the Pd salt with $NaBH_4$ [54]. TEM characterization evidenced a good dispersion of Pd NPs on the NFs, which was attributed to the complexation between Pd(II) and the free amine groups of PEI. The average diameter of the produced Pd NPs was 3.4 nm, some aggregation being noticed. The HRP/AgNPs/PVA/PEI NFs/GCE biosensor, obtained after adsorption of HRP, exhibited higher CV response to H_2O_2 than the HRP/AgNPs/PVA/PEI NFs/GCE biosensor, demonstrating that Pd NPs play a key and positive role in the electron transfer between the redox-active site of H_2O_2 and the electrode. Jose et al. [55] proposed to combine the advantages of electrospinning with those of both CNTs and Au NPs to produce an amperometric glucose biosensor of enhanced performances. In this approach, PAN NFs were first decorated with Au NPs using a seed-mediated electroless deposition method. Carboxylated MWCNTs were further coated onto the Ag NPs/PAN NFs by electrophoretic deposition. SEM images revealed a complete and uniform coating of the fibers surface with MWCNTs. The carboxylated MWCNTs provided the anchor for covalent immobilization of GOX. The direct electron transfer between GOx and the electrode surface was demonstrated. The LOD for glucose was as low as 4 μM.

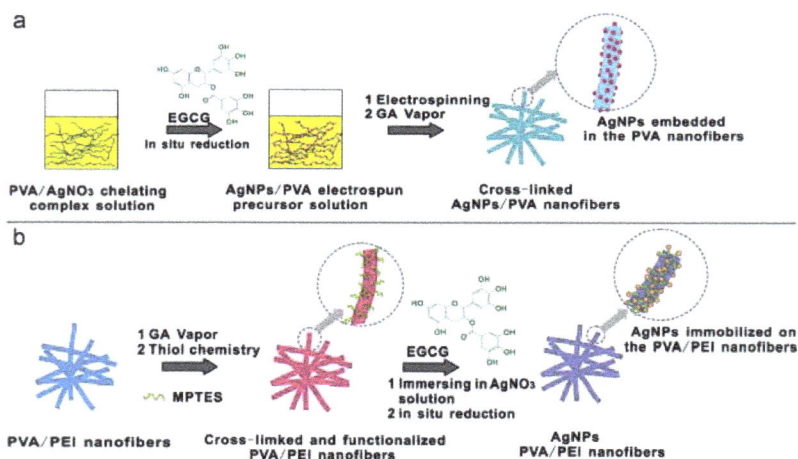

Figure 6. Fabrication process of the (**a**) AgNPs embedded in the PVA water-stable nanofibers and (**b**) AgNPs immobilized on the functionalized PVA/PEI water-stable nanofibers Reprint with permission from [53]. Copyright 2013 Elsevier.

Conducting Polymers in the Fabrication of Electrospun NFs

Conducting polymers (CPs), such as polyaniline (PANI), polypyrrole (PPy) or poly(3,4-ethylenedioxythiophene) (PEDOT) are of particular interest in the elaboration of electrochemical biosensors due to their unique electrical properties. They possess fascinating chemical and physical properties, such as intrinsic conductivity, derived from their conjugated π-electron system and so they have been used to enhance the speed, sensitivity and versatility of many biosensors. Direct electrospinning of CPs would therefore be a facile and rapid way to create conducting NFs. However, processing intrinsically conducting polymers has always represented a challenge. Indeed, most of them are insoluble and infusible due to the stiffness of their all-conjugated aromatic backbone structures, which renders them hardly electrospinnable.

To overcome this issue, different approaches may be used. The first strategy consists in blending the CP with a non-conducting spinnable polymer. The latter serves as a carrier to improve the CP spinnability. Gladish et al. [56] created polymer NFs based platforms for enzymes immobilization by electrospinning blends of PAN and highly conductive sulfonated PANIs on ITO electrodes. Apart from sulfonic acid groups, the polymers exhibit carboxylic acid groups which were used for covalent immobilization of the pyrroloquinoline quinone-dependent glucose dehydrogenase (PQQ-GDH). The modified electrodes demonstrated high catalytic current responses to glucose and a wide range of detection (2.5 μM–1 mM) could be achieved.

In a second approach, electrospun NFs are covered with a CP. Fu et al. [57] prepared hierarchical PANI/carboxymethyl cellulose (CMC)/cellulose NFs on GCE electrodes by in situ polymerization of aniline on the CMC-modified cellulose NFs (Figure 7). Highly dense PANI nanorods (60 nm × 180 nm) grown onto the surface of CMC/cellulose NFs could be visualized by TEM (Figure 8). The NFs exhibited an average diameter of 310 ± 8 nm. A Nafion/Lac solution was dropped at the NFs surface for Lac fixation. The fabricated Lac/PANI/CMC/cellulose/GCE exhibited a highly sensitive detection toward catechol with a broad linear range and low detection limit (0.374 μM).

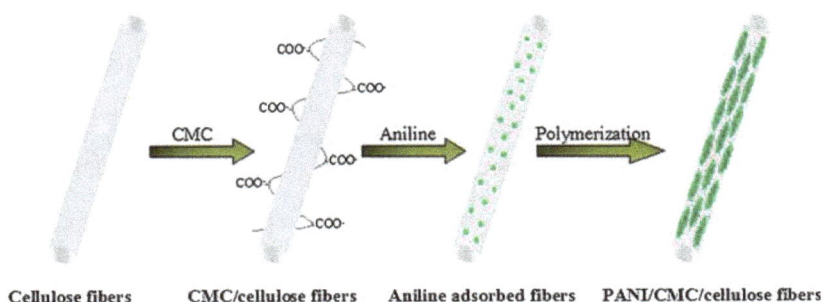

Cellulose fibers   **CMC/cellulose fibers**   **Aniline adsorbed fibers**   **PANI/CMC/cellulose fibers**

Figure 7. Preparation of PANI/CMC/cellulose nanofibers. Reprinted with permission from [57]. Copyright 2015 Elsevier.

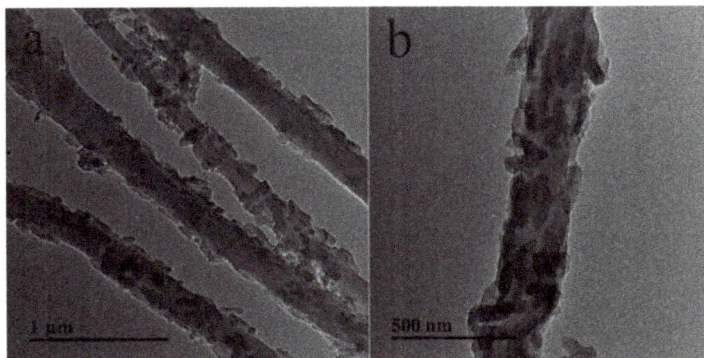

Figure 8. TEM images of PANI/CMC/cellulose nanofibers at different magnifications. Reprinted with permission from [57]. Copyright 2015 Elsevier.

In a similar approach, electrospun poly(L-lactide) (PLLA) NFs were produced onto Pt microelectrode arrays and covered with GOx-PEDOT films by electropolymerization of EDOT monomer in presence of poly(sodium-*p*-styrene sulfonate) and GOx [58]. The analytical performances of the NFs modified electrodes were compared to those of Pt electrodes covered with PEDOT/GOx films of similar thickness (around 330 nm). In addition to improved sensitivity, the PEDOT NFs-GOx biosensors exhibited a lower LOD for amperometric glucose detection (0.12 mM at +700 mV) than the PEDOT film-GOx biosensors (0.45 mM). Guler et al. [72] also reported the fabrication of poly(ε-caprolactone) (PCL) NFs onto an ITO-PET electrode followed by in situ chemical polymerization of pyrrole with Fe^{3+} as an oxidant and Cl^- as a dopant. The PCL/PPy NFs were functionalized with calf thymus ssDNA, used as model, by physical immobilization. Different techniques evidenced the effective incorporation of PPy and ssDNA on PCL NFs, and demonstrated the possibility of using the PCL/PPy NFs as electrochemical DNA biosensor after immobilization of a specific probe DNA onto the fibers.

To improve the electrochemical properties of the final nanofibrous membranes, some authors proposed to make the electrospun polymer conductive by integrating CNTs into the electrospun solution. Uzun et al. [59] fabricated a glucose sensor by preparing electrospun nylon 6,6 NFs incorporating MWCNTs, which were further coated with a conducting polymer named PBIBA [poly-4-(4,7-di(thiophen-2-yl)-1*H*-benzo[d]imidazol-2-yl)benzaldehyde] to covalently attach GOx on the surface of the fibers through the free aldehyde groups of the conducting polymer. The resulting novel amperometric glucose biosensor revealed higher stability and sensitivity in presence of MWCNTs. The linear response for glucose detection is in the range of 0.01 mM to 2 mM with a LOD of 9 μM. In the same way, MWCNTs-doped nylon 6,6 composite NFs were prepared using electrospinning, and served as backbone for pyrrole electropolymerization [73]. The functional composite was used to immobilize wild type p53 ssDNA for hybridization detection of a specific mutation in the p53 tumor suppressor gene. Ekabutr et al. [60] reported the fabrication of PAN-MWCNTs hybrid NFs on a screen-printed carbon electrode. The NFs were covered with PPy using an original approach by vapour phase polymerization (VPP) of pyrrole using Fe(III) *p*-toluenesulfonate as oxidizing agent. GOx, used as a model enzyme, was subsequently adsorbed on the modified electrode and the biosensor was evaluated for amperometric detection of glucose. The LOD was 0.98 mM. The VPP approach was also employed in another study by Jun et al. [75] to cover carbon NFs scaffold decorated with ZnO nanonodules with carboxylated PPy. The platelet-derived growth factor-B (PDGF-B) binding aptamer was conjugated to the NFs at the surface of a FET transducer via PPy carboxyle groups. The biosensor was highly sensitive (5 fM) and extremely selective for isoforms of PDGFs.

3.2. Enzyme Entrapment into the NFs

When enzymes are attached to NFs after electrospinning, only the external surface of the NFs may be used for immobilization, without taking full advantage of the internal volume of fibers that can protect enzyme molecules from harsh conditions. Enzyme loading is therefore rather limited. Therefore, another way to proceed consists in electrospinning a blend of enzymes and water-soluble polymer. This approach might be also used to entrap antibodies, but has been rarely investigated due to the high amount of biomolecules required and their high cost [31]. In contrast to the protocols described in the previous sections, the entrapment does not involve any bond between the enzyme and the NFs surface that helps preserving their biological activity [77]. The major drawback of the approach is that some enzymes may be shielded inside the NFs and are not accessible to the substrate. A summary of the enzymatic biosensors prepared this way may be found in Table 1.

PVA is the most common water-soluble polymer employed for enzyme entrapment into NFs as it is a non-toxic, hydrophilic and highly biocompatible synthetic polymer. However, data reported in the literature are not easy to compare. Indeed, PVA displays various degrees of polymerization and hydrolysis (DP and DH, respectively). These parameters highly affect the viscosity and conductivity of PVA solutions, and therefore the electrospinning conditions and the morphology of electrospun NFs produced. Ren et al. [78] proposed an amperometric glucose biosensor prepared by electrospinning a mixture of PVA and GOx in PBS followed by cross-linking using GA in solution to form water-insoluble NFs. The NF diameters obtained in this work were somewhat heterogeneous, ranging from 70 nm to 250 nm in the absence of GOx, as observed by SEM. In the presence of GOx, the fibers became irregular and were interspersed with shuttle-shape beads. The presence of GOx in the NFs was evidenced by IR spectroscopy. The amperometric response to 0.5 mM glucose was more rapid (1 s) than the response observed for PVA/GOx films (5.2 s) under the same conditions. The enzymes remained active when entrapped in the NFs and a LOD of 0.05 mM was measured. Oriero et al. [79] immobilized tyrosinase onto an ITO-coated glass substrate modified with silica-tyrosinase-PVA NFs for amperometric detection of catechol, phenol, and *p*-cresol. The biosensor sensitivity decreased in the order catechol > phenol > *p*-cresol, with a LOD of about 10 μM for each compound and a linear response up to 100 mM.

To enhance the conductivity of PVA NFs, the incorporation of carbon NMs, mostly graphene oxide (GO) and MWCNTs, or Au NPs has been proposed. Su et al. [80] developed a novel biosensing platform for glucose detection, by electrospinning a solution containing PVA, Chit, GO and GOx as model enzyme to fabricate NFs. After electrospinning, the Pt electrode modified with electrospun Chit-GOx-GO-PVA NFs was placed in glutaraldehyde vapor for cross-linking to form water-insoluble NFs. Then, a thin layer of Nafion was deposited on the surface of the matrix, and the prepared electrode was used for glucose amperometric detection. The electrode exhibited high sensitivity, good stability, low detection limit (5 μM) and wide linear range (5 μM–4 mM). Glucose concentrations determined in human serum samples by the NF-based biosensor were in good agreement with the values obtained using a standard clinical assay. The same methodology was used to entrap choline oxidase. The choline biosensor exhibited excellent analytical features, demonstrating the versatility of the proposed method. A GOx-GO-PVA NFs modified Pt electrode was also constructed for glucose detection [81]. The highest sensitivity was obtained by incorporating 20 ppm GO in the electrospun solution. The biosensor response was very rapid, the steady-state current being achieved only 10–14 s after glucose injection.

Our group has proposed two strategies to produce enzyme NF-based electrochemical biosensors from enzymes-polymer blends [82]. GOx was chosen as model enzyme. In a first approach, we reported the one-step and facile fabrication of water-stable NFs from blends of the photochemically cross-linkable polyvinyl alcohol styrylpyridinium polymer (PVA-SbQ), MWCNT and GOx. MWCNTs were functionalized with carboxyl groups to improve their compatibility with PVA and PVA/GOx aqueous solutions The NFs were stabilized by UV-cross-linking of PVA-SbQ, which enabled a fast, easy and soft cross-linking step without any added chemicals. The enzyme conformation and activity could be preserved. The addition of MWCNT resulted in a significant decrease in the average fiber diameter, from 350 ± 20 nm to 250 ± 50 nm, as evidenced by SEM. In addition, embedded nanotubes, well dispersed in the polymer matrix and aligned along the NFs direction, could be detected by TEM for the lowest MWCNT loadings (Figure 9b,c). MWCNTs tended to agglomerate at the highest MWCNTs loading, and clusters of coiled/bundles nanotubes were observed (Figure 9d). The combination of MWCNT and PVA-SbQ polymer improved electron transfer ability of the generated nanofibrous mats. The resulting enzyme voltammetric biosensor was linear in a wide range of glucose concentration (up to at least 4 mM) and a very low LOD (2 μM) was achieved.

In a second study [83], aqueous solutions of GOx-PEI-PVA were electrospun on Au electrodes and the NFs were decorated with Au NPs. It was shown that adhesion of the fibers mat onto gold electrodes could be improved by modifying the surface with a self-assembled monolayer of 4-ATP bearing thiol groups for covalent binding to the gold surface and amine groups to react with the amine groups of PEI in a subsequent cross-linking step. This step also helped improving the water stability of the produced NFs. A significant enhancement of the NFs' conductive properties, as characterized by cyclic voltammetry, was achieved by decorating the NFs with AuNPs. The highest density of particles was observed using Au colloidal solutions of pH 5, which could be attributed to hydrogen bondings and ionic interactions between the amine groups of PEI and the carboxylic groups of citrate-stabilized Au NPs (Figure 10). Glucose could be successfully detected by impedimetry, the response being linear in the 10–200 μM range. A very low limit of detection (0.9 μM) could be achieved and the biosensor exhibited good operational and storage stabilities.

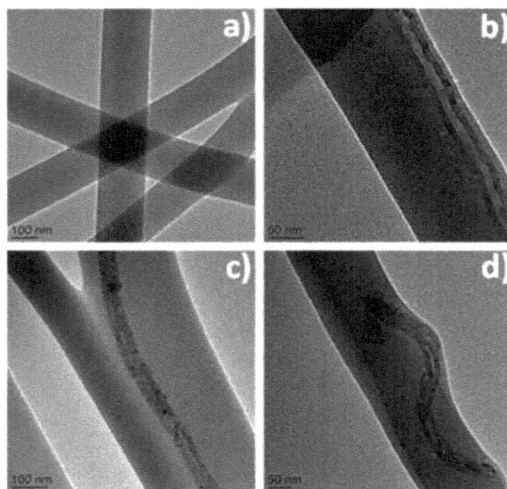

Figure 9. TEM images of (**a**) pure PVA-SbQ NFs, (**b**) PVA-SbQ/MWCNTs (1 wt %) NFs, (**c**) PVA-SbQ/MWCNTs (5 wt %) NFs, (**d**) PVA-SbQ/MWCNTs (10 wt %). Reprinted with permission from [82]. Copyright 2015 The Electrochemical Society

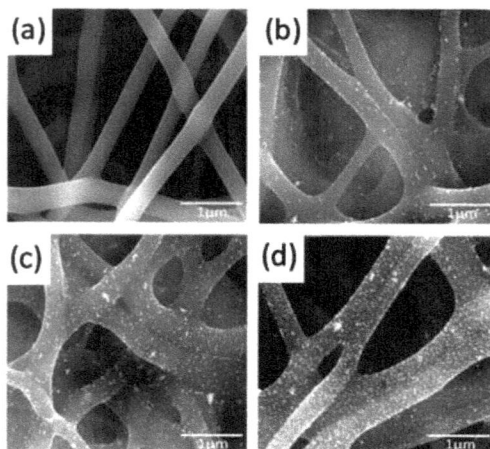

Figure 10. SEM images of water-stable electrospun PVA/PEI NFs before immersion in the Au NPs solution (**a**), after immersion in colloidal Au NPs solutions of pH 7.0 (**b**), pH 6.0 (**c**), pH 5.0 (**d**). Reprinted with permission from [83]. Copyright 2016 Elsevier.

4. Electrospun NFs in ELISA-Type Bioassays and Point-of-Care (POC) Devices

Owing to their high surface area and their ability to bind biomolecules, electrospun NFs have also been exploited to the elaboration of improved Enzyme-Linked ImmunoSorbent Assays (ELISAs) and of advanced point-of-care (POC) devices. The analytical performances of these devices of course depends strongly on the immobilization of the affinity elements (mostly antibodies) to the assay substrate. Random adsorption is by far the most common immobilization technique, but bioaffinity attachment, using for example the specific binding of biotin to avidin and streptavidin, is also relatively widespread [89]. Covalent attachment has been also investigated to limit antibodies leaching and loss of functionality that physical adsorption could induce, and to achieve a better distribution of the biomolecules onto the surface. Both electrochemical and optical detection are employed, with a predominance of the latter.

ELISA is perhaps the most well-known and widely used immunoassay for the detection of proteins (antibodies, biomarker proteins), pathogens (bacteria, virus) and smaller molecules of toxicological or environmental interest. However, it suffers from several major weaknesses. The conventional test is time-consuming and laborious, requires large amounts of sample and exhibits low sensitivity, which impedes early diagnosis of diseases. In addition, it is a single target analyte test, whereas it is known, for example, that the combined measurement of several biomarkers is required to increase the accuracy of disease diagnosis. Finally, samples must be purified before the test in order to avoid interferences. Different strategies have been proposed to overcome these different issues, resulting in the development of improved ELISA-type assays and advanced POC devices, among which lab-on-a-chip/lab-on-a-disc devices, lateral flow and microfluidic paper-based bioassays [90–94]. The following sections will show how the integration of electrospun NFs in such systems can contribute to improve their performances.

4.1. ELISA-Type Bioassays

Conventional ELISA relies on two-dimensional surfaces of planar substrates to immobilize the capture molecules. Due to their large surface area-to-volume ratio, the three-dimensional electrospun NF mats have the capacity to immobilize larger amounts of capture molecules, resulting in higher expected sensitivities. Wang et al. [95] recently proposed a high-throughput immunoassay based on electrospun PS NFs and compared its performance with conventional PS substrate. A plasma treatment was applied to make PS NFs hydrophilic, facilitating infiltration of antibodies into the NFs scaffold. Three different cancer biomarkers (AFP, CEA, VEGF) were further used to compare the analytical performance of the two substrates using a sandwich approach and fluorescence detection. A 300-fold decrease of the LOD was achieved by replacing the planar PS substrate by electrospun PS NFs.

Other polymers of various levels of hydrophilicity and functionalities have been investigated to evaluate their ability to generate sensitive and more rapid ELISA. Sadir et al. [96] produced NF mats from two polymers: poly-L-lactic acid (PLLA) and cellulose acetate (CA). The CA NFs were treated with GA to enhance their water stability. Electrospun NF membranes were placed in a microwell plate and the sensitivity of the two bioassay platforms were compared for the detection of CRP cardiac biomarker using a colorimetric sandwich ELISA approach. The LODs were 13 pg mL^{-1} and 53 pg mL^{-1} for PLLA and CA, respectively, lower than that of the conventional ELISA. Moreover, the total analysis time was reduced by a factor 2. It was shown that the amount of antibodies adsorbed on PLLA was higher than on CA and that the proteins were more stable. In another work, polyhydroxybutyrate (PHB) NF membranes were produced by electrospinning and were coated with poly(methylmethacrylate-co-methacrylic acid) co-polymer, poly(MMA-coMAA) [97], which was synthetized with different molar ratios of the monomers, to achieve nanofiber membranes combining the high surface area of PHB NFs with functional carboxylic groups inherited from MMA segments of the copolymer. The hydrophobicity of PHB NFs gradually decreased and the number of functional groups increased as the number of MAA segments introduced via poly(MMA-co-MAA) coating increased. The coated and uncoated PHB NF membranes were placed into a PS ELISA well-plate for subsequent colorimetric detection of dengue virus using a sandwich approach. Two modes of primary antibodies immobilization were tested, i.e., physical adsorption and covalent binding through EDC/NHS chemistry method. Covalent binding generally yielded lower signals than physical adorption but higher than conventional ELISA. The coated NFs with 9:1 MMA:MAA molar ratio of exhibited the lowest LOD and highest specificity. In a recent work, Mahmoudifard et al. [98] reported the fabrication of electrospun NF mats from the hydrophobic polysulfone polymer. Oxygen plasma treatment enabled the creation of carboxylic functional groups at the surface of the nanofibrous membrane and monoclonal IgG antibodies were further immobilized covalently or by physical adsorption to perform a simple colorimetric sandwich ELISA in well-plate format. Covalent attachment resulted in higher signals compared to those obtained with adsorption approach. In addition, it was shown that the proposed platform integrating electrospun NFs was more sensitive than the standard ELISA.

Polymers with biocompatible and antifouling properties have been also proposed to reduce the background noise coming from nonspecific adsorption of proteins in ELISA. Using electrospun NFs prepared from a functional polymer bearing both phosphorylcholine and active ester groups, Chantasiricot et al. [99] developed an ELISA for human IgG detection. The LOD was improved by a factor 4.6 compared to the conventional ELISA performed on PS substrate and 1.8 compared to the test performed with a dip-coated polymer substrate. A much wider linear range could be also achieved. In addition, the blocking step could be omitted, resulting in a significant decrease of the total time of analysis. Hersey et al. [100] synthetized a set of water insoluble poly(oxanorbornene) derivatives with biotin (bioactive) and triethylene glycol (antifouling) functionalities. Electrospun NF meshes were generated from these copolymers with the ability to specifically bind streptavidin while minimizing the nonspecific binding of other proteins. The protein binding capabilities of NF meshes were evaluated using a simplified colorimetric ELISA for a model target protein, mouse IgG, under both static (26 h) and flow conditions (1 h). Under flow conditions, the LOD was 2.2-fold lower than the LOD of the conventional ELISA in streptavidin coated plates.

Recent works have also been devoted to the development of original microarray-based immunoassays using either hydrogel-micropatterned PS-based NFs [101] or droplets immobilized on polycaprolactone NFs [102].

4.2. Lab-on-a Chip/Lab-on-a-Disc POC Devices

The integration of bioaffinity tests into more compact, rapid, simple and easy-to-use devices usable on-site by non-qualified persons have been the subject of tremendous research works over the past decades, resulting in the development of POC devices. In these devices, minimal human intervention should be achieved, ideally only for sample introduction, signal read-out and interpretation. The consumption of minimal volumes of sample is also targeted. Lab-on-a chip [90] or lab-on-a-disc [91] devices integrate sample preparation and detection in one system. They can address the requirement of providing a fast and multiplexed diagnostic at the point-of-need.

Electrospun NFs can contribute to the improvement of biosensing lab-on-chip devices in several ways. It has been shown recently that they can be very useful as passive mixers, with capacities comparable to or better than many passive mixers already reported [26]. Jin et al. [103] developed a capillary flow microfluidic bio-chip for chemiluminescence detection of *E. coli 0157* cells combining lateral flow assay and microfluidic technology, demonstrating the applicability of water soluble NFs for on-chip reagent delivery. In this work, PVP NFs containing HRP tagged antibodies were produced and incorporated into the chip and the delivery of binding reagents, i.e., antibodies, was achieved by dissolution of PVP NFs. Electrospun NFs can be also used to specifically capture and concentrate the analytes for enhanced purification and detection. One major difficulty comes from the integration of NFs into the microchannel of the lab-on-a-chip system. Cho et al. [104] fabricated patterned aligned PVA NFs using gold microelectrodes array into PMMA microchannels. The same group later evaluated the approach for specific liposomes retention [24]. Very recently, they proposed a more simple fabrication method, where thin layer of NFs were first prepared, then manually peeled off the grounded collector and stacked together to create multilayer mats with more homogeneous morphologies [25]. The mats were finally transferred to PMMA substrates, cut into small strips and incorporated into the microchannels. Positively and negatively charged PVA NFs were prepared. The negatively charged PVA mats bearing carboxylate groups was used to attach anti-*E. coli* K12 antibodies via EDC/NHS chemistry. Positively charged NFs enabled 87% retention of negative *E. coli* K12 bacterial cells through non-specific electrostatic interactions, while antibody-functionalized negatively mats were capable of the specific capture of 72% of the cells while reducing non-specific analyte retention due to the repealing effect of the polymer. This work therefore demonstrated the potential of the method for efficient capture and concentration of microbial cells. Lee et al. [105] exploited these attractive properties to immobilize high amounts of suitable antibodies on TiO_2 NFs integrated in a lab-on-a-disc device. The TiO_2 NFs had been previously treated with O_2 plasma and functionalized for antibodies binding. The chemiluminescence biosensing device enabled the sensitive and rapid (30 min) detection of two cardiac biomarkers (CRP and cTnI). LODs were 0.8 pg/mL and 37 pg/mL, respectively, and the method required only 10 μL of sample. Ali et al. [106] reported the development of a label-free electrochemical immuno-biochip detection of breast cancer biomarkers. The biosensor utilized a uniquely structured working electrode made of porous hierarchical graphene foam modified with electrospun carbon-doped TiO_2 NFs inserted in the microchannel of the PDMS microfluidic device. Antibodies were immobilized covalently onto the electrode. High sensitivity and specificity could be achieved.

4.3. Paper-Based POC Devices

Due to their low-cost and various easy-to-use formats, paper-based devices have appeared for many years as very appealing and cheaper alternatives to lab-on-a-chip systems. The porous structure of paper allows fluid flow by capillary action without the need of additional pumping systems. The paper-based devices exist under various formats, from the basic dipsticks and lateral flow assays (LFAs) to the more advanced and complex forms, i.e., microfluidic paper-based analytical devices, with possible multiplexing [92–94].

LFA format typically consist of four segments, including the sample pad, the conjugate pad, reaction membrane and an adsorbent, all being enclosed in a plastic cassette [93]. Luo et al. [107] reported the fabrication of a LFA using nitrocellulose nanofibrous membranes functionalized with antibodies directed against bovine viral diarrhea virus (BVDV) as the capture pad. The capillary action of the membrane was enhanced using oxygen plasma treatment. LFA was very rapid (8 min) and a good LOD (10^3 CCID/mL) was obtained. Reinholt et al. [108] also demonstrated that PLA-based electrospun nanofibers could be successfully incorporated as platform in paper-based biological LFAs. Antibodies were attached to the NFs by adsorption and a colorimetric enzymatic sandwich immunoassay was developed for the detection of *E. coli* O157:H7.

5. Conclusions and Future Prospects

A number of electrospun NFs-based biosensors have been developed, using a large variety of approaches, polymers, integrating or not organic or inorganic fillers (e.g., conducting NMs) and post-treatment processes to generate robust biosensing platforms with enhanced functionalities and analytical performances. There is no doubt that the design and fabrication of NFs-based biosensors is an expanding area of research but it is still in an early stage of academic study in laboratory.

At present, most of the biosensors reported use enzymes as sensing elements and electrochemical transduction. Electrospun NFs are excellent supports for enzyme immobilization, providing large surface areas, porosity, functionalized surfaces for attachment or entrapment, favorable environments around the biomolecules to improve enzyme stability and activity. In addition, most enzymatic biosensors based on electrospun NFs have been designed and evaluated using a model enzyme, GOx. The high versatility of electrospinning process and of the methodologies developed open the way to their application to the immobilization of a wider range of biomolecules, i.e., other enzymes, antibodies, DNA or aptamers, and other transduction modes, e.g., optical. Table 2 shows that very good analytical performances, at least in terms of linear range and LOD, can be achieved for electrochemical detection of glucose using NF-modified electrodes. However, only a very limited number of them has been applied to real samples analysis and no commercial devices are yet available. Electrospun NFs have also demonstrated high potential for the elaboration of advanced POC devices, acting as efficient supports for antibodies capture, analyte concentration, signal amplification, with interesting capillary properties for integrating in paper-based devices.

Table 2. Enzymatic electrospun NFs-based biosensors for glucose detection.

Transduction	NFs/Transducer	Linear Range (mM)	LOD (μM)	Selectivity Test	Real Samples	References
Immobilization of enzymes after electrospinning						
Amperometry	L-Cys-GOx/PVA/ZnO NFs/Au electrode	0.25–19	1	Chol,L-Cys, AA, urea	No	[35]
	Chit/GOx/TiO$_2$ NFs/Pt electrode	0.01–7	10	NR	No	[36]
	GOx-Nafion/Mn$_2$O$_3$-Ag NFs/GCE	up to 1.1	1.73	UA, AA	No	[38]
	Nafion/GOx/NCNS-ECNFs/GCE	0.012–1	2	UA, AA, DA	No	[44]
	GOx/Mesoporous ECNFs/SPE	Up to at least 20	NR	UA, sucrose	No	[45]
	GOx-Chit/PB/ECNFs/GCE	0.02–12	0.5	UA, AA, DA, mannose, galactose, fructose, lactate, BSA, L-Cys	No	[47]
	Nafion/GOx/MWCNTs(PDDA)-PMMA NFs/ITO	0.02–15	1	UA, AA	Human blood serum	[48]
Amperometry	GOx/MWCNTs-PANCAA NFs /Pt electrode	0.67–7	670	NR	No	[49]
	MWCNTs/Au NPs/PAN NFs/Au electrode	Up to 30	4	NR	No	[55]
	GOx-PEDOT-PSS/PLLA NFs/Pt microelectrodes	Up to 5 mM	120 (+700 mV)	NR	No	[58]
	GOx/PBIBA/MWCNTs-Nylon 6,6 NFs/GCE	0.01–2	9	AA, urea, oxalic acid	Beverages	[59]
	GOx/PPy/MWCNTs-PAN NFs/SPCE	0.125–7 (+0.36 mV)	980	NR	No	[60]
Luminescence	GOx/BSA/Ir complex-PS NFs	3 10^{-7}–0.13	10^{-4}	8 substances among which lactose, sucrose, fructose	Human blood serum	[61]
Electrospinning of enzymes-polymer blends						
Amperometry	GOx-PVA NFs/Au electrode	1–10	50	NR	No	[78]
	Nafion/GO-Chit-PVA NFs/Pt electrode	0.005–3.5	5	AA, UA, lactose, sucrose	Human blood serum	[80]
	GOx-Graphene-PVA NFs/Pt electrode	Up to 10	NR	NR	No	[81]
CV	GOx-MWCNT-PVASbQ NFs/Au electrode	0.005–4	2	NR	No	[82]
EIS	Au NPs/GOx-PVA-PEI NFs/Au electrode	0.01–0.2	0.9	AA, UA	No	[83]

AA: ascorbic acid, Chol: cholesterol, DA: dopamine, ECNFs: electrospun carbon nanofibers, GCE: glassy carbon electrode, GOx: glucose oxidase, ITO: indium tin oxide, MWCNTs: Multiwall carbon nanotubes, NR: not reported, PBIBA: poly-4-(4,7-di(thiophen-2-yl)-1*H*-benzo[d]imidazol-2-yl)benzaldehyde, PDDA: poly(diallyldimethylammonium chloride), PEDOT: poly(3,4-ethylenedioxythiophene), PLLA: poly(L-lactide), PMMA: poly(methylmethacrylate), PPy: polypyrrole, PS: poly(styrene), PSS: poly(sodium-*p*-styrene sulfonate), SPCE: screen-printed carbon electrode, SPE: screen-printed electrode, UA: uric acid.

Although significant progress has been made in the past decade in the field, there is still some areas where further improvements and refinements are required, i.e., a better control of the NF production, diameter and arrangement on the transducers, as well as of the immobilization of the biomolecules and of the physico-chemical and biological processes occurring at the NF-based biointerfaces. More fundamental and modeling approaches will be very helpful to better understand and control the interactions between the polymers, fillers and biomolecules (both before and after electrospinning), and the effect of the various electrospinning/immobilization parameters on the size, morphology and electrochemical/optical properties of the NF-based biohybrid materials produced. It will be also interesting to explore new strategies of producing NFs with enhanced electronic/optical properties or functionalities. An original approach has been recently proposed, which combines electrospinning and atomic layer deposition technique to generate controlled ZnO nanocristalline layers and ZnO/Al$_2$O$_3$ nanolaminates at the surface of PAN electrospun NFs used as template. The enhanced photolelectronic and photoluminescence features of the NFs allows applications in optical and electrochemical biosensors [109,110]. Gonzales et al. [111] reported the successful one-step production of a biotin surface functionalized hydrophilic non-water soluble PLA NFs. Incorporation of PLA-*b*-PEG copolymer together with biotin in the spinning solution significantly increased the amount of biotin available at the fibers surface able to bind avidin. The functional hydrophilic NFs proposed in this work are really promising for use in POC devices.

Finally, the conventional electrospinning process, used in most of the works reported, generates NFs of random orientation, limiting the repeatability of biosensors fabrication and therefore hampering a future reliable and consistent production at the industrial scale. Some strategies and advanced electrospinning set-ups have been recently proposed for a more accurate deposition of various patterns of polymer NFs with controlled orientation and spacing [10,84]. The development of new devices, such as coaxial set-ups, which enable the production of core-shell NFs, is also very promising in the biosensor area. Using this approach, Ji et al. [85] recently reported the fabrication of a colorimetric ready-to-use glucose test strip based on GOx and HRP co-immobilized with two commonly used chromogenic agents, in the hollow chamber or shell of polyurethane hollow NFs.

Acknowledgments: The authors thank the French government for E. Sapountzi PhD grant and NATO's Public Diplomacy Division for its financial support in the framework of "Science for Peace" (project CBP.NUKR SFP 984173).

Conflicts of Interest: The authors declare no conflict of interest.

References

1. Karimian, N.; Moretto, L.M.; Ugo, P. Nanobiosensing with arrays and ensembles of nanoelectrodes. *Sensors* **2017**, *17*, 65. [CrossRef] [PubMed]
2. Lopez, G.A.; Estevez, M.-C.; Solera, M.; Lechuga, L.M. Recent advances in nanoplasmonic biosensors: Applications and lab-on-a-chip integration. *Nanophotonics* **2017**, *6*, 123–136. [CrossRef]
3. Kim, J.; Kumar, R.; Bandodkar, A.J.; Wang, J. Advanced materials for printed wearable electrochemical devices: A review. *Adv. Electron. Mater.* **2017**, *3*, 1–15. [CrossRef]
4. Chen, C.; Zhao, X.-L.; Li, Z.-H.; Zhu, Z.-G.; Qian, S.-H.; Flewitt, A.J. Current and emerging technology for continuous glucose monitoring. *Sensors* **2017**, *17*, 182. [CrossRef] [PubMed]
5. Othman, A.; Karimi, A.; Andreescu, S. Functional nanostructures for enzyme based biosensors: Properties, fabrication and applications. *J. Mater. Chem. B* **2016**, *4*, 7178–7203. [CrossRef]
6. Zhang, Y.; Wei, Q. The role of nanomaterials in electroanalytical biosensors: A mini review. *J. Electroanal. Chem.* **2016**, *781*, 401–409. [CrossRef]
7. Arduini, F.; Micheli, L.; Moscone, D.; Palleschi, G.; Piermarini, S.; Ricci, F.; Volpe, G. Electrochemical biosensors based on nanomodified screen-printed electrodes: Recent applications in clinical analysis. *Trends Anal. Chem.* **2016**, *79*, 114–126. [CrossRef]
8. Walcarius, A.; Minteer, S.D.; Wang, J.; Lin, Y.; Merkoçi, A. Nanomaterials for bio-functionalized electrodes: Recent trends. *J. Mater. Chem. B* **2013**, *1*, 4878–4908. [CrossRef]

9. Holzinger, M.; Le Goff, A.; Cosnier, S. Nanomaterials for biosensing applications: A review. *Front. Chem.* **2014**, *2*, 1–10. [CrossRef] [PubMed]

10. Sun, B.; Long, Y.Z.; Zhang, H.D.; Li, M.M.; Duvail, J.L.; Jiang, X.Y.; Yin, H.L. Advances in three-dimensional nanofibrous macrostructures via electrospinning. *Prog. Polym. Sci.* **2014**, *39*, 862–890. [CrossRef]

11. Agarwal, S.; Greiner, A.; Wendorff, J.H. Functional materials by electrospinning of polymers. *Prog. Polym. Sci.* **2013**, *38*, 963–991. [CrossRef]

12. Stepanyan, R.; Subbotin, A.; Cuperus, L.; Boonen, P.; Dorschu, M.; Oosterlinck, F.; Bulters, M. Fiber diameter control in electrospinning. *Appl. Phys. Lett.* **2014**, *105*, 173105. [CrossRef]

13. Pelipenko, J.; Kocbek, P.; Kristl, J. Critical attributes of nanofibers: Preparation, drug loading, and tissue regeneration. *Int. J. Pharm.* **2015**, *484*, 57–74. [CrossRef] [PubMed]

14. Sahay, R.; Kumar, P.S.; Sridhar, R.; Sundaramurthy, J.; Venugopal, J.; Mhaisalkar, S.G.; Ramakrishna, S. Electrospun composite nanofibers and their multifaceted applications. *J. Mater. Chem.* **2012**, *22*, 12953–12971. [CrossRef]

15. Quirós, J.; Boltes, K.; Rosal, R. Bioactive applications for electrospun fibers. *Polym. Rev.* **2016**, *56*, 631–667. [CrossRef]

16. Khalf, A.; Madihally, S.V. Recent advances in multiaxial electrospinning for drug delivery. *Eur. J. Pharm. Biopharm.* **2017**, *112*, 1–17. [CrossRef] [PubMed]

17. Khorshidi, S.; Solouk, A.; Mirzadeh, H.; Mazinani, S.; Lagaron, J.M.; Sharifi, S.; Ramakrishna, S. A review of key challenges of electrospun scaffolds for tissue-engineering applications. *J. Tissue Eng. Regen. Med.* **2016**, *10*, 715–738. [CrossRef] [PubMed]

18. Ray, S.S.; Chen, S.-S.; Li, C.-W.; Nguyenac, N.C.; Nguyenac, H.T. A comprehensive review: Electrospinning technique for fabrication and surface modification of membranes for water treatment application. *RSC Adv.* **2016**, *6*, 85495–85514. [CrossRef]

19. Lu, X.; Wang, C.; Favier, F.; Pinna, N. Electrospun nanomaterials for supercapacitor electrodes: Designed architectures and electrochemical performance. *Adv. Energy Mater.* **2017**, *7*, 1–43. [CrossRef]

20. Sun, G.; Sun, L.; Xie, H.; Liu, J. Electrospinning of nanofibers for energy applications. *Nanomaterials* **2016**, *6*, 129. [CrossRef] [PubMed]

21. Luzio, A.; Canesi, E.V.; Bertarelli, C.; Caironi, M. Electrospun polymer fibers for electronic applications. *Materials* **2014**, *7*, 906–947. [CrossRef] [PubMed]

22. Reyes-Gallardo, E.M.; Lucena, R.; Cardenas, S. Electrospun nanofibers as sorptive phases in microextraction. *Trends Anal. Chem.* **2016**, *84*, 3–11. [CrossRef]

23. Moheman, A.; Alam, M.S.; Mohammad, A. Recent trends in electrospinning of polymer nanofibers and their applications in ultra thin layer chromatography. *Adv. Colloid Interface Sci.* **2016**, *229*, 1–24. [CrossRef] [PubMed]

24. Matlock-Colangelo, L.; Cho, D.; Pitner, C.L.; Frey, M.W.; Baeumner, A.J. Functionalized electrospun nanofibers as bioseparators in microfluidic systems. *Lab Chip* **2012**, *12*, 1696–1701. [CrossRef] [PubMed]

25. Matlock-Colangelo, L.; Coon, B.; Pitner, C.L.; Frey, M.W.; Baeumner, A.J. Functionalized electrospun poly(vinyl alcohol) nanofibers for on-chip concentration of *E. coli* cells. *Anal. Bioanal. Chem.* **2016**, *408*, 1327–1334. [CrossRef] [PubMed]

26. Matlock-Colangelo, L.; Colangelo, N.W.; Fenzl, C.; Frey, M.W.; Baeumner, A.J. Passive Mixing Capabilities of Micro- and Nanofibres When Used in Microfluidic Systems. *Sensors* **2016**, *16*, 1238. [CrossRef] [PubMed]

27. Ding, B.; Wang, M.; Wang, X.; Yu, J.; Sun, G. Electrospun nanomaterials for ultrasensitive sensors. *Mater. Today* **2010**, *13*, 16–27. [CrossRef]

28. Su, Z.; Ding, J.; Wei, G. Electrospinning: A facile technique for fabricating polymeric nanofibers doped with carbon nanotubes and metallic nanoparticles for sensor applications. *RSC Adv.* **2014**, *4*, 52598–52610. [CrossRef]

29. Mondal, K.; Sharma, A. Recent advances in electrospun metal-oxide nanofiber based interfaces for electrochemical biosensing. *RSC Adv.* **2016**, *6*, 94595–94616. [CrossRef]

30. Zhang, M.; Zhao, X.; Zhang, G.; Wei, G.; Su, Z. Electrospinning design of functional nanostructures for biosensor applications. *J. Mater. Chem. B* **2017**, *5*, 1699–1711. [CrossRef]

31. Matlock-Colangelo, L.; Baeumner, A.J. Biologically Inspired Nanofibers for Use in Translational Bioanalytical Systems. *Annu. Rev. Anal. Chem.* **2014**, *7*, 23–42. [CrossRef] [PubMed]

32. Senthamizhan, A.; Balusamy, B.; Uyar, T. Glucose sensors based on electrospun nanofibers: A review. *Anal. Bioanal. Chem.* **2016**, *408*, 1285–1306. [CrossRef] [PubMed]
33. Zander, N.E. Hierarchically Structured Electrospun Fibers. *Polymers* **2013**, *5*, 19–44. [CrossRef]
34. Zhang, C.-L.; Yu, S.-H. Nanoparticles meet electrospinning: Recent advances and future prospects. *Chem. Soc. Rev.* **2014**, *43*, 4423–4448. [CrossRef] [PubMed]
35. Ahmad, M.; Pan, C.; Luo, Z.; Zhu, J. A single ZnO nanofiber-based highly sensitive amperometric glucose biosensor. *J. Phys. Chem. C* **2010**, *114*, 9308–9313. [CrossRef]
36. Tang, H.; Yan, F.; Tai, Q.; Chan, H.L.W. The improvement of glucose bioelectrocatalytic properties of platinum electrodes modified with electrospun TiO_2 nanofibers. *Biosens. Bioelectron.* **2010**, *25*, 1646–1651. [CrossRef] [PubMed]
37. Mondal, K.; Ali, M.A.; Agrawal, V.V.; Malhotra, B.D.; Sharma, A. Highly Sensitive biofunctionalized mesoporous electrospun TiO_2 nanofiber based interface for biosensing. *ACS Appl. Mater. Interfaces* **2014**, *6*, 2516–2527. [CrossRef] [PubMed]
38. Huang, S.; Ding, Y.; Liu, Y.X.; Su, L.; Filosa, R., Jr.; Lei, Y. Glucose biosensor using glucose oxidase and electrospun Mn_2O_3-Ag nanofibers. *Electroanalysis* **2011**, *23*, 1912–1920. [CrossRef]
39. Mao, X.; Tian, W.; Hatton, T.A.; Rutledge, G.C. Advances in electrospun carbon fiber based electrochemical sensing platforms for bioanalytical applications. *Anal. Bioanal. Chem.* **2016**, *408*, 1307–1326. [CrossRef] [PubMed]
40. Inagaki, M.; Yang, Y.; Kang, F. Carbon Nanofibers Prepared via Electrospinning. *Adv. Mater.* **2012**, *24*, 2547–2566. [CrossRef] [PubMed]
41. Zhang, L.; Aboagye, A.; Kelkar, A.; Lai, C.; Fong, H. A review: Carbon nanofibers from electrospun polyacrylonitrile and their applications. *J. Mater. Sci.* **2014**, *49*, 463–480. [CrossRef]
42. Fu, J.; Qiao, H.; Li, D.; Luo, L.; Chen, K.; Wei, Q. Laccase biosensor based on electrospun copper/carbon composite nanofibers for catechol detection. *Sensors* **2014**, *14*, 3543–3556. [CrossRef] [PubMed]
43. Li, D.; Luo, L.; Pang, Z.; Ding, L.; Wang, Q.; Ke, H.; Huang, F.; Wei, Q. Novel phenolic biosensor based on a magnetic polydopamine-laccase-nickel nanoparticle loaded carbon nanofiber composite. *ACS Appl. Mater. Interfaces* **2014**, *6*, 5144–5151. [CrossRef] [PubMed]
44. Zhang, X.; Liu, D.; Li, L.; You, T. Direct electrochemistry of glucose oxidase on novel freestanding nitrogen-doped carbon nanospheres@carbon nanofibers composite film. *Sci. Rep.* **2015**, *5*, 1–11. [CrossRef]
45. Bae, T.-S.; Shin, E.; Im, J.S.; Kim, J.G.; Lee, Y.-S. Effects of carbon structure orientation on the performance of glucose sensors fabricated from electrospun carbon fibers. *J. Non-Cryst. Solids* **2012**, *358*, 544–549. [CrossRef]
46. Cui, K.; Song, Y.; Guo, Q.; Xu, F.; Zhang, Y.; Shi, Y.; Wang, L.; Houb, H.; Li, Z. Architecture of electrospun carbon nanofibers-hydroxyapatite composite and its application act as a platform in biosensing. *Sens. Actuators. B Chem.* **2011**, *160*, 435–440. [CrossRef]
47. Wang, L.; Ye, Y.; Zhu, H.; Song, Y.; He, S.; Xu, F.; Hou, H. Controllable growth of Prussian blue nanostructures on carboxylic group functionalized carbon nanofibers and its application for glucose biosensing. *Nanotechnology* **2012**, *23*, 1–10. [CrossRef] [PubMed]
48. Manesh, K.M.; Kim, H.T.; Santhosh, P.; Gopalan, A.I.; Lee, K.-P. A novel glucose biosensor based on immobilization of glucose oxidase into multiwall carbon nanotubes-polyelectrolyte-loaded electrospun nanofibrous membrane. *Biosens. Bioelectron.* **2008**, *23*, 771–779. [CrossRef] [PubMed]
49. Wang, Z.G.; Wang, Y.; Xu, H.; Li, G.; Xu, Z.K. Carbon nanotube-filled nanofibrous membranes electrospun from poly(acrylonitrile-*co*-acrylic acid) for glucose biosensor. *J. Phys. Chem. C* **2009**, *113*, 2955–2960. [CrossRef]
50. Numnuam, A.; Thavarungkul, P.; Kanatharana, P. An amperometric uric acid biosensor based on chitosan-carbon nanotubes electrospun nanofiber on silver nanoparticles. *Anal. Bioanal. Chem.* **2014**, *406*, 3763–3772. [CrossRef] [PubMed]
51. Bourourou, M.; Holzinger, M.; Bossard, F.; Hugenell, F.; Maaref, A.; Cosnier, S. Chemically reduced electrospun polyacrilonitrile-carbon nanotube nanofibers hydrogels as electrode material for bioelectrochemical applications. *Carbon* **2015**, *87*, 233–238. [CrossRef]
52. Devadoss, A.; Han, H.; Song, T.; Kim, Y.-P.; Paik, U. Gold nanoparticle-composite nanofibers for enzymatic electrochemical sensing of hydrogen peroxide. *Analyst* **2013**, *138*, 5025–5030. [CrossRef] [PubMed]

53. Zhu, H.; Du, M.L.; Zhang, M.; Wang, P.; Bao, S.Y.; Wang, L.N.; Fu, Y.Q.; Yao, J.M.; Zhu, H.; Du, M.L.; et al. Facile fabrication of AgNPs/(PVA/PEI) nanofibers: High electrochemical efficiency and durability for biosensors. *Biosens. Bioelectron.* **2013**, *49*, 210–215. [CrossRef] [PubMed]

54. Wang, P.; Zhang, M.; Cai, Y.; Cai, S.; Du, M.; Zhu, H.; Bao, S.; Xie, Q. Facile fabrication of palladium nanoparticles immobilized on the water-stable polyvinyl alcohol/polyehyleneimine nanofibers via In-Situ reduction and their high electrochemical activity. *Soft Mater.* **2014**, *12*, 387–395. [CrossRef]

55. Jose, V.; Marx, S.; Murata, H.; Koepsel, R.R.; Russell, A.J. Direct electron transfer in a mediator-free glucose oxidase-based carbon nanotube-coated biosensor. *Carbon* **2012**, *50*, 4010–4020. [CrossRef]

56. Gladisch, J.; Sarauli, D.; Schäfer, D.; Dietzel, B.; Schulz, B.; Lisdat, F. Towards a novel bioelectrocatalytic platform based on "wiring" of pyrroloquinoline quinonedependent glucose dehydrogenase with an electrospun conductive polymeric fiber architecture. *Sci. Rep.* **2016**, *6*, 1–10. [CrossRef] [PubMed]

57. Fu, J.; Pang, Z.; Yang, J.; Huang, F.; Cai, Y.; Wei, Q. Fabrication of polyaniline/carboxymethyl cellulose/cellulosenanofibrous mats and their biosensing application. *App. Surf. Sci.* **2015**, *349*, 35–42. [CrossRef]

58. Yang, G.; Kampstra, K.L.; Abidian, M.R. High-performance conducting polymer nanofiber biosensors for detection of biomolecules. *Adv. Mater.* **2014**, *26*, 4954–4960. [CrossRef] [PubMed]

59. Uzun, S.D.; Kayaci, F.; Uyar, T.; Timur, S.; Toppare, L. Bioactive surface design based on functional composite electrospun nanofibers for biomolecule immobilization and biosensor applications. *ACS Appl. Mater. Interfaces* **2014**, *6*, 5235–5243. [CrossRef] [PubMed]

60. Ekabutr, P.; Chailapakul, O.; Supaphol, P. Modification of disposable screen-printed carbon electrode surfaces with conductive electrospun nanofibers for biosensor applications. *J. Appl. Polym. Sci.* **2013**, 3885–3893. [CrossRef]

61. Zhou, C.; Shi, Y.; Ding, X.; Li, M.; Luo, J.; Lu, Z.; Xiao, D. Development of a fast and sensitive glucose biosensor using iridium complex-doped electrospun optical fibrous membrane. *Anal. Chem.* **2013**, *85*, 1171–1176. [CrossRef] [PubMed]

62. Paul, K.B.; Kuma, S.; Tripathy, S.; Vanjari, S.R.K.; Singh, V.; Singh, S.G. A highly sensitive self-assembled monolayer modified copper doped zinc oxide nanofiber interface for detection of *Plasmodium falciparum* histidine-rich protein-2: Targeted towards rapid, early diagnosis of malaria. *Biosens. Bioelectron.* **2016**, *80*, 39–46. [CrossRef] [PubMed]

63. Ali, M.A.; Mondal, K.; Singh, C.; Malhotra, B.D.; Sharma, A. Anti-epidermal growth factor receptor conjugated mesoporous zinc oxide nanofibers for breast cancer diagnostics. *Nanoscale* **2015**, *7*, 7234–7245. [CrossRef] [PubMed]

64. Paul, B.; Singh, V.; Vanjari, S.R.K.; Singh, S.G. One step biofunctionalized electrospun multiwalled carbon nanotubes embedded zinc oxide nanowire interface for highly sensitive detection of carcinoma antigen-125 K. *Biosens. Bioelectron.* **2017**, *88*, 144–152. [CrossRef] [PubMed]

65. Zhang, N.; Deng, Y.; Tai, Q.; Cheng, B.; Zhao, L.; Shen, Q.; He, R.; Hong, L.; Liu, W.; Guo, S.; et al. Electrospun TiO$_2$ nanofiber-based cell capture assay for detecting circulating tumor cells from colorectal and gastric cancer patients. *Adv. Mater.* **2012**, *24*, 2756–2760. [CrossRef] [PubMed]

66. Li, Q.; Liu, D.; Xu, L.; Xing, R.; Liu, W.; Sheng, K.; Song, H. Wire-in-tube IrOx architectures: Alternative label-Free amperometric immunoassay toward α-Fetoprotein. *ACS Appl. Mater. Interfaces* **2015**, *7*, 22719–22726. [CrossRef] [PubMed]

67. Mondal, K.; Ali, M.A.; Srivastava, S.; Malhotra, B.D.; Sharma, A. Electrospun functional micro/nanochannels embedded in porous carbon electrodes for microfluidic biosensing. *Sens. Actuators B Chem.* **2016**, *229*, 82–91. [CrossRef]

68. Gupta, P.K.; Gupta, A.; Dhakate, S.R.; Khan, Z.H.; Solanki, P.R. Functionalized polyacrylonitrile-nanofiber based immunosensor for *Vibrio cholerae* detection. *J. Appl. Polym. Sci.* **2016**, *44170*, 1–9. [CrossRef]

69. Netsuwan, P.; Mimiya, H.; Baba, A.; Sriwichai, S.; Shinbo, K.; Kato, K.; Kaneko, F.; Phanichphant, S. Long-range surface plasmon resonance immunosensor based on water-stable electrospun poly(acrylic acid) fibers. *Sens. Actuators B Chem.* **2014**, *204*, 770–776. [CrossRef]

70. Sun, M.; Ding, B.; Lin, J.; Yu, J.; Sun, G. Three-dimensional sensing membrane functionalized quartz crystal microbalance biosensor for chloramphenicol detection in real time. *Sens. Actuators B Chem.* **2011**, *160*, 428–434. [CrossRef]

71. Tripathy, S.; Vanjari, S.R.K.; Singh, V.; Swaminathan, S.; Singh, S.G. Electrospun manganese (III) oxide nanofiber based electrochemical DNA nanobiosensor for zeptomolar detection of dengue consensus primer. *Biosens. Bioelectron.* **2017**, *90*, 378–387. [CrossRef] [PubMed]

72. Guler, Z.; Erkoc, P.; Sarac, A.S. Electrochemical impedance spectroscopic study of single-stranded DNA-immobilized electroactive polypyrrole-coated electrospun poly(ε-caprolactone) nanofibers. *Mater. Express* **2015**, *5*, 269–279. [CrossRef]

73. Wang, X.; Wang, X.; Wang, X.; Chen, F.; Zhu, K.; Xu, Q.; Tang, M. Novel electrochemical biosensor based on functional composite nanofibers for sensitive detection of p53 tumor suppressor gene. *Anal. Chim. Acta* **2013**, *765*, 63–69. [CrossRef] [PubMed]

74. Kim, S.G.; Lee, J.S.; Jun, J.; Shin, D.H.; Jang, J. Ultrasensitive bisphenol A Field-Effect Transistor sensor using an aptamer-modified multichannel carbon nanofiber transducer. *ACS Appl. Mater. Interfaces* **2016**, *8*, 6602–6610. [CrossRef] [PubMed]

75. Jun, J.; Lee, J.S.; Shin, D.H.; Jang, J. Aptamer-functionalized hybrid carbon nanofiber FET-type electrode for a highly sensitive and selective platelet-derived growth factor biosensor. *ACS Appl. Mater. Interfaces* **2014**, *6*, 3859–13865. [CrossRef] [PubMed]

76. Wang, Z.-G.; Wan, L.-S.; Liu, Z.-M.; Huang, X.-J.; Xu, Z.-K. Enzyme immobilization on electrospun polymer nanofibers: An overview. *J. Mol. Catal. B* **2009**, *56*, 189–195. [CrossRef]

77. Tang, C.; Saquing, C.D.; Morton, S.W.; Glatz, B.N.; Kelly, R.M.; Khan, S.A. Cross-linked polymer nanofibers for hyperthermophilic enzyme immobilization: Approaches to improve enzyme performance. *ACS Appl. Mater. Interfaces* **2014**, *6*, 11899–11906. [CrossRef] [PubMed]

78. Ren, G.; Xu, X.; Liu, Q.; Cheng, J.; Yuan, X.; Wu, L.; Wan, Y. Electrospun poly(vinyl alcohol)/glucose oxidase biocomposite membranes for biosensor applications. *React. Funct. Polym.* **2006**, *66*, 1559–1564. [CrossRef]

79. Oriero, D.A.; Gyan, I.O.; Bolshaw, B.W.; Cheng, I.F.; Aston, D.E. Electrospun biocatalytic hybrid silica-PVA-tyrosinase fiber mats for electrochemical detection of phenols. *Microchem. J.* **2015**, *118*, 166–175. [CrossRef]

80. Su, X.; Ren, J.; Meng, X.; Ren, X.; Tang, F. A novel platform for enhanced biosensing based on the synergy effects of electrospun polymer nanofibers and graphene oxides. *Analyst* **2013**, *138*, 1459–1466. [CrossRef] [PubMed]

81. Wu, C.M.; Yu, S.A.; Lin, S.L. Graphene modified electrospun poly(vinyl alcohol) nanofibrous membranes for glucose oxidase immobilization. *Express Polym. Lett.* **2014**, *8*, 565–573. [CrossRef]

82. Sapountzi, E.; Braiek, M.; Farre, C.; Arab, M.; Chateaux, J.-F.; Jaffrezic-Renault, N.; Lagarde, F. One-step fabrication of electrospun cross-linkable polymer nanofibers incorporating multiwall carbon nanotubes and enzyme for biosensing. *J. Electrochem. Soc.* **2016**, *162*, B275–B281. [CrossRef]

83. Sapountzi, E.; Braiek, M.; Vocanson, F.; Chateaux, J.-F.; Jaffrezic-Renault, N.; Lagarde, F. Gold nanoparticles assembly on electrospun poly(vinyl alcohol)/poly(ethyleneimine)/glucose oxidase nanofibers for ultrasensitive electrochemical glucose biosensing. *Sens. Actuators B Chem.* **2017**, *238*, 392–401. [CrossRef]

84. Luo, G.; Teh, K.S.; Liu, Y.; Zang, X.; Wen, Z.; Lin, L. Direct-write, self-aligned electrospinning on paper for controllable fabrication of three-dimensional structures. *ACS Appl. Mater. Interfaces* **2015**, *7*, 27765–27770. [CrossRef] [PubMed]

85. Ji, X.; Su, Z.; Wang, P.; Ma, G.; Zhang, S. "Ready-to-use" hollow nanofiber membrane-based glucose testing strips. *Analyst* **2014**, *139*, 6467–6473. [CrossRef] [PubMed]

86. Mateo, C.; Palomo, J.M.; Fernandez-Lorente, G.; Guisan, J.M.; Fernandez-Lafuente, R. Improvement of enzyme activity, stability and selectivity via immobilization techniques. *Enzyme Microb. Technol.* **2007**, *40*, 1451–1463. [CrossRef]

87. Dai, Y.; Liu, W.; Formo, E.; Sun, Y.; Xia, Y. Ceramic nanofibers fabricated by electrospinning and their applications in catalysis, environmental science and energy technology. *Polym. Adv. Technol.* **2011**, *22*, 326–338. [CrossRef]

88. Both Engel, A.; Bechelany, M.; Fontaine, O.; Cherifi, A.; Cornu, D.; Tingry, S. One-Pot Route to Gold Nanoparticles Embedded in Electrospun Carbon Fibers as an Efficient Catalyst Material for Hybrid Alkaline Glucose Biofuel Cells. *ChemElectroChem* **2016**, *3*, 629–637. [CrossRef]

89. Gubala, V.; Harris, L.F.; Ricco, A.J.; Tan, M.X.; Williams, D.E. Point of Care Diagnostics: Status and Future. *Anal. Chem.* **2012**, *84*, 487–515. [CrossRef] [PubMed]

90. Schumacher, S.; Nestler, J.; Otto, T.; Wegener, M.; Ehrentreich-Forster, E.; .Michel, D.; Wunderlich, K.; Palzer, S.; Sohn, K.; Weber, A.; et al. Highly-integrated lab-on-chip system for point-of-care multiparameter analysis. *Lab Chip* **2012**, *12*, 464. [CrossRef] [PubMed]

91. Burger, R.; Amato, L.; Boisen, A. Detection methods for centrifugal microfluidic platforms. *Biosens. Bioelectron.* **2016**, *76*, 54–67. [CrossRef] [PubMed]

92. Dincer, C.; Bruch, R.; Kling, A.; Dittrich, P.S.; Urban, G.A. Multiplexed Point-of-Care Testing—xPOCT. *Trends Biotechnol.* **2017**, *35*, 728–742. [CrossRef] [PubMed]

93. Yetisen, A.K.; Akram, M.S.; Lowe, C.R. Paper-based microfluidic point-of-care diagnostic devices. *Lab Chip* **2013**, *13*, 2210–2251. [CrossRef] [PubMed]

94. Mahato, K.; Srivastava, A.; Chandra, P. Paper based diagnostics for personalized health care: Emerging technologies and commercial aspects. *Biosens. Bioelectron.* **2017**, *96*, 246–259. [CrossRef] [PubMed]

95. Wang, J.; Kang, Q.-S.; Lv, X.-G.; Song, J.; Zhan, N.; Dong, W.-G.; Huang, W.-H. Simple Patterned Nanofiber Scaffolds and Its Enhanced Performance in Immunoassay. *PLoS ONE* **2013**, *8*, e82888. [CrossRef] [PubMed]

96. Sadir, S.; Prabhakaran, M.P.; Wicaksono, D.H.B.; Ramakrishna, S. Fiber based enzyme-linked immunosorbent assay for C-reactive protein. *Sens. Actuators B Chem.* **2014**, *205*, 50–60. [CrossRef]

97. Hosseini, S.; Azari, P.; Farahmand, E.; Gan, S.N.; Rothan, H.A.; Yusof, R.; Koole, L.H.; Djordjevic, I.; Ibrahim, F. Polymethacrylate coated electrospun PHB fibers: An exquisite outlook for fabrication of paper-based biosensors. *Biosens. Bioelectron.* **2015**, *69*, 257–264. [CrossRef] [PubMed]

98. Mahmoudifard, M.; Soudi, S.; Soleimani, M.; Hosseinzadeh, S.; Esmaeili, E.; Vossoughi, M. Efficient protein immobilization on polyethersolfone electrospun nanofibrous membrane via covalent binding for biosensing applications. *Mater. Sci. Eng. C* **2016**, *58*, 586–594. [CrossRef] [PubMed]

99. Chantasirichot, S.; Ishihara, K. Electrospun phospholipid polymer substrate for enhanced performance in immunoassay system. *Biosens. Bioelectron.* **2012**, *38*, 209–214. [CrossRef] [PubMed]

100. Hersey, J.S.; Meller, A.; Grinstaff, M.W. Functionalized Nanofiber Meshes Enhance Immunosorbent Assays. *Anal. Chem.* **2015**, *87*, 11863–11870. [CrossRef] [PubMed]

101. Lee, Y.; Lee, H.J.; Son, K.J.; Koh, W.-G. Fabrication of hydrogel-micropatterned nanofibers for highly sensitive microarray-based immunosensors having additional enzyme-based sensing capability. *J. Mater. Chem.* **2011**, *21*, 4476–4483. [CrossRef]

102. Nicolini, A.M.; Fronczek, C.F.; Yoon, J.-Y. Droplet-based immunoassay on a 'sticky' nanofibrous surface for multiplexed and dual detection of bacteria using smartphones. *Biosen. Bioelectron.* **2015**, *67*, 560–569. [CrossRef] [PubMed]

103. Jin, S.; Dai, M.; Ye, B.-C.; Nugen, S.R. Development of a capillary flow microfluidic *Escherichia coli* biosensor with on-chip reagent delivery using water-soluble nanofibers. *Microsyst. Technol.* **2013**, *19*, 2011–2015. [CrossRef]

104. Cho, D.; Matlock-Colangelo, L.; Xiang, C.; Asiello, P.J.; Baeumner, A.J.; Frey, M.W. Electrospun nanofibers for microfluidic analytical systems. *Polymer* **2011**, *52*, 3413–3421. [CrossRef]

105. Lee, W.S.; Sunkara, V.; Han, J.-R.; Parka, Y.-S.; Cho, Y.K. Electrospun TiO₂ nanofiber integrated lab-on-a-disc for ultrasensitive protein detection from whole blood. *Lab Chip* **2015**, *15*, 478–485. [CrossRef] [PubMed]

106. Ali, M.A.; Mondal, K.; Jiao, Y.; Oren, S.; Xu, Z.; Sharma, A.; Dong, L. Microfluidic Immuno-Biochip for Detection of Breast Cancer Biomarkers Using Hierarchical Composite of Porous Graphene and Titanium Dioxide Nanofibers. *ACS Appl. Mater. Interfaces* **2016**, *8*, 20570–20582. [CrossRef] [PubMed]

107. Luo, Y.; Nartker, S.; Miller, H.; Hochhalter, D.; Wiederoder, M.; Wiederoder, S.; Setterington, E.; Drzal, L.T.; Alocilj, E.C. Surface functionalization of electrospun nanofibers for detecting *E. coli* O157:H7 and BVDV cells in a direct-charge transfer biosensor. *Biosens. Bioelectron.* **2010**, *26*, 1612–1617. [CrossRef] [PubMed]

108. Reinholt, S.J.; Sonnenfeldt, A.; Naik, A.; Frey, M.W.; Baeumner, A.J. Developing new materials for paper-based diagnostics using electrospun nanofibers. *Anal. Bioanal. Chem.* **2014**, *406*, 3297–3304. [CrossRef] [PubMed]

109. Viter, R.; Iatsunskyi, I.; Fedorenko, V.; Tumenas, S.; Balevicius, Z.; Ramanavicius, A.; Balme, S.; Kempiński, M.; Nowaczyk, G.; Jurga, S.; et al. Enhancement of Electronic and Optical Properties of ZnO/Al₂O₃ Nanolaminate Coated Electrospun Nanofibers. *J. Phys. Chem. C* **2016**, *120*, 5124–5132. [CrossRef]

Sensors **2017**, *17*, 1887

110. Iatsunskyi, I.; Vasylenko, A.; Viter, R.; Kempinski, M.; Nowaczyk, G.; Jurga, S.; Bechelany, M. Tailoring of the electronic properties of ZnO-polyacrylonitrile nanofibers: Experiment and theory. *Appl. Surf. Sci.* **2017**, *411*, 494–501. [CrossRef]

111. González, E.; Shepherd, L.M.; Saunders, L.; Frey, M.W. Surface Functional Poly(lactic Acid) Electrospun Nanofibers for Biosensor Applications. *Materials* **2016**, *9*, 47. [CrossRef] [PubMed]

Review

France's State of the Art Distributed Optical Fibre Sensors Qualified for the Monitoring of the French Underground Repository for High Level and Intermediate Level Long Lived Radioactive Wastes

Sylvie Delepine-Lesoille [1,*], **Sylvain Girard** [2], **Marcel Landolt** [1], **Johan Bertrand** [1], **Isabelle Planes** [1,2], **Aziz Boukenter** [2], **Emmanuel Marin** [2], **Georges Humbert** [3], **Stéphanie Leparmentier** [1,3], **Jean-Louis Auguste** [3] **and Youcef Ouerdane** [2]

[1] National Radioactive Waste Management Agency (Andra), F-92298 Chatenay-Malabry, France; marcel.landolt@andra.fr (M.L.); johan.bertrand@andra.fr (J.B.); isabelle.planes@univ-st-etienne.fr (I.P.); stephanie.leparmentier@xlim.fr (S.L.)

[2] Laboratoire Hubert Curien CNRS UMR 5516, University of Lyon, F-42000 Saint-Etienne, France; sylvain.girard@univ-st-etienne.fr (S.G.); aziz.boukenter@univ-st-etienne.fr (A.B.); emmanuel.marin@univ-st-etienne.fr (E.M.); ouerdane@univ-st-etienne.fr (Y.O.)

[3] XLIM Research Institute, UMR 7252 CNRS/University of Limoges, 123 Avenue Albert Thomas, F-87060 Limoges, France; georges.humbert@xlim.fr (G.H.); auguste@xlim.fr (J.-L.A.)

* Correspondence: sylvie.lesoille@andra.fr; Tel.: +33-1-4611-8071

Academic Editors: Nicole Jaffrezic-Renault and Gaelle Lissorgues
Received: 31 March 2017; Accepted: 4 May 2017; Published: 13 June 2017

Abstract: This paper presents the state of the art distributed sensing systems, based on optical fibres, developed and qualified for the French Cigéo project, the underground repository for high level and intermediate level long-lived radioactive wastes. Four main parameters, namely strain, temperature, radiation and hydrogen concentration are currently investigated by optical fibre sensors, as well as the tolerances of selected technologies to the unique constraints of the Cigéo's severe environment. Using fluorine-doped silica optical fibre surrounded by a carbon layer and polyimide coating, it is possible to exploit its Raman, Brillouin and Rayleigh scattering signatures to achieve the distributed sensing of the temperature and the strain inside the repository cells of radioactive wastes. Regarding the dose measurement, promising solutions are proposed based on Radiation Induced Attenuation (RIA) responses of sensitive fibres such as the P-doped ones. While for hydrogen measurements, the potential of specialty optical fibres with Pd particles embedded in their silica matrix is currently studied for this gas monitoring through its impact on the fibre Brillouin signature evolution.

Keywords: optical fibres; optical fibre sensors; Raman; Brillouin; Rayleigh scatterings; temperature; strain; radiation effects; hydrogen

1. Introduction

In the Cigéo project, the French underground repository for high-level and intermediate level long-lived radioactive wastes program, monitoring in the civil engineering structures used and host rock is planned in order to confirm the long-term safety evaluation and recoverability of wastes during the exploitation period. Work is in progress in order to select the exact parameters to monitor. Andra, the French national agency for radioactive waste management, supports sensor developments and qualification tests for the most relevant processes that will be developed during Cigéo implementation, namely thermal, hydraulic, mechanical, chemical and radioactive (T-H-M-C-R).

To ensure long-term monitoring despite harsh conditions, several and diverse monitoring technologies are considered and would be combined inside the repository cells: both non-destructive evaluation and embedded sensors, as well as electronic and optical sensors, including direct and indirect measurements of parameters of interest, mixing traditional and innovative sensing technologies.

Optical fibre sensors (OFS) are found to be exceptional tools, as they enable distributed measurements, thus providing data over the entire structure. Monitoring with a single fibre can provide information all along the structure behavior, and thus overcome limitations of traditional sensors, whose information is restricted to local effects. Moreover, optical fibre's small size enables one to reduce invasiveness. Remote sensing would enable the maintenance of the optoelectronic devices during the facility lifetime; only the optical fibre, that is known to be more resistant than electronics, can be exposed to the harsh conditions during the operating period. This explains why, since 2006, Andra has deployed a large qualification process on distributed optical fibre sensing systems in collaboration with several companies and research laboratories.

Optical fibres are known and widely spread in industry to provide strain and temperature measurements. Yet they must be designed to resist to the specific environmental conditions of Cigéo: presence of hydrogen, relatively high temperature (100 °C) and gamma rays. This article presents the target application in a first part (Section 2), and the qualification methodology in a second part (Section 3). State of the art temperature sensing is presented in Section 4, while strain sensing is detailed in Section 5, hydrogen and radiation sensing are reported in Sections 6 and 7, respectively. The updated knowledge is quoted in Technology Readiness levels (TRL, [1]). This overview includes developments supported by Andra in France or reported in literature by other French labs.

2. The Target Application: Monitoring the Underground Repository of High Level and Intermediate Level Long-Lived Radioactive Wastes

2.1. Repository Cells

Cigéo is designed to isolate the radioactive waste from humans and the biosphere and to confine it within a deep geological formation to prevent dissemination of the radionuclides contained in this waste. In the French design, the disposal takes placed in a clay formation, an argillaceous rock (clay rock) known as the Callovo-Oxfordian formation, which is approximately 155 million years old and lies at a depth of 400 to 600 m.

As sketched in Figure 1, the underground facility has separate areas designed for high-level waste (HLW) and intermediate-level long-lived waste (ILW-LL) in order to limit phenomenological interactions between these different waste categories. The cells are spherical or ovoid underground excavations on a horizontal axis or slight slope dug out of the Callovo-Oxfordian formation. There are two categories of cells. HLW micro-tunnel cells would contain only one disposal package per cell section; with main dimensions of 0.7 m diameter, 100 m long, and a metallic liner. ILW-LL tunnel cells would contain several disposal packages per cell section (in the order of 9 m size diameter). Their mechanical stability during operation is guaranteed by a concrete liner, left in place at closure. Length would be on the order of 600 m.

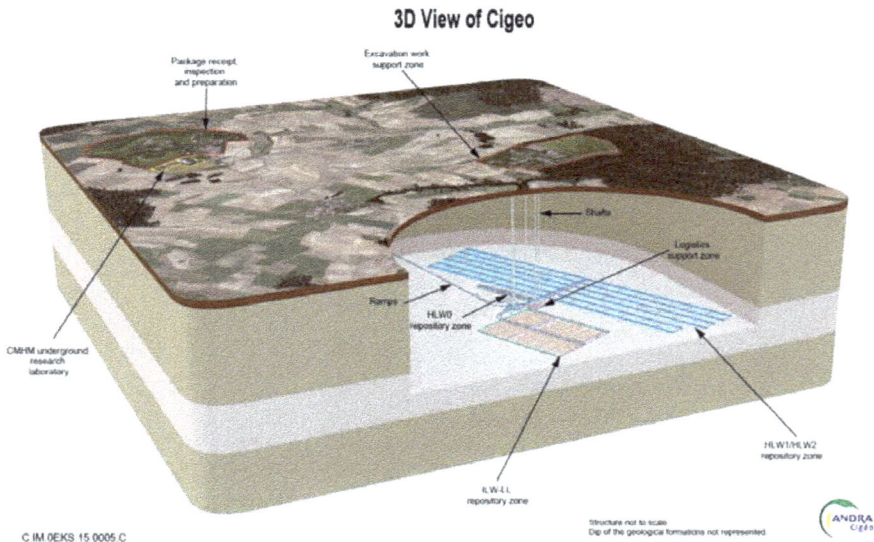

Figure 1. Preliminary design of Cigéo, the underground repository for high level and intermediate level long-lived radioactive wastes.

2.2. Monitoring Objectives

An appropriate level of monitoring and control will be applied to Cigéo during its construction and its operation, to ensure the protection and preservation of the passive post-closure safety features. Observation and surveillance parameters are not fully stated yet. A global analysis based on safety analysis, technology readiness level [1] of monitoring technologies and risks is currently performed within Andra, and also at the European level within the MoDeRn and MoDeRn2020 European projects [2,3].

OFS are mainly envisioned for temperature and mechanical evaluations. Temperature monitoring contributes to the assessment confirmation basis for long term safety, by quantifying margins of prior predictions for the duration of the thermal period and the expected thermal peaks. Mechanical parameters are also required, to confirm that the waste recoverability is possible, if decided; indeed repository reversibility is stated by the French Planning Act No. 2006-739 dated 28 June 2006 on the sustainable management of radioactive waste.

To provide accurate and verified measurements, it is also important to characterize all parameters influencing the sensing chain performances. This is why hydrogen and radiation also appear in the target monitoring parameters of Table 1. Gamma radiation dose rate ranges from 1 to 10 $Gy(SiO_2)/h$ depending on the waste category. Over the target lifetime of 100 years, the sensors could experience ionizing doses up to 10 $MGy(SiO_2)$. The atmosphere will become hydrogen-rich because H_2 emissions originating from anoxic corrosion of metallic materials or some specific radioactive IL waste release hydrogen (cf. [2]). Measuring either hydrogen or oxygen (related to corrosion speed) or both would assess chemical evolutions.

Possible target monitoring parameters, expected sensing range and uncertainties are detailed in Table 1. As most phenomena kinetics are slow, their corresponding measurement frequencies can be selected on a daily basis.

As the parameters are not homogeneously distributed around the repository cells, it is important to develop sensors able to perform truly distributed sensing, to characterize a spatial mapping distribution.

Table 1. Main parameters to be monitored in the underground repository.

Parameters	Typical Range	Target Sensitivity	Spatial Homogeneity
Temperature	[20–90 °C]	±0.1 °C	20 cm
Displacement	+0.5 mm/m to 2.5 mm/m	1 µm/m	10 cm
Strain in concrete	10 µm/m	3 µm/m	10 cm
Concrete Crack	Threshold for openings: 200 µm		10 cm
Gap evolution inside the cell	10 mm (in 100 years)	0.5 mm	1 m
Hydrogen	[0–4%] sensitivity of 500 ppm [4–10%] sensitivity of 1%	100 ppm <1%	3 m (ILW-LL waste package size) ~1.5 m (HL waste package)
Gamma radiation	0.1–1 Gy/h Total cumulated dose 10 MGy (100 years)	50 mGy	~1.5 m (HL waste package)

2.3. Envisionned Implementation of OFS in Repository Cells

In the reference monitoring system design, sensors are embedded in concrete liners of Intermediate—level (IL) repository cells, as sketched in Figure 2, or on the external surface of the metallic liner of High-Level (HL) repository cells. Regarding the application, more details can be found in [2].

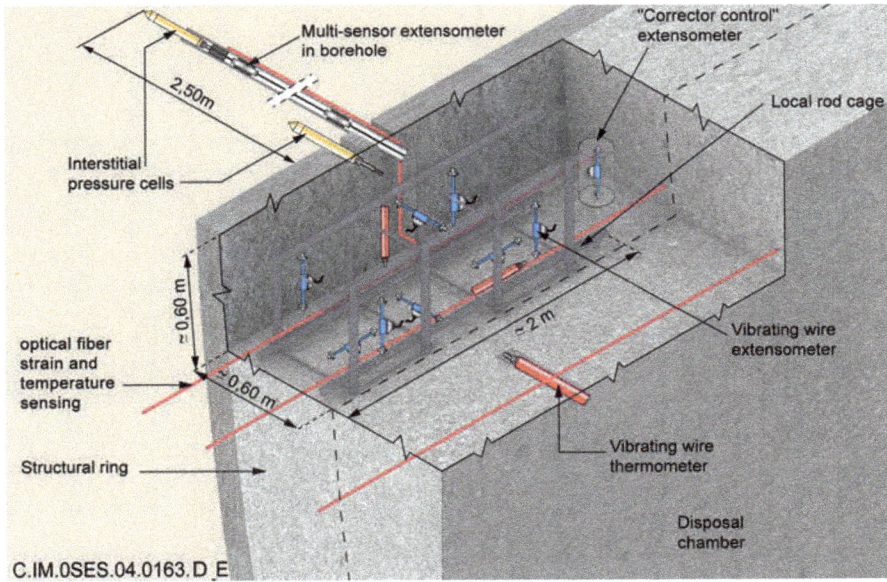

Figure 2. Envisioned monitoring system for ILW-LL repository cells.

2.4. Environmental Conditions in the Repository Cells

Improving and hardening the sensor radiation tolerance is a major issue for the design of reliable, long-term equipment. Worst case will be considered for the qualification process performed by Andra, namely the vicinity of HLW repository cell. At the extreme conditions, humidity will reach 100%, hydraulic pressure 6 MPa, lithostatic pressure 12 MPa, dose rate 1 Gy/h and temperature 90 °C.

3. Materials and Methods

3.1. Qualification Method

To fulfill these challenging application requirements, several monitoring technologies are considered and would be used in combination inside the repository cells. For each sensing chain, a qualification method in several steps is deployed. First we select the most promising technologies which are tested in controlled situations. If these tests are successful, the sensor can be implemented into real structures, to quantify its impact, integration and response in representative environments. A last step is devoted to the harsh environments vulnerability study: accelerated aging tests, for instance radiation exposure, are organized. This methodology is summarized in Figure 3.

Figure 3. Sensing chain qualification methodology.

In the present review, we will summarize and point out developments realized in France by Andra and its research partners to qualify a sensing chain and provide truly distributed measurements of temperature, strain, hydrogen and radiation dose levels.

3.2. Choice of Sensing Principle—Scatterings in Optical Fibres

Distributed sensing provides a versatile and powerful monitoring tool. The term distributed optical fibre sensor designates the case in which the silica-based material becomes a sensor. It is thus no longer necessary to implement anticipated sensor positions since measurements are being performed all along the optical fibre connected to the probe/reading device (as well as within the extension cables). Truly distributed measuring systems for temperature and strain sensing were transposed from labs to industrial applications in 2000 s. Instruments combine (i) a sensitive phenomenon based on Brillouin, Rayleigh or Raman scatterings with (ii) a localization process, usually Optical Time Domain Reflectometry (OTDR), Optical Frequency Domain Reflectometry (OFDR, as in Luna-OBR device mentioned later) or coherent probe-pump techniques (as in the Neubrescope device mentioned later).

Rayleigh and Brillouin scatterings are sensitive to both strain and temperature whereas Raman scattering depends only on temperature. More precisely, the Raman Anti-Stokes intensity is significantly more sensitive to temperature than the Stokes component. Most commercial devices determine the optical fibre temperature from the spectral analysis of the scattered light, by computing the ratio between the Anti-Stokes and Stokes intensities from the Equation (1) [4]:

$$\frac{I_{AS}}{I_S} = \frac{(\nu_0 + \Delta\nu_{Raman})^4}{(\nu_0 - \Delta\nu_{Raman})^4} \cdot e^{-\frac{h \cdot \Delta\nu_{Raman}}{K_B \cdot T}} \tag{1}$$

where I_{AS} and I_S are the Anti-Stokes and Stokes intensities respectively, h is Planck's constant (given as $6.62607004 \times 10^{-34} \cdot m^2 \cdot kg \cdot s^{-1}$), $\Delta\nu_{Raman}$ is the Raman frequency shift (13.2 THz for silica) and ν_0 is the probe laser pulse frequency (in Hz), K_B is the Boltzmann constant (given as $1.38064852 \times 10^{-23} \, m^2 \cdot kg \cdot s^{-2} \cdot K^{-1}$) and T the absolute temperature (in Kelvin, K).

The Brillouin frequency shift $\Delta\nu_B$ is known to be proportional to temperature and strain (ε) variations [5], so it can be defined as:

$$\Delta\nu_B = C_\varepsilon{}^B \varepsilon + C_T{}^B \Delta T \tag{2}$$

where $C_T{}^B$ and $C_\varepsilon{}^B$ are the sensitivity coefficients for both temperature and strain, respectively.

If measured at a single wavelength, Rayleigh scattering provides only a detection of events along an optical fibre. To supply temperature and strain measurements, commercial devices exploit the temperature and strain sensitivities of the Rayleigh backscattered intensity over a spectral window. Two measurements should be acquired and correlated, this technique being able to follow the strain or temperature changes occurring with respect to an initial condition considered as a reference. The scattered profiles from the two datasets are cross-correlated along the perturbed portion of the fiber to obtain the spectral shift in this part of the fiber caused by temperature and /or strain change. We call this change of the correlation peak position "Rayleigh spectral shift", $\Delta \upsilon^R$. It is related to both strain and temperature via calibration coefficients C_T^R and C^R_ε [6]:

$$\Delta \upsilon^R = C_\varepsilon^R \varepsilon + C_T^R \, \Delta T, \tag{3}$$

Detailed description of s truly distributed sensing scheme is available in [7] and more recently in [8]. Performances are summarized in Table 2 for comparison. Values are orders of magnitude; performances may significantly change with the operation wavelength, the device supplier, the sensing optical fibre type... Raman device performances should reach 0.1 °C uncertainty over 30 km, with a length resolution of 1m. Brillouin systems provide 1 °C or 20°·μm/m uncertainties, over 30 km, with a spatial resolution of 1 m. Rayleigh scattering sensing devices exhibit higher performances than Brillouin based systems. Depending on the localization process, there is a trade-off to find between the maximal sensing range and the spatial resolution. For OFDR [9] (and tunable-OTDR, respectively [10]) the measurement distance range is 70 m (or 20 km, respectively), and spatial resolution in the order of 1 cm (resp. 20 cm). It is worth noticing that both Rayleigh and Brillouin instruments are using single-mode fibres (SMF) and operate at wavelength around 1.55 μm. For distance range smaller than 20 km, most commercially-available Raman devices use multi-mode fibres (MMF) at around 1064 nm. It implies that the three device types cannot be paired with the same optical fibre. Raman singlemode devices exist but are less efficient on short distance consistently with [11] (and more expensive).

Table 2. Typical performances of distributed temperature and sensing systems based on Rayleigh, Brillouin and Raman scatterings in optical fibres.

Scattering	Rayleigh		Brillouin		Raman
Process	Elastic		Inelastic		Inelastic
Optical fibre type	Single-mode		Single-mode		Mainly multi-mode
Measuring principle	OFDR	TW-COTDR	BOTDR	BOTDA	R-OTDR
Acces to fibre	Single end		Single end	Loop configuration	Single end
Maximal distance range	70 m	20 km	30 km		30 km
Best spatial resolution	10 mm	20 cm	1 m	10 cm	1 m
Temperature sensitivity	$C_T^R = -1.5$ GHz/°C		1 MHz/°C		0.1 °C
Strain sensitivity	$C^R_\varepsilon = -0.15$ GHz/με		0.05 MHz/με		Not sensitive
Measurement Uncertainty	0.1 °C	0.5 °C	5 °C	1 °C	0.01 °C
Measurement duration	10 s	10 min	10 min		1 min
Optical budget	70 dB	10 dB	10 dB		10 dB

Performances summarized in Table 2 are obtained with optical fibres in laboratories and would be significantly degraded in real application. More precisely, although Andra's application is limited to the kilometer range, with the high excess of optical losses induced by radiations, optical budget would reach 10 dB after a propagation length of 1 km only. One should wonder whether short distance with high losses equals long distance and low-loss, on the sensing performance point of view.

Because of the complementary performances, of these various techniques for strain and temperature distributed sensing, qualification procedure has been realized on the sensors exploiting the three scattering processes.

Influences of hydrogen and radiations presence on strain and temperature distributed sensing were poorly known in 2008 when Andra started its qualification process. The first studies were devoted to the vulnerability of those technologies to radiations. It was demonstrated that the sensors based on the three scattering processes are significantly influenced by: (i) radiation; (ii) hydrogen diffusion into the optical fibre and (iii) their coupled effects. The main results of these analyses as the outcomes of the next projects devoted to the hardening of these technologies are discussed in the following sections of this review (Sections 3.8 and 3.9). Recent works and related methodologies for radiation and hydrogen sensing are detailed in sections Sections 6 and 7, respectively.

3.3. Optical Fibre Choice/Design: Dopant and Primary Coatings

Claimed performances are obtained with standard telecommunication fibres, usually single-mode G652 type for temperature sensing, G657 type for strain sensing based on Brillouin or Rayleigh scatterings, graded-index 50 µm core multi-mode fibre for temperature sensing based on Raman scatterings.

These properties might not be preserved under harsh environmental conditions of Cigéo. Optical fibres darken under radiation through a phenomenon called Radiation-Induced Attenuation (RIA). The amplitudes and kinetics of these excess losses depend on many parameters including the fibre composition and the operation wavelength. Standard telecommunication fibres usually possess Ge-doped core and various possible cladding compositions. Both core and cladding dopants strongly impacts the fibre radiation response [12]. They are known to handle low gamma irradiation doses. For harsh environments such as Cigéo, pure silica core fibres with F dopants in the optical claddings have been developed [13]. On the opposite, P and Al dopants are used to obtain radiation sensitive fibres [14]. This is why Andra's qualification procedure has been implemented with this variety of optical fibres, from the highly sensitive fibre to the most resistant one. More precisely, Andra's partners, Laboratoire Hubert Curien, (LabHC—Université de Lyon, CNRS UMR 5516), selected a large variety of optical fibres. Samples whose results are presented in the article are detailed in Table 3.

Table 3. Optical fibers selected for the ANDRA qualification process.

Type	Core	Cladding	Coating	Names
SM-Ge	Ge-SiO$_2$ (28 wt %)	Pure silica	Acrylate	CMS
SM-F	F-SiO$_2$ (0.2 wt %)	F-SiO$_2$ (1.8 wt %)	Acrylate	SIO2/F
MMF	Pure Silica Core	F-SiO$_2$	Acrylate	Fibre I
MMF	P-SiO$_2$	Doped-SiO$_2$	Carbon	HDR MMF

Regarding primary coatings, acrylate is used for telecommunication fibres and it is able to withstand temperatures up to 80 °C. As the 20–100 °C range is expected for Cigéo application, it appears mandatory to consider alternative coating materials able to resist to higher temperatures. For such conditions, High-Temperature (HT) acrylate, polyimide coatings and metallic ones (Al, Au, Cu) can be used. Carbon layers deposited between the fibre glass and its coating also endure 100 °C and are incorporated to prevent the hydrogen diffusion from the surrounding environment into silica-based optical fibres [15]. As regards the various fibre prices, availability and optical performances, Andra selected HT-acrylate, polyimide and carbon-coated fibres. Au coating has been tested in France by EDF for Raman sensing in the 300–350 °C temperature range [16].

Samples were either purchased from the suppliers of Draka (Douvrin, France) and Fibercore (Southampton, UK) commercial off the shelf samples (COTS in the following), or ad hoc samples manufactured by the French iXBlue Photonics company (Lannion, France), allowing us to have a better control on the preform composition and its drawing conditions.

3.4. Optical Fibre Sensing Cables

On top of measuring principle and optical fibre type, selection of the sensing cable is of crucial importance. The external sheath ensures protection of the optical fibre (mechanical, chemical, hydraulic) and drives the strain transfer from the structure to sense/supervise to the optical fibre where sensing is performed. These two requirements are somehow opposite.

Electricité de France (EDF, the French Electricity Company) and Andra have selected and tested several sensing cables, for soil monitoring, as illustrated in Figure 4 [17]. Similar comparative tests can be also performed for concrete and metallic structure monitoring [18,19]. The selected cables present advantages and drawbacks regarding handling (flexible and easy to install) and durability (metallic reinforced cable).

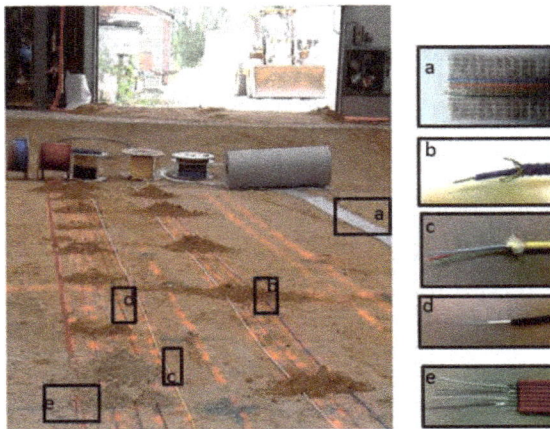

Figure 4. (Left) Picture of a one-to-one scale mock-up of dyke where several strain sensing cables (whose zooms are provided on the right) were tested.

3.5. Strain and Temperature Metrological Benches

Several metrological benches were developed in order to impose controlled solicitation on optical fibre samples with several meter lengths (longer than the 1 m spatial resolution of most optoelectronic devices).

3.5.1. Temperature Metrology

In collaboration with Andra and EDF, the Laboratoire National d'Essai (LNE, in Trappes near Paris, France) has built a dedicated facility in order to evaluate and to qualify distributed temperature sensing chains. Spatial resolution of the instrument is typically 1 m; the furnace had thus to be longer than 10 m for ensuring a spatially resolved measurement. As described in extended details in [20] and sketched in Figure 5, the horizontal furnace has a tubular configuration arranged as an assembly of five concentric stainless steel tubes. The central tube (internal diameter of 18 mm) can host 25 m of unrolled optical fibre. This furnace enables to perform temperature measurements with optoelectronic devices, by controlling the stability and homogeneity of the temperature around the fibre and by avoiding any mechanical stress. The fibre is fully free to move inside the tube, and free from any mechanical constraints, besides its own weight. The furnace temperature is controlled and stabilized by a water jacket. Enclosure is insulated to ensure a homogenous temperature field.

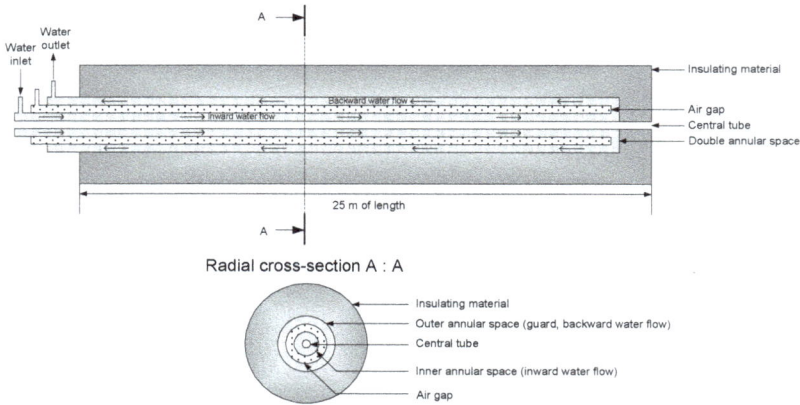

Figure 5. Schematic view of the 25 m horizontal furnace. Adapted with permission from [20]. Copyright 2016 IOP Publishing.

The 25 m long bench enables one to evaluate temperature sensing chain performances with temperature homogeneity and temperature stability over a duration of 20 h better than 0.1 °C and 0.05 °C, respectively [20].

3.5.2. Strain Metrology

Andra and LabHC use two different techniques to evaluate strain sensitivities of optical fibres. The first design is based on pulleys [21]. Several meter-long fibre is supported by several pulleys, fixed at the two extremities; and strained over a length section L by applying a stepwise displacement ΔL using a micrometer screw gauge (Figure 6, right). The strain applied along the fibre is given by the relation: $\varepsilon = \Delta L/L$, where L and ΔL are the length of stretched fibre and the displacement respectively. The total sample length was 8 m while the strain step was 100 µε in our case. As illustrated in Figure 6 left, the bench enables to create several controlled and homogeneous strain levels over more than one meter of fibre, between the fixation points.

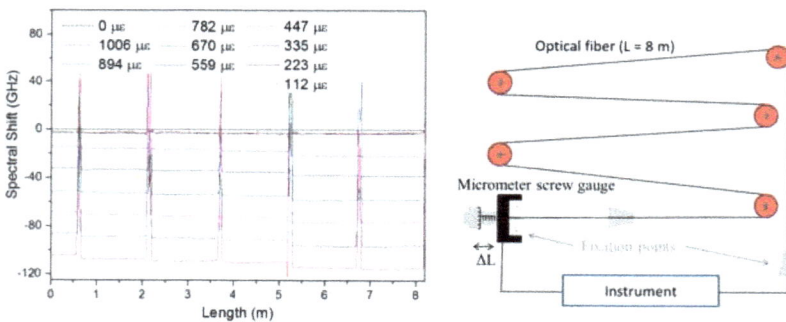

Figure 6. (Left) Example of distributed strain measurement obtained with TW-OTDR while a fibre is pulled at 10 different stress levels on the strain sensing bench (sketched on the right).

Another solution is to coil the Fibre Under Test (FUT) around a borosilicate [21] or a quartz [22] tube, by applying different controlled strains along the sample length. An illustration is reported in Figure 7 with a step strain every 15 m. Borosilica has a thermal expansion coefficient (TEC) identical to that of silica. As a consequence, such a setup permits one to verify in situ the impact of radiation on the fibre strain sensitivity coefficient. However, to transform weight into strain, the optical fibre Young modulus is required. It is proved to change with the optical fibre type and its influence is not negligible. This is why both strain calibration methods are useful.

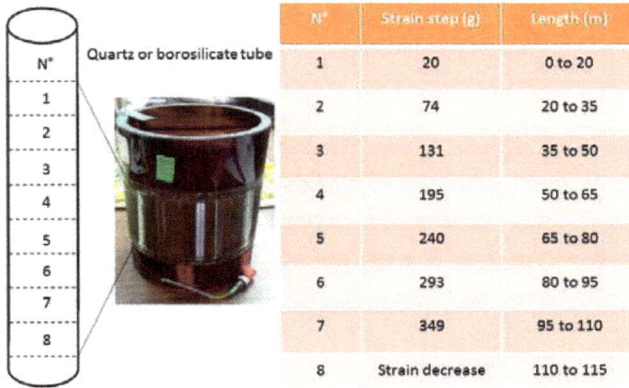

N°	Strain step (g)	Length (m)
1	20	0 to 20
2	74	20 to 35
3	131	35 to 50
4	195	50 to 65
5	240	65 to 80
6	293	80 to 95
7	349	95 to 110
8	Strain decrease	110 to 115

Figure 7. Schematic representation of the fibre coiling and picture of fibre sample strain steps. Borosilicate holder darkened under irradiation.

Finally, more robust bench designs must be used when sensing cables (opposite to optical fibres in primary coating), have to be characterized. Previous studies [23] demonstrated that the fibre strain coefficients of the Rayleigh and Brillouin responses are affected by the fibre packaging. Andra has used the testing bench of the Marmota company (Zurich, Switzerland), illustrated in Figure 8 to characterize the coefficients of the fibre cable. The fixation is designed in such way that it fixes the cable properly with a specific attention of no slippage effect between the cable and the clamp. The total length of the sample is about 5 m; the applied strain step can be controlled down to 100 µε while the room temperature is regulated.

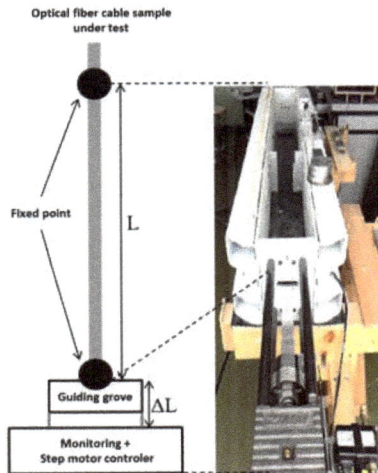

Figure 8. Schematic representation (**left**) of the bench used to determine fibre optic cables' sensitivities to strain and pictures (**right**) from Marmota Company. Adapted with permission from [23]. Copyright 2016 Optical Society of America Publishing.

3.6. Outdoor Tests

Andra has created an Underground Research Laboratory (URL, see Figure 9) at the beginning of 2000 in the town of Bure in Meuse district, at −490 m under the surface inside the Callovo-Oxfordian clay rock targeted for Cigéo project. Several one-to-one scale demonstrators of repository cells have been realized. Civil engineering designs are similar to those of the target application, but there is no radiation.

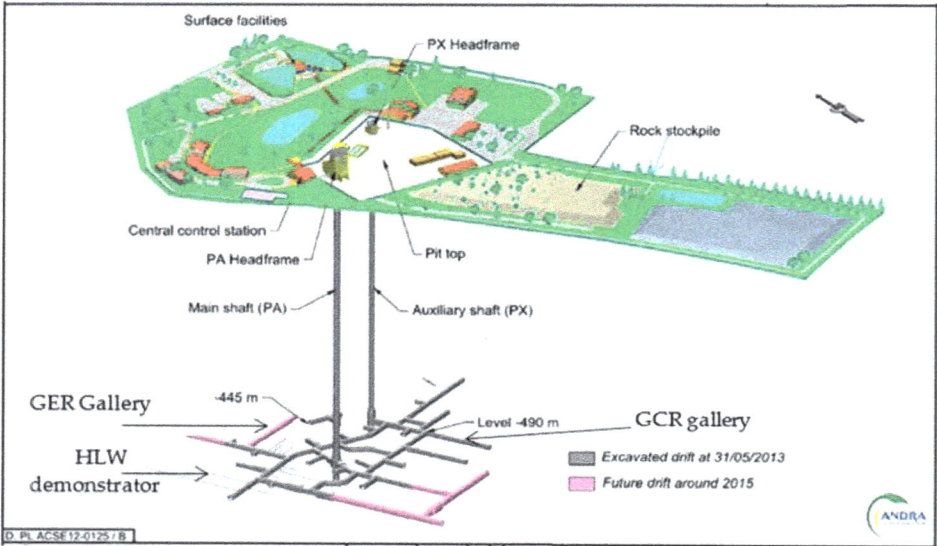

Figure 9. Schematic representation of the Andra's URL and instrumented galleries (GER and GCR).

Two ILW-LL cell monitoring demonstrators were realized [24,25]. They are tunnels of 5 m external diameter, with a shotcrete layer covering the host rock, a layer of poured concrete covering the shotcrete, basement construction, then slab and finally vault. In 2011 (respectively 2015), the "GCR" (resp. "GER") gallery was dug according to the major (resp. minor) horizontal stress field. In these two galleries, several instrumented sections were implemented (see Figure 10), equipped with both classical and innovative sensor technologies (i) Fibre Optic Sensors (FOS), Vibrating Wire Extensometers (VWE), total pressure cells in the shotcrete; (ii) FOS, displacement sensors and pore water pressure in boreholes; (iii) FOS, VWE, Time Domain Reflectometry sensors (TDR) for water content measurement, permeability pulse sensor, total pressure cell. Such structures allow to test temperature and strain measurement solutions in real operating conditions. An instrumented HLW-LL demonstrator had also been realized and is depicted in details in [26].

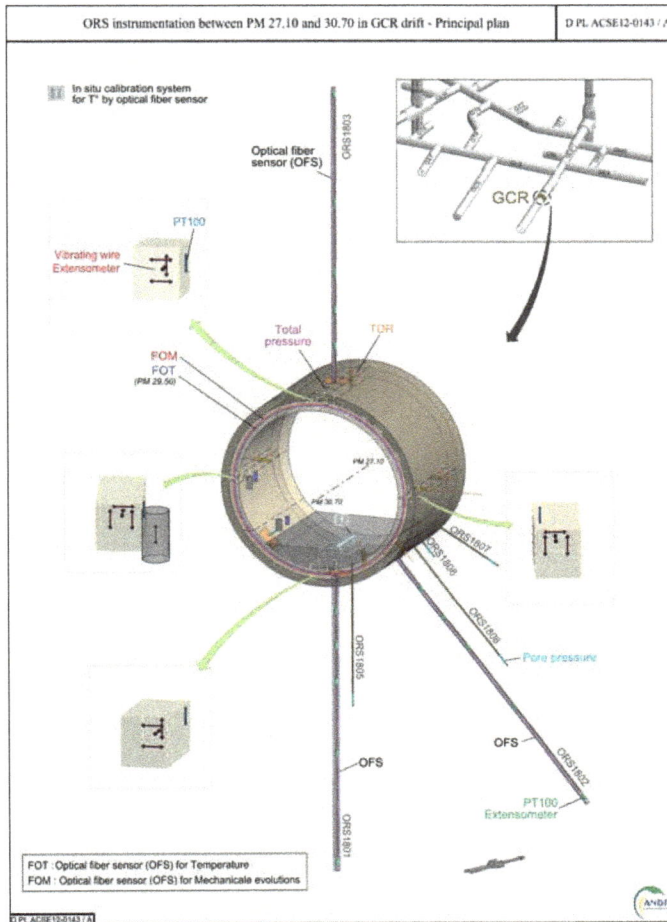

Figure 10. Schematic view of the monitoring section in the GCR gallery and pictures of implemented sensors, including optical fibre sensors.

3.7. Aging Tests

3.7.1. Methodology for Aging Tests under Radiation

As it is usually the case for radiation environments, it is not possible to qualify the radiation response of the developed technologies in conditions representative of the application. At the external surface of the metallic liner of HLW repository cell, the total ionizing dose of 10 MGy will be deposited on the sensor in one century. As a consequence, it is mandatory to perform accelerated tests on the sensors and to extrapolate the expected radiation response of the system in the targeted environment. The strategy selected by Andra and LabHC consists in a three stage qualification. First the optical fibres have been irradiated passively (without monitoring) at the maximal dose expected for Cigéo: 10 MGy and the irradiated samples have been characterized post irradiation to evaluate the levels of permanent damages after such a huge dose. In a second step, the radiation response of the OFS are characterized online at a dose rate representative of the application but for a limited period (typically one or two weeks), providing information on the impact of such dose rate on their performances. Finally, a last campaign is achieved where the OFS performances are measured in situ at higher dose rate than the application and up to its maximal dose (here 10 MGy). This last campaign is considered as a worst case scenario. Indeed, it is well known that the degradation of the fibre in terms of RIA depends on both dose and dose rate and that for passive optical fibres, the degradation levels increase at higher dose rate and at fixed doses [27]. To perform this vulnerability study, Andra used the IRMA ^{60}Co. source of the Nuclear Safety and Radiation Protection Institute (IRSN, Fontenay-aux-Roses, France) as this irradiation facility offers dose rates representative of the ones expected for the Cigéo project. The second used facility is the Big Radius Installation under Gamma Irradiation for Tailoring and Testing Experiments (BRIGITTE) ^{60}Co facility of SCK-CEN (MOL, Belgium) that allows to obtain MGy doses representative of the total dose after one century of monitoring in Cigéo. These two irradiators are illustrated in Figure 11.

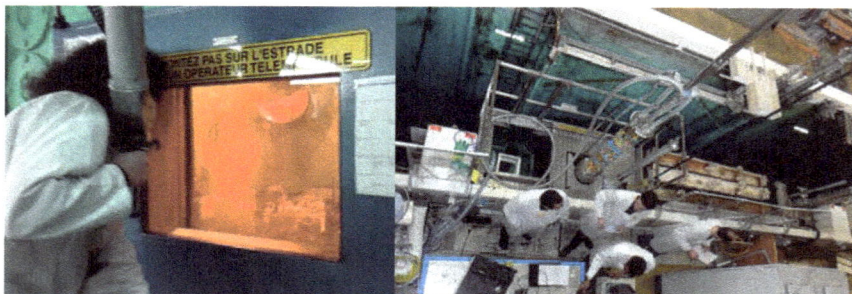

Figure 11. Picture of IRMA (**left**) and BRIGITTE (**right**) facilities used for irradiation of optical fibre sensors.

3.7.2. Methodology for Aging under Hydrogen

The impact of the presence of hydrogen into optical fibres has been widely studied in the past, as loading germanosilicate fibres with this gas allows to increase its photosensitivity to ultraviolet laser light, opening the way to the photo-inscription of Fibre Bragg Grating [28]. As a consequence, tools exist allowing to saturate the fibres with hydrogen (or sometimes deuterium) gas in a few days, the kinetics of loading being adjusted by the pressure and temperature of the gas tank used for the treatment.

To investigate the effect of the presence of hydrogen molecules in the fibre core and cladding on the various technologies of OFS, the following procedure has been used. First, the various types of optical fibres (single-mode, multi-mode, different coatings . . .) have been treated under several conditions of temperature and pressure to reach different hydrogen saturation states. As soon as the fibre is removed from the gas tank, desorption process starts and the concentration of gas in the fibre core decreases with time. For a given fibre structure, known temperature and pressure conditions, the kinetics of gas absorption, the saturation level and the gas desorption kinetics can be calculated ([29] Figure 12).

The presence of H_2 molecules can be monitored (as well as an estimation of its concentration) through the well-known absorption bands of the hydrogen in the IR part of the spectrum [30].Then the impact of the hydrogen (or other gas) on the performances of an OFS can be evaluated for the whole range of hydrogen concentration (from zero to maximal initial concentration) by continuously monitoring the OFS performance during the long desorption period (typically a few day or weeks). In the framework of Andra research, existing tools and facilities of LabHC in Saint-Etienne, and iXBlue in Lannion have been used to perform such hydrogen loading in various experimental conditions.

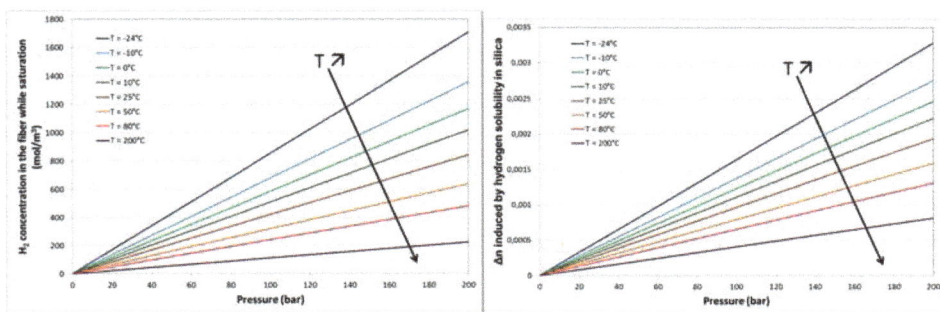

Figure 12. Theoretical values of the pressure and temperature dependence of the H_2 saturation concentration in the fibre core center (**left**) and the related refractive index variation (**right**).

3.8. Methodology for Radiation Sensing

Most of the studies about radiation effects on optical fibres are done to improve their radiation hardness to ensure the systems exploiting these fibres will be able to survive to the harsh environment of interest. However those studies also permit to identify radiation-sensitive optical fibres and it has been shown that their radiation response can be diversely exploited to monitor either the TID (total ionizing dose) and/or the dose rate. Optical fibres can then act as dosimeters. For some applications, the thermoluminescence or optically-stimulated luminescence (OSL) properties of optical fibres have been shown to be very attractive for post-irradiation dose measurements of various irradiation types with performances exceeding those of COTS Thermo-Luminescent Dosimeters (TLDs) [31]. Even more attractive, some fibres emit light (radioluminescence, RIL) when exposed to radiation, the RIL intensity being for adequate composition related to the dose rate or particle flux [32,33]. Most of the COTS scintillating fibres are based on polymer matrix [34], able to resist to low TID but recent progresses were done to achieve highly sensitive fibres in highly radiation resistant silica-based matrixes. However, these dosimeters exploiting the radiation induced emission of light are today mainly punctual sensors and more work is needed to design distributed measurements based on this RIE phenomenon in an unique fibre. Another class of fibre-based dosimeters exploit the dose dependence of the radiation-induced attenuation (RIA) in sensitive fibres such as those containing phosphorus, aluminium ... Using reflectometry technique (OTDR, OFDR), the spatial distribution of the RIA can be monitored along a fibre exposed to irradiation and for each portion of the fibre, the deposited TID can be recovered from the RIA value (supposing the RIA vs dose dependence is known). Among the most promising fibres for such distributed dose measurements are the Phosphorus-doped ones as it was shown that the RIA levels at specific wavelength increase almost linearly with the dose and is quite insensitive to dose rate or temperature fluctuations [35].

3.9. Methodology for Hydrogen Sensing

Many types of H_2 optical fibre sensors based on H_2-sensitive materials such as WO_3 [36], yttrium oxide (Y_2O_3) [37] or Pd compounds have been studied for several years. In contact with H_2 gas, Pd forms an hydride PdH_x (with x is a functions of the gas rate) leading to a variation of both the material refractive index [38] and lattice cell volume [38,39]. Efficient hydrogen sensing was achieved by using standard optical fibres (like G652 fibres used for telecommunications) with (i) Pd films deposited on the fibre cross-section [40] (ii) laid around the fibre [39,41]; (iii) around the core [42] or (iv) around tapered fibre [43]. However, these sensors are mainly dedicated to local gas detection, and they suffer from untimely deterioration in harsh environments and poor robustness [44]. They are mainly dedicated to local gas detection only.

Recently, Andra has demonstrated that H_2 rich atmosphere leads to a very large shift of the backscattered Brillouin spectrum (around 21 MHz) in H_2-loaded standard single-mode fibres [45]. Andra and Xlim intent to take advantage of this sensitivity for developing distributed hydrogen sensors. In this prospect, the Brillouin frequency shift induced by H_2 diffusion in optical fibres need to be studied and finely quantified for dissociating H_2 variation to others ones (temperature, pressure ...). In contrast to most H_2 sensing application that requires fast response time, slow variation of H_2 concentration has to be monitored in Cigéo project which is compatible with the kinetic of H_2 diffusion in silica fibre.

In this context, the methodology developed for sensing H_2 is based on continuously recording simultaneous Brillouin and Rayleigh backscattering measurements during the H_2 desorption process at ambient temperature and atmosphere. Following the methodology for hydrogen influence evaluation (cf. Section 3.7.2), optical fiber samples are H_2 saturated in gas tank (at the iXBlue company in Lannion, France). Then, the fiber samples are sent to the Xlim Research Institute for desorption measurements.

Rayleigh and Brillouin backscattering measurements are realized simultaneously, on H_2-loaded and pristine fibre samples with the instrument Neubrescope NBX-7020F (Neubrex Co. Ltd., Kobe, Japan) by tunable wavelength optical time domain reflectometry (resolution 1 GHz) and Brillouin optical time domain analysis (resolution 1 MHz) methods respectively [46]. Furthermore, a white light source and an optical spectra analyzer (ANDO AQ-6315A, Yokogawa, Tokyo, Japan) are used for measuring the attenuation of the peak at 1.245 μm induced by hydroxyl formation in the optical fibre [28]. This measure gives additional information on H_2 concentration in the fibre during the desorption process. Contributions from acoustic velocity and refractive index variations induced H_2 diffusion in the optical fibres are dissociated from Rayleigh and Brillouin backscattering measurements. To confirm the measurement of the refractive index variation, a fibre sample composed of a Bragg grating could be also inserted into the gas tank during H_2 loading. An ASE light source and an optical spectra analyzer are then used for measuring the shift of the Bragg wavelength during H_2 desorption, with a resolution of 50 pm. Experiments are realized with different H_2 concentration at saturation that are obtained by H_2 loading fibre samples with different temperature and pressure. All measurements in desorption are realized at room temperature (22 $\pm$ 1 °C), atmospheric pressure, and the fibre samples have been kept loose to minimize bias measurements.

Nevertheless, this methodology requires to send the H_2 loaded fibre samples from the two French research partners, namely from Lannion to Limoges, which limits accurate measurements at the first desorption time, even if the desorption kinetic is strongly reduced by packing the fibre samples with ice blocks during the travel. In this context, the hydrogen setup developed by Mons University in Belgium (through MODERN2020 EU-project) that enables in situ measurements during H_2 loading and desorption (with pressure and temperature continuously monitored) offers great prospects for pursuing our developments.

4. Results on Distributed Temperature Measurements

Andra plans to use Rayleigh and Brillouin scatterings for strain sensing after temperature characterization through Raman sensors. Based on its unique properties, Raman-based sensing is the Andra reference solution for temperature measurements.

4.1. Distributed Temperature Measurements Based on Raman Scattering

4.1.1. Metrological Evaluation

The implemented methodology enabled to demonstrate the importance of wording, namely "spatial resolution" versus "sampling". DTS response to a temperature variation step over one meter (spatial resolution typically claimed by the manufacturers) of sensing optical fibre corresponds to only 90% of the temperature step magnitude, whereas the full DTS response is obtained in fact for 10 m (the practical spatial resolution) of sensing optical fibre solicited by this temperature step variation [20]; depending on the instrument under test it can be 90% over 50 cm and 97% over 1 m [16]. Within METRODECOM project [47], Andra, EDF and LNE are performing a benchmark. Up to now, five commercially-available instruments were tested and compared and the results will soon be released [48]. Temperature uncertainty reaches 0.1 °C and degrades with distance along fibre, improves with measurement duration and spatial resolution (see Figure 13). Each instrument presents advantages and drawbacks to be confronted to the target application. For Cigéo monitoring, Andra presently prefers instrument specialized for short distance range, high spatial resolution and high Mean-Time Between Failure (MTBF).

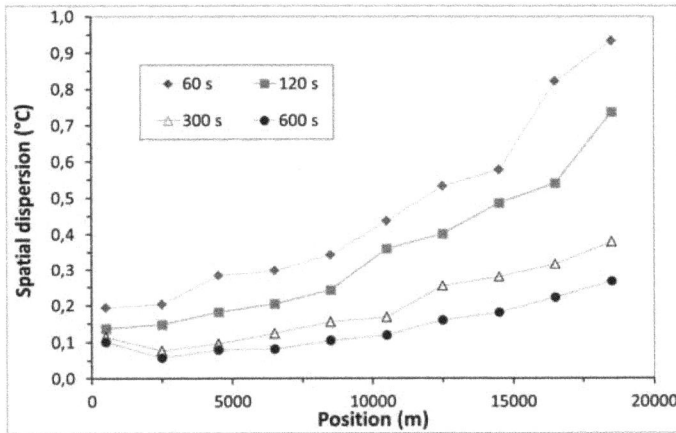

Figure 13. Measured temperature uncertainty (°C) as a function of the optical fibre length (m) at different averaging measurement times over at 23 °C and 1m spatial resolution.

4.1.2. On-Site Evaluation

OFS were implemented in 2011 in the "GCR" gallery of the Andra URL. Fibre-optic Raman temperature monitoring systems were installed in three boreholes and in the concrete liner. The instrument has been working continuously for 5 years and must have been replaced in 2017. The second Raman device hold by Andra also failed after 5 years, despite exclusive use in laboratory conditions and several transportations. Similarly, Andra partners faced durability limitation: in the Mont Terri underground laboratory, Nagra had to improve the dust protection after failure of the Raman device during tunnel construction [49]. This is why MTBF is now an important criteria in device selection.

If the OFS cables are compressed or pinched, losses are created and measurement noise increases. What is more, as the bending losses are different at the Stokes and Anti-Stokes wavelengths, intensity ratio exploited to determine the temperature will be false (Equation (1)). Special tools were designed to limit the puncturing of the optical fibre, at both extremities of the borehole, that is to say the sealing and the far extremity (see Figure 14 [49]).

To enable high-quality measurement provided by loop configurations (also required for radiation tolerance), the sensing fibre was installed with a U-shape. However, borehole diameter (8 cm and 13 cm) was smaller than the minimum curvature radius of the optical fibre. We took the risk and developed a curvature guide to optimise the position of the fibre. Measurements errors were noted anyway, and could be totally compensated subtracting the first measurement taken after the installation. This is why we conclude that the installation was successfully accomplished and Raman measurements are efficient as a relative temperature measuring system.

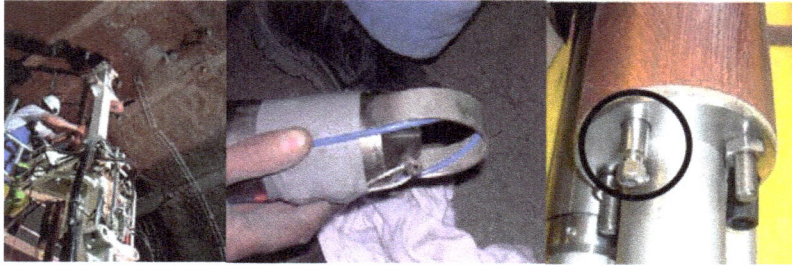

Figure 14. Pictures of distributed temperature system implementation inside vertical borehole (**left**). Example for a curvature guide (**middle**) and connector (**right**) to get the sensing cable through packers.

Another difficulty that must be faced is how to extract the interesting measurements among the whole sensing line. From this point of view, distributed measurements are a powerful tool which might rapidly become time-consuming if not carefully configured. The device provides temperature as a function of a location in the optical fibre, which is never placed straight in the structure. To match Euclidien distance and curvilinear abscissa, it is of utmost important to realize a dynamic a map after installation, for instance, by heating or cooling remarkable locations. More industrial reference points can also be included in the sensing lines [50]. Under these precautions, as illustrated in Figure 15, distributed Raman systems produced good quality data over several hundred meters. This measurement was acquired during the concrete liner pouring and reveals heat propagation in the clay rock. Resolution and accuracy are satisfactory and in the same range as standard platinum probe temperature sensors (0.1 °C) placed nearby. The distributed temperature measurements along the fibre-optic cable reveal detailed insights into the spatio-temporal varying temperature field in the rock around the gallery. During the last four years, the most fragile parts proved to be the connectors on the multiplexor, which must be cleaned occasionally, unlike the sensing cables placed in the rock that resist very well [49].

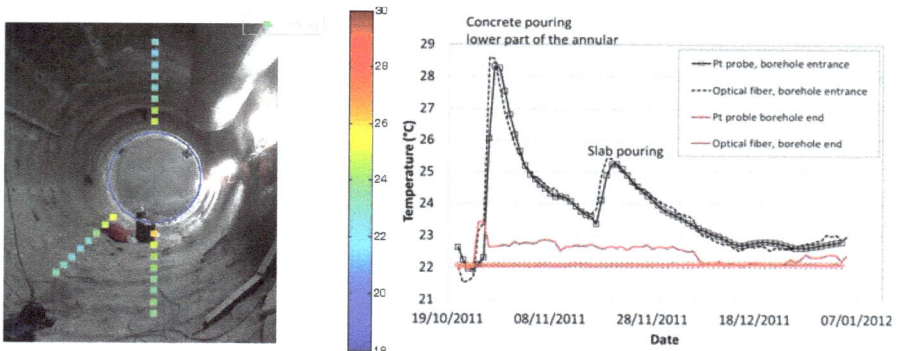

Figure 15. Distributed temperature measurement obtained with Raman OFS during concrete liner pouring (**left**) and comparison with collocated platinum probes as a function of time during concrete liner pouring (**right**).

4.1.3. Accelerated Aging Tests: Hydrogen Influence

As detailed in [51,52], large temperature errors were observed along fibre samples exposed to hydrogen (80 °C with 202 bars of pressure for 62 h) then removed from the hydrogen tank, while the ratio between Stokes and Anti-Stokes Raman scattering (see Figure 16) was measured and interpreted in temperature following Equation (1). We also verify (Figure 16 right) that the presence of a carbon layer is efficient (and mandatory) to prevent hydrogen diffusion and degradation of the Raman distributed temperature sensors.

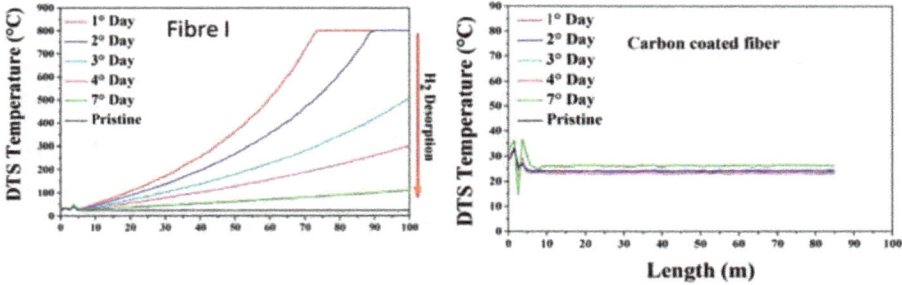

Figure 16. Influence of hydrogen exposure on temperature sensing based on Raman scattering for standard (**left**) and carbon (**right**) primary coating. Adapted with permission from [52]. Copyright IEEE 2015.

4.1.4. Accelerated Aging Tests: Radiation Influence

Regarding radiation influence on temperature measurement based on Raman scattering, dramatic influence is observed (Figure 17): up to 50 °C error on a 100 m sample. Error increases with distance. Transient response significantly differs from post-mortem measurements, which enhances the importance of our methodology, with representative dose rate and cumulated dose, both post-mortem and on-line tests. Appropriate composition for the fibre (F-doped) allows reducing the amplitude of the temperature errors due to permanent effects of radiations but this is not sufficient to obtain acceptable resolution. The pre-treatments of the fiber, i.e., the ex situ pre-irradiation at 10 MGy(SiO$_2$) reduces temperature measurement error ($\leq$2 °C) [52].

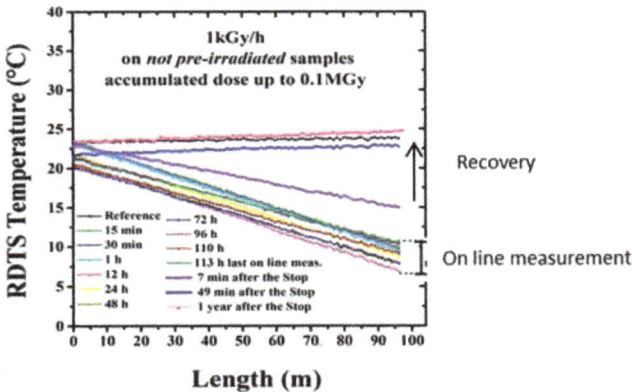

Figure 17. Distributed temperatures measurement performed, on Fiber I. Results obtained both during and after the γ-irradiation at 1 kGy(SiO$_2$)/h dose rate. The red line indicates the first measurement from the start (our reference) and the dot line designates the room temperature (RT) at the beginning of the irradiation run. Copyright IEEE 2015.

Clear explanation of this phenomenon has been provided [51–53]. It is due to the differential attenuation at the Stokes and Anti-Stokes wavelengths created by RIA. α_S and α_{AS} differ rapidly under radiation, and the factor in Equation (1) will become unneglectable. To solve this wrong signal analysis, double-end measurements are promoted in literature [54]. New devices based on double-Raman probes are also to be considered [55]. Measuring RIA along the sensing device at S and AS wavelengths is another solution.

4.1.5. Accelerated Aging Tests: Coupled Hydrogen and Radiation Influence

Finally, it is worth notifying that hydrogen and radiation aging influence should not be considered independently. As illustrated in Figure 18 and detailed in [56], radiation impact is different whether the fibre has been previously exposed to hydrogen. We observe a meaningful negative effect of γ-rays on the sample H_2-loaded two months before irradiation campaign and left to desorb naturally at ambient condition. Indeed, the response of the sensor is worse than without the H_2 pre-treatment. We observe a temperature error exceeding the one of the 10 MGy irradiated sample (not reported in the figure). At the opposite, the sample irradiated in presence of H_2 during the irradiation is only slightly affected by this combined treatment one month after the end of the irradiation.

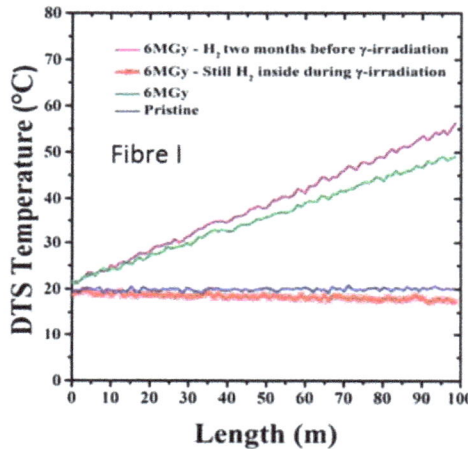

Figure 18. Distributed Temperature measurement performed on Fiber I, at room temperature. Results of pristine (light blue line), γ-irradiated at 6 MGy (red line), γ-irradiated at 6 MGy with H_2 inside (green line) and γ-irradiated at 6 MGy two months after H_2-loading (blue line) samples are shown.

4.1.6. Conclusion for Temperature Measurements Based on Raman Scattering

Studies have shown that carbon coating is mandatory and fully satisfactory to perform temperature measurements in hydrogen rich atmosphere. Concerning radiation hardening, transient effects proved to be very important: post-mortem analysis cannot replace on-line testing. F-doped optical fibre (eventually pre-irradiated) is mandatory to limit RIA value and maximize the possible distance sensing range but is not efficient enough to reduce temperature measuring errors based on single-ended RDTS A solution is to implement other Raman scattering measuring configurations, such as double-end measurements (but sensing range is limited by a factor of 2) or to use innovative RDTS architectures.

We have shown that hydrogen and radiation effects are coupled. In Cigéo, there is also temperature to take into account. Andra has recently launched evaluation of coupled temperature influence (100 °C for Cigéo monitoring) on aging tests. Results will be released soon. Recent French study shows it should minimize radiation impact [16].

Regarding temperature measurements based on Raman scattering, TRL is presently estimated at 6 for Cigéo application: with all the listed precautions, OFS are ready to be implemented into the Pilot Phase.

4.2. Distributed Temperature Measurements Based on Loose Tube

Rayleigh or Brillouin scatterings are attractive for temperature sensing because both measurements can be performed in a single single-mode fibre. It fastens on-site implementation and thus reduces cost; a single device paired with a multiplexor could be used for both temperature and strain measurements, reducing costs.

Because of strain sensitivity, temperature measurement with these scatterings requires to isolate the sensing fibre from mechanical influence of the structure. This is called a loose optical fibre sensing cable. Andra has conducted several tests on "loose tubes" in the past, which revealed difficulties to isolate an optical fibre over very long distances [18]. Friction was unneglectable.

Overlength must carefully be selected as a function of the application and the maximal temperature range. Implementation procedure must also be carefully adapted to ensure free dilatation. Meanwhile, the cable has to walk along and around the structure to provide a 3D characterization, but it is necessary to respect the radii of curvature to reduce friction. As detailed in [57], during a fire test performed at Efectis (Maizières-lès-Metz, France), to reduce the influence of differential thermal expansion, the cable was disconnected from the demonstrator. The selected process relied on curved copper tubes fixed to the structure. The sensing cables were threaded through these tubes. With such precautions, we managed to perform distributed temperature measurements with Brillouin and Rayleigh scattering acquired in a 50 m long loose tube. Brillouin and Rayleigh measurements were compared to Raman temperature measurements and to reference probes. Accordance was very good: Strain influence could not be distinguished. Slight discrepancies were observed on few locations (see for instance Figure 19). They attributed to the different acquiring spatial resolution (5 cm for Brillouin and Rayleigh versus 25 cm for Raman), while spatial gradients were important.

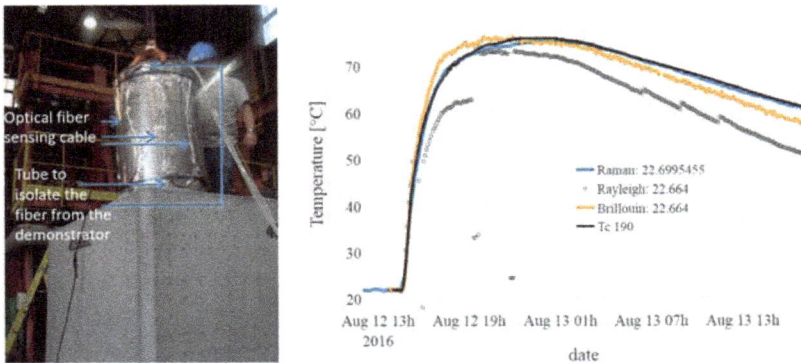

Figure 19. Loose tube implementation and test. (**left**) Picture of the emplacement of a demonstrator of primary waste package, instrumented with optical fibres (**right**) measured temperature on the waste demonstrator, obtained with Raman Rayleigh and Brillouin devices versus reference sensors.

As a conclusion, distributed temperature based on Brillouin or Rayleigh scattering is now validated up to step 3 (over 4) of the qualification procedure for Cigéo monitoring. Compared to Raman scattering, implementation is more difficult, this is why it is not the reference solution. Aging tests will be presented in the following part, dedicated to strain sensing.

5. Results on Distributed Strain Measurements

5.1. Metrological Evaluation

From 2008 to 2012, Andra, EDF, Telecom-ParisTech and IFSTTAR have characterized many different single-mode optical fibres to evaluate their Brillouin scattering properties, both spectrum and: thermal and mechanical sensitivities. The goal was to evaluate whether the several Brillouin peaks could have significant sensitivities, in order to separate the two parameters to sense. Numerical modeling was also implemented; the influence of draw tension on the Brillouin peak position (up to -20 MHz/100 g) and its linewidth was quantified [58,59].

More recently, the strain sensitivity of the sensing cables were characterized. As detailed in [23], they may differ from the optical fibre' sensitivity. For instance, following the methodology described in Section 3.2.5, Rayleigh strain coefficients of 0.79 $\mu\varepsilon^{-1}$ for the optical fibre versus 0.77 $\mu\varepsilon^{-1}$ for the fibre cable were obtained. For Brillouin scattering paired on two different strain sensing cable, we measured 0.0394 MHz $\mu\varepsilon^{-1}$ for the soft cable and 0.0465 MHz $\mu\varepsilon^{-1}$ for the reinforced cable [25].

These values were obtained on laboratory. Once embedded or attached to the structure to monitor, strain profiles measured in the optical fibre may differ from the actual strain in the structure, due to the shear transfer through the intermediate material layers between the optical fibre and the host material (i.e., in the protective coating of the sensing cable and in the adhesive).

The influence of the external sheath, both the geometry and the constitutive materials, has driven extensive evaluation in France these last years. In [60,61], EDF, in collaboration with Andra, and IFSTTAR have developed a methodology for the qualification of strain sensing cables in the host material. It relies on a numerical modelling of the cable, in which the mechanical parameters are calibrated from experiments. Once the transfer function is determined, measured strain into the optical fibre can be transformed into the strain in concrete. This work has been performed on a soft sensing cable, meant for embedment into concrete, illustrated in Figure 4. As detailed in [60], a four-point bending test was realized on a meter-long instrumented concrete beam. Crack openings were detected and quantified before it could be seen on external surface. It has also been shown the sensing cable could be attached on concrete rebars, there were no difference if placed between rebars. This eases and fastens real site implementation. Later, EDF developed an algorithm to automatically analyze crack evolution [62]. This process provided a precise map of cracks with the evolution of their amplitudes. Recent work focused on the transfer function of a sensing cable attached on the surface of structures [63]. It is of prior importance for old structure where sensors cannot be implemented inside materials anymore (which is not the case for Cigéo however Andra has application for the monitoring of short-lived waste repositories presently in exploitation).

5.2. On-Site Implementation

Truly distributed optical fibres strain sensing systems have been implemented in France in many civil engineering structures, including dykes [64], bridges, roads, tunnels and the nuclear power plant under construction, the EPR [65].

Andra and its industrial partners (IDIL Fibres optiques, EGIS, Cementys, Marmota, Solexperts and GTC) have implemented distributed strain sensing systems into concrete slab, during a building construction on surface [18], two concrete tunnels in the Andra URL (similar to a ILLL repository cell) and one horizontal tunnel with metallic liner (representing a HL cell demonstrator).

Obtained performances were reported in [19,24,25]. Specific one-site implementation difficulties were detailed in [49]. Similarly with temperature sensing, a key issue with OFS implementation was the pressure induced by clamps. Pinches generated high losses along the optical lines and poor signal to noise ratio. Unlike temperature sensing cables, where the fibre is protected from the host material, strain sensing optical lines suffered during the tunnel construction. The weakest part proved to be the interface between the sensing cables and communication cables. They could be repaired at the expense of total length change. Such slight changes (in the order of 20 cm of 500 m) in the total distance range and in the sensing area location proved to be very difficult to manage; On-line references are a promising solution [50].

At this stage we showed that in the concrete liner, the tested strain optical fibre cables were sufficiently robust to tolerate construction conditions (with several dB losses in only few hundred meters). In GCR gallery, after 6 years more than 95% of the sensors still provide measurements, without any drift. What is more, strain measurements obtained in concrete liners with optical sensing cables are very in good accordance with collocated reference sensors, namely VWE [19,24]. In GER gallery, for FOS placed in the poured concrete layer, survival rate is 100%, after more than 1.5 year and good accordance is also obtained [25] (cf. Figure 20). Colocated optical lines were implemented, either spliced or equipped with connectors. Both survived to installation and provide accurate measurements. Since connected sensors allow faster operation than on-site splices, such configuration results much cheaper and will be promoted for future monitoring system implementation.

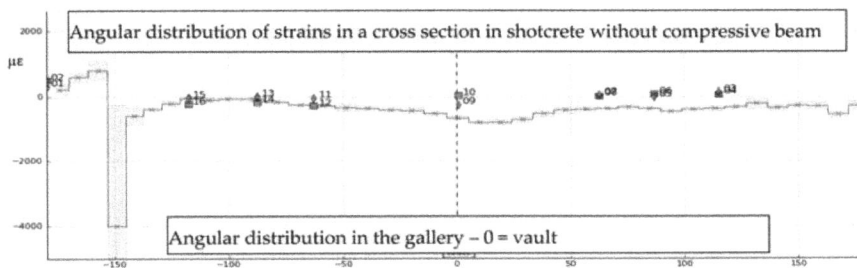

Figure 20. Strains measured by FOS (blue segments) and VWE (red squares for intrados location and green diamonds for extrados) around the cross section of the GER gallery (0° in vault and 180° in the bottom). FOS measured is the average stain over segments of 0.5 m. Blue line represents the moving window average of 12 h, with 95% confidence interval.

Feasibility of implementation in the shotcrete layer has been demonstrated in the GER gallery. FOS distributed strain measurements showed a good correlation with VWE's [23]. However, as expected due to high strain in the shotcrete, the optical loop became too noisy after the first days, forbidding the use of stimulated Brillouin measurements. Strain measurements using Rayleigh instruments remained possible several weeks but also failed rapidly. New sensing cable that would tolerate larger strain and local pressure are under design in ITN-FINESSE project [66].

Finally, in Bure we also implemented strain optical fibre sensors in the HLW demonstrator [26]. An armored cable was placed at the interface between a metallic liner and the Callovo-Oxfordian clay. The cable endured very large stresses during liner pushing operations. However, the furthest extremity broke: at this location, the sensing cable was open inside a connection box, where the two optical fibres were spliced to provide a loop. It enhances the importance of the measuring device to be able to perform measurements on open loops and well as connected loops. Strain sensing cables implemented in the inner surface of the liner have been providing strain measurements for the last five years. Regarding durability of strain sensing devices [49], in Bure, a Brillouin device is running continuously since October 2011. We faced two returns to supplier over the three last years (one hard disk's breakdown). The Rayleigh device test failed three times in Bure (contrary to several successful laboratory experiments and tests in HADES). We suspect the instrument to be totally incompatible with vibrations, created by permanent digging in several galleries of the Bure URL.

5.3. Aging Tests: Hydrogen Influence on Strain Measurements

Similarly with temperature sensing, single-mode fibres were hydrogenated. H_2 rich atmosphere leads to large shifts of the backscattered Brillouin spectrum (around 21 MHz) in single-mode fibers loaded under 200 bars and 80 °C [45]. This corresponds to huge strain measuring errors, on the order of 400 με.

Optical fiber dopant type slightly changes the kinetics, not the Brillouin shift value (see Figure 21 and [67]). Carbon primary coating proved to be efficient to prevent hydrogen migration into silica [45]. This is why this primary coating is mandatory for strain sensing in hydrogen rich atmosphere, such as several repository cells of Cigéo (depending on radioactive waste type).

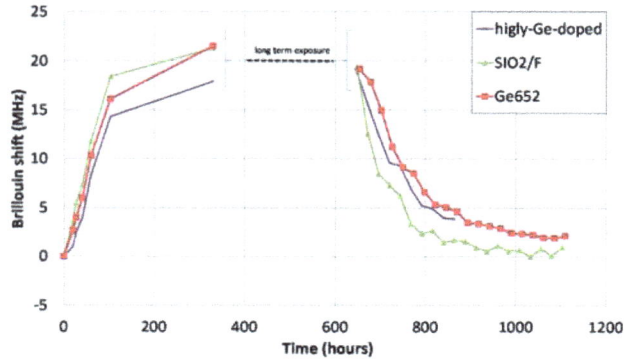

Figure 21. Brillouin frequency shift during the sorption and desorption period for CMS fiber (purple), Fluorine (green) and the most standard commercial G652 type fibre (red).

5.4. Aging Tests: Influence of Radiations

Radiations are known to influence the optical properties of silica-based optical fibres through ionizing or displacement damages. Depending on the irradiation conditions (nature of particles, dose, dose rate, temperature of irradiation), the observed changes in the optical and structural properties of the pure and doped amorphous silica of the fibre core can strongly differ. Regarding sensing applications, all OFS will be affected by the radiation-induced attenuation (RIA) that by decreasing the transparency of the fibre glass limits the sensing distance. Depending on the physical process used to functionalize the optical fibre as a sensor and of the sensor architecture, other phenomena can alter the OFS performance, such as radiation-induced refractive index change, glass compaction ...

5.4.1. On the Single-Mode Fibre Attenuation

The amplitudes and kinetics of the RIA depend on many parameters that have been previously reviewed [68]. Among them, a crucial one is the composition of the fibre core and cladding that can totally change the RIA levels. Some compositions have to be avoided in radiation environments, such as the phosphorus doped or codoped optical fibres (except for dosimetry). Telecom-grade germanosilicate optical fibres present an intermediate radiation response that is acceptable for steady state kGy dose levels in the IR part of the spectrum but RIA is usually too high for applications at the MGy dose levels of CIGEO project, implying to select so-called radiation-hardened optical fibres that are usually designed either with fluorine or pure-silica core and fluorine-doped claddings [13]. Orders of magnitudes of the room temperature stable γ-ray RIA are provided in Figure 22 [69] at the wavelength of 1550 nm for a radiation hardened pure-silica core fibre and two germanosilicate fibres with moderate or high levels of Ge in their cores. If before irradiation, those fibres present attenuation below a few dB/km at this wavelength, the losses increase up to 400 dB/km after a 10 MGy dose. An important outcome of these studies is that at high radiation doses, the 1.3 μm RIA becomes below the one at 1.55 μm [69] but even by optimizing the fibre choice and operating wavelength, the OFS range will be reduced from several kilometres before irradiation to hundreds of meters after MGy irradiation.

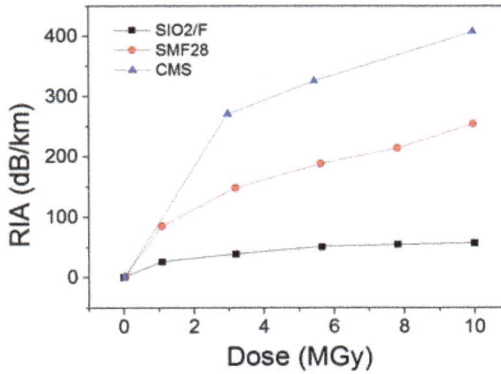

Figure 22. Measured RIA at 1.55 μm after different doses of irradiation from 1 to 10 MGy for three single-mode optical fibres: pure silica core fibre (SiO₂/F), Telecom-grade germanosilicate (SMF28) and highly Ge-doped (CMS).

5.4.2. Radiation Impact on Brillouin Scattering

The radiation effects on Brillouin based sensors have been studied in the recent years in the frameworks of PhD thesis between Andra and LabHC. Radiations affect the performances of those sensors by two mechanisms. First, as for other technologies, the RIA limits the possible sensing range, reducing it to hundredths of meters at MGy doses for radiation-hardened optical fibres. Furthermore, radiations also induce a Brillouin frequency shift (RIBFS, see Figure 23) that causes a direct error on the measurement of the strain and or the temperature. Our studies shows that the amplitudes and dose dependences of this RIBFS depend on the fibre composition. Indeed, Figure 23 compares the RIBFS in a germanosilicate and a pure-silica core optical fibres. An outcome of our work is that selection of a pure or fluorine doped optical fibres appears the best choice to reduce simultaneously the RIA issues and the RIBFS. Doing this, Brillouin sensing remains possible for Cigéo application with limited expected errors on temperature (below 2 °C) and strain (below 40 μɛ) [70].

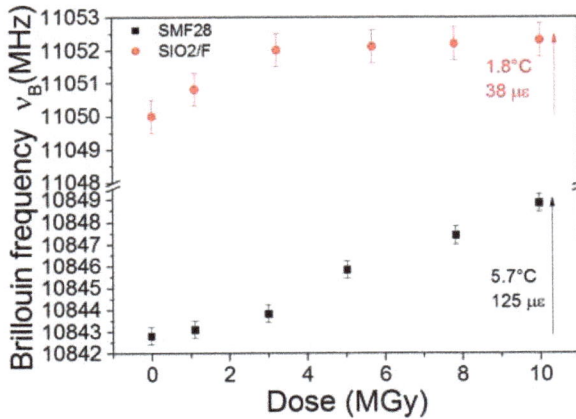

Figure 23. Measured Brillouin frequency shift for standard fibre with Ge dopants and F-doped fibre [69].

More recently, Andra and LabHC have also checked with on-line tests that transient effects were not important for Brillouin sensing, at least for moderate doses [21,51,71]. Based on promising results from [72], N-doped fibre has also been tested but it revealed less adapted to Cigéo than the qualified F-doped fiber [21].

5.4.3. Radiation Impact on Rayleigh Scattering

The radiation effects on Rayleigh-based sensors have been studied in France in the recent years in the frameworks of PhD theses between Andra, LabHC and AREVA [21,73–75]. The same methodology as described in Section 3.7.1 was applied, either with OBR instrument [9] or TW-OTDR [10]. It was shown that the Rayleigh signature of various types of optical fibres (from radiation sensitive to radiation hardened) was not affected by irradiation up to 10 MGy, the temperature and strain coefficients being stable at 5%; this percentage can be improved by appropriate pre-treatment of the fibre coating (80 °C and/or pre-irradiation up to 3 MGy) to stabilize it for large irradiation tolerance. Furthermore, online testing up to 1 MGy dose and at various temperatures (from −40 °C to 75 °C) demonstrates the feasibility of temperature monitoring in such harsh conditions.

In conclusion, Rayleigh scattering seems then poorly sensitive to gamma rays. This is promising for Cigéo application, as the main remaining issue will be the RIA and its induced limitation in terms of sensing range.

5.5. Conclusion on Strain Sensing

Laboratory and in-situ tests were successful. Based on the presented aging tests, we are confident that Brillouin scattering performed in a carbon-coated F-doped fibre should handle 100 years of use under both radiation and hydrogen exposure. This conclusion has been reached based on the hypothesis that: (i) measuring device turns from 1.55 μm to 1.3 μm; (ii) post-mortem measurements at higher doses are similar or a worst case compared with the on-line influence at lower dose rate (iii) carbon primary coating remains efficient after radiation exposure. In a near future, on-line measurements must be performed to check that these results remain valid in case of on-line irradiation (the case of Cigéo application). It will also be important is to evaluate possible coupled impacts of both temperature, radiation and hydrogen on Brillouin and Rayleigh scatterings, similarly with the phenomenon observed for Raman scattering. Tests are on-going and results will be published soon.

Strain sensing will also be possible with Rayleigh measurements; on-line irradiations at high doses are on-going to check that the transient effects remain low. With these two qualified measuring methods (Brillouin and Rayleigh scatterings), a perspective is to perform both temperature and strain measurements in a single sensing cable, linked with the structure.

For Cigéo, the main perspective is to insert the qualified optical fiber into the qualified strain sensing cable (see Section 5.2); this work is on-going within MODERN2020 project. TRL is presently estimated at 5 for distributed strain sensing in the Cigéo application.

6. Results on Distributed Hydrogen Measurements

The Xlim research institute (CNRS and Limoges University, France) in collaboration with Andra, exploits hydrogen diffusion into silica and induced effects (see Figure 21) to realize distributed hydrogen measurements. This work initiated in 2014 was in first focused in understanding the diffusion mechanism of hydrogen inside silica optical fibre and its dependency to conditions (both temperature and pressure of the tank), and the geometry of the optical fibre. As illustrated in Figure 12, H_2 concentration (in the fibre core) at saturation increases with larger pressure and lower temperature. This theoretical study is currently experimentally tested. It is of prior importance because it links optical properties such as variation of the refractive index, acoustic velocity) with external hydrogen quantities, the real sensing parameter.

This theoretical work has allowed us to investigate the physical mechanisms that provokes Brillouin frequency shift when H_2 diffuses into silica. We have experimentally demonstrated that the Brillouin frequency shift is not only due to refractive index change with H_2 concentration, but also to a variation of the acoustic velocity that increases with H_2 concentration [29]. For example, we have measured a Brillouin frequency shift of 21 MHz induced by 1.7 %mol of H_2 in silica that was obtained by exposing, until saturation state, a sample of standard single mode fibre (G652-type) to pure H_2 gas at 25 °C with a pressure of 150 bars [45]. By decreasing the concentration of H_2 at saturation in the fibre, we have measured a smaller shift of the Brillouin frequency (19 MHz for 1.14 %mol of H_2 in silica) confirming the influence of H_2 concentration on Brillouin frequency and its interest for H_2 sensing applications. This smaller concentration at saturation was mainly obtained by increasing the temperature of the gas chamber (from 25 °C to 80 °C) as theoretically predicted and illustrated in Figure 12. Furthermore, the association of Brillouin and Rayleigh backscattered measurements allowed us to measure, during H_2 outgassing period (13 days), a variation of the fibre core refractive index of the fibre core of 12×10^{-4} leading to a variation of the acoustic velocity by +6 m/s at saturation state.

In addition to these studies, specialty optical fibres have been developed to enhance hydrogen sensitivity. Andra and Xlim proposed to introduce Pd particles into the silica cladding of optical fibres in order to protect the sensing metal from harsh environments and for enabling distributed sensing of H_2 gas along long lengths. Embedding Pd into fibres might improve the sensitivity and the response time of the distributed fibre gas sensor, by exploiting the mechanical strain induced by the crystal lattice expansion of Pd particles in contact with H_2 gas. The fabrication feasibility of this kind of optical fibre has been demonstrated by exploiting an original fabrication process (based on powder technology) developed at Xlim [76].

Different optical fibre topologies with Pd particles embedded into the silica cladding are presented in Figure 24. The fibres have been fabricated with lengths of several hundred meters and PdO concentration ranging from 0.01% to 5% mol (in addition to silica). Structural and microstructural characterizations of the preforms and fibres demonstrate that the fabrication process yields homogenous long lengths of fibres with metallic Pd particles randomly spread concentrically in the external cladding. However, the interest of Pd particules embedded into the silica cladding to enhance H_2 sensitivity remains to be proved. Up to now, fibres were exposed to hydrogen. Then removed from exposure and measured when hydrogen was diffusing out of the fibre. It was difficult to start measurements at saturation, that is to say as soon as the fibres were removed from the hydrogen tank. The novel setup developed by Mons University will allow us to continuously measure Brillouin frequency shift during H_2 loading and desorption. These on-line measurements will be performed in the coming months. Presently, we rate these developments at TRL 3 (laboratory stage only) level.

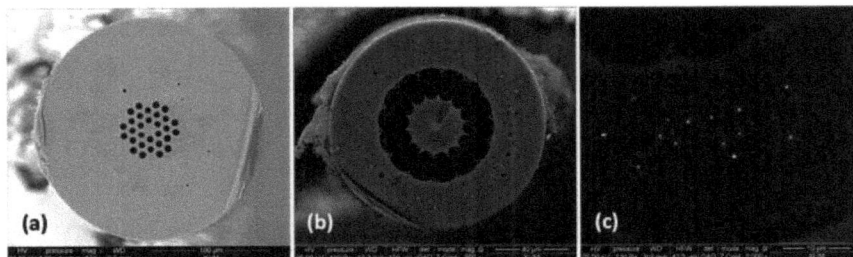

Figure 24. SEM (in the backscattered configuration) pictures of cross-sections of two optical fibres with Pd particles embedded into the silica cladding realized by the powder in tube process: (**a**) a pure silica core microstructured fibre; (**b**) a SiO_2-GeO_2 step index core with a microstructured cladding fibre; (**c**): zoom-in the cladding region of the fibre.

7. Results on Distributed Radiation Sensing

Several studies have been conducted to investigate the potential of OTDR or OFDR measurements of the radiation-induced attenuation (RIA) to monitor the distribution of TID along the optical fibre. Usually, those studies focused on the well-known radiation sensitive optical fibre such as those containing the Al or P dopants [14], but recently more radiation sensitive fibres ($\times 10$ times) have been reported in the framework of the DROID project from Universities of Perpignan and Nice [77]. A recent work was achieved by CERN on this topic combining commercial multimode P-doped fibres with COTS OTDR [78]. The authors demonstrate their ability to follow the dose distribution, with meter scale spatial resolutions and dose resolution down to 10–15 Gy, into the CERN High energy AcceleRator Mixed field facility, CHARM facility, that has an environment similar than the one of the Large Hadron Collider. One of the difficulties with such COTS solution is the trade-off between the mandatory high sensitivity to be able to detect the low dose RIA over typically one meter of the fibre and the range of measurements limited by the dynamics of interrogators. Preliminary study has been performed in Mons to investigate the potential of OFDR measurements of RIA offering higher spatial resolution (a few mm) than OTDR solution [14]. The authors demonstrate that such sensor can be used to monitor the dose increase up to 9 kGy with 15 cm resolution over about 1 m fiber length, the sensing length can be increased but only if lower dose levels are considered. In addition to these crucial issues, the integration of such solutions into facilities implies to calibrate the fibre RIA in function of not only the dose but also the temperature, the dose rate, the photobleaching effects that affect the RIA dose dependence [31]. Such a work is currently under progress in the framework of a PhD thesis between CERN and LabHC with the recent deployment of an OTDR-based dosimetry system exploiting the RIA at 1.55 μm in a P-doped multimode fiber at the Proton Synchroton Booster at CERN [79]. TRL is in the order of 3 (feasibility demonstrated in laboratory).

As a perspective, in 2017, the SURFIN project (a consortium between the Universities of Lille, Nice, Clermont Ferrand and Saint-Etienne (LabHC)) has been awarded for a grant by Andra within the framework of the French State "d'Investissements d'Avenir" program to investigate the feasibility of innovative dosimetry techniques to perform first punctual dose and dose rate measurements over a large range of doses, second to identify how the radioluminescence (RIL) or optically-stimulated luminescence can be exploited in a spatially-resolved dose and dose measurement and third how fiber materials can be adapted to discriminate between ionizing and non-ionizing radiations. Preliminary results have been obtained by University of Lille and LabHC with Cu and Ce-doped fiber materials, demonstrating their capability to monitor through RIL, X-ray [32] and proton [33] flux well lower than 1 Gy/s whereas being able to resist to more than 10 kGy accumulated doses.

8. Conclusions

We have presented the French state of the art distributed sensing systems, based on optical fibres, developed and qualified for the French Cigéo project, the underground repository for high level and intermediate level long-lived radioactive wastes. Four main parameters, namely strain, temperature, radiation and hydrogen concentration are currently investigated by optical fibre sensors, as well as the tolerances of selected technologies to the unique constraints of the Cigéo's severe environment (temperature, hydrogen and radiation). A qualification method in four steps is being deployed: sensing chains are tested in laboratory, implemented into real structures, then evaluated in accelerated aging tests.

The performances of Raman, Brillouin and Rayleigh scattering have been studied for many optical fiber types, such as F-doped, Ge-doped, N-doped fibers and P-doped fibers, and several sensing cables. Succeful on-site implementation were reported; feedback enables to list several recommandations for tunnel monitoring with distributed OFS. The influences of both representative gamma radiation dose rates and total dose were estimated, as well as hydrogen influence. Cross-influences of hydrogen and radiations proved to be important. F-doped fiber (eventually pre-irradiated) and carbon primary-coatings are mandatory to limit RIA value and maximize the possible distance sensing range. The additional temperature influence on aging tests is still under evaluation, with the final objective to insert the selected optical fibre into the qualified strain sensing cable.

Strain and temperature distributed measurements based on Raman, Brillouin and Rayleigh scattering signatures now reach high TRL, suitable for implementation into the repository cells of radioactive wastes in the Cigéo Pilot Phase. Double-ended Raman-based sensing is the Andra reference solution for temperature measurements, with TRL6. Brillouin scattering proved to be appropriate for strain sensing (TRL5) but Rayleigh scattering is also studied as a back-up solution.

Future work will turn towards methodologies for fine measurement interpretations (from strain measurements to the real target parameters, such as tunnel convergence for instance). Metrology of optoelectronics is also driving much attention. Several optoelectronic devices endure failure or require maintenance after 5 years; methodology to characterize MTBF and interoperability of instruments are becoming a major issue.

Regarding distributed hydrogen and radiations sensing, developments presently remain at the laboratory stage. Promising solutions are proposed based on Radiation Induced Attenuation (RIA) responses of sensitive fibres such as the P-doped ones. While for hydrogen measurements, the potential of specialty optical fibres with Pd particles embedded in their silica matrix is currently studied for this gas monitoring through its impact on the fibre Brillouin signature evolution.

Acknowledgments: Authors acknowledge the financial support of the European projects MoDeRn, MoDeRn2020, Metrodecom. Investigation was partially done as part of the COST Action TD1001 "Novel and Reliable Optical Fibre Sensor Systems for Future Security and Safety Applications" (OFSeSa).

Conflicts of Interest: The authors declare no conflict of interest.

References

1. Mankins, J. *Technology Readiness Levels, a White Paper*; NASA Advanced Concepts Office: Washington, DC, USA, April 1995.
2. MoDeRn European Project 7th Framework n° 232598. Monitoring Developments for Safe Repository Operation and Staged Closure. Deliverable 4.1 "Case Studies Final Report". Available online: http://www.modern-fp7.eu/publications/project-reports/index.html (accessed on 7 June 2017).
3. MoDeRn2020 Euratom Horizon 2020 Program, no. 662177. Development and Demonstration of Monitoring Strategies and Technologies for Geological Disposal. Available online: http://cordis.europa.eu/project/rcn/196921_en.html (accessed on 7 June 2017).
4. Hausner, M.; Suárez, F.; Glander, K.E.; van de Giesen, N.; Selker, J.S.; Tyler, S.W. Calibrating Single-ended fiber-optic Raman spectra distributed temperature sensing data. *Sensors* **2011**, *11*, 10859–10879. [CrossRef] [PubMed]
5. Parker, T.R.; Farhadiroushan, M.; Handerek, V.A.; Rogers, A.J. Temperature and strain dependence of the power level and frequency of spontaneous Brillouin scattering in optical fibers. *Opt. Lett.* **1997**, *22*, 787–789. [CrossRef] [PubMed]
6. Kreger, S.; Gifford, D.; Froggatt, M.; Soller, B.; Wolfe, M. High resolution distributed strain or temperature measurements in single-and multi-mode fiber using swept-wavelength interferometry. In Proceedings of the Optical Fiber Sensors Conference, Cancun, Mexico, 23–27 October 2006.
7. Lanticq, V.; Gabet, R.; Taillade, F.; Delepine-Lesoille, S. Distributed Optical Fibre Sensors for Structural Health Monitoring: Upcoming challenges. In *Optical Fibre, New Developments*; Lethien, C., Ed.; In-The: Rijeka, Croatia, 2010; Chapter 9, pp. 177–199.
8. Barrias, A.; Casas, J.; Villalba, S. Review of Distributed Optical Fibre Sensors for Civil Engineering Applications. *Sensors* **2016**, *16*, 748. [CrossRef] [PubMed]
9. Product Optical Backscatter Reflectometer OBR4600. Available online: http://lunainc.com/product/sensing-solutions/obr-4600/ (accessed on 3 March 2017).
10. Product Neubrescope NBX-7020. Available online: http://lunainc.com/product/sensing-solutions/obr-4600/ (accessed on 3 March 2017).
11. Farahani, M.A.; Gogolla, T. Spontaneous Raman scattering in Optical fibers with modulated probe light for distributed temperature Raman remote sensing. *J. Lightwave Tech.* **1999**, *17*, 8. [CrossRef]
12. Girard, S.; Keurinck, J.; Ouerdane, Y.; Meunier, J.P.; Boukenter, A. Gamma-rays and pulsed X-ray radiation responses of germanosilicate single-mode optical fibres: Influence of cladding codopants. *J. Lightwave Technol.* **2004**, *22*, 1915. [CrossRef]

13. Brichard, B.; Tomashuk, A.L.; Bogatyrjov, V.A.; Fernandez, A.F.; Klyamkin, S.N.; Girard, S.; Berghmans, F. Reduction of the radiation-induced absorption in hydrogenated pure silica core fibres irradiated in situ with γ-rays. *J. Non-Cryst. Solids* **2007**, *353*, 466–472. [CrossRef]

14. Faustov, A. Advanced Fibre Optics Temperature and Radiation Sensing in Harsh Environments. Ph.D. Thesis, Université de Mons, Mons, Belgium, 2014.

15. Lemaire, P.J.; Lindholm, E.A. *Hermetic Optical Fibres: Carbon Coated Fibres in Specialty Optical Fibres Handbook*; Mendez, A., Morse, T.F., Eds.; New York Academic: New York, NY, USA, 2007; pp. 453–490.

16. Lecomte, P. Mesure Haute Température en Environnement Irradié par Fibre Optique Utilisant L'Effet Raman. Ph.D. Thesis, Université de Perpignan via Domitia, Perpignan, France, April 2017.

17. Blairon, S.; Delepine-Lesoille, S.; Vinceslas, G. Truly distributed optical fibre extensometers for geomechanical structure monitoring (dikes and underground repository): Influence of sensor external coating. In Proceedings of the 8th International Symposium on Field Measurements in GeoMechanics, Berlin, Germany, 12–16 September 2011.

18. Dubois, J.-P.; Delepine-Lesoille, S.; Tran, V.-H.; Buschaert, S.; Mayer, S.; Henault, J.-M.; Salin, J.; Moreau, G. Raman versus Brillouin optical fibre distributed temperature sensing: An outdoor comparison (metallic beam and concrete slab). In Proceedings of the Conference SHMII-4, Zurich, Switzerland, 22–24 July 2009.

19. Farhoud, R.; Delepine-Lesoille, S.; Buschaert, S.; Righini-Waz, C. Monitoring System Design of Underground Repository for Radioactive Wastes—In Situ Demonstrator. *IACSIT Int. J. Eng. Technol.* **2015**, *7*, 484–489. [CrossRef]

20. Failleau, G.; Beaumont, O.; Delepine-Lesoille, S.; Plumeri, S.; Razouk, R.; Beck, Y.-L.; Henault, J.-M.; Bertrand, J.; Hay, B. Development of facilities and methods for the metrological characterization of distributed temperature sensing systems based on optical fibres. *Meas. Sci. Tech.* **2017**, *28*, 015009. [CrossRef]

21. Planes, I.; Girard, S.; Boukenter, A.; Marin, E.; Delepine-Lesoille, S.; Marcandella, C.; Ouerdane, Y. Steady γ-ray effects on the performances of PPP-BOTDA and TW-COTDR fibre sensing. *Sensors* **2017**, *17*, 396. [CrossRef] [PubMed]

22. Pheron, X.; Ouerdane, Y.; Girard, S.; Marcandella, C.; Delepine-Lesoille, S.; Bertrand, J.; Taillade, F.; Merliot, E.; Mamdem, Y.S.; Boukenter, A. In situ radiation influence on strain measurement performance of Brillouin sensors. In Proceedings of the 21st International Conference on Optical Fibre Sensors, Ottawa, ON, Canada, 15–19 May 2011.

23. Planes, I.; Girard, S.; Boukenter, A.; Marin, E.; Delepine-Lesoille, S.; Farhoud, R.; Zghondi, J.; Fischli, F.; Ouerdane, Y. Distributed Strain Monitoring in Tunnel Shotcrete (with and without yieldable concrete wedges) by fibre-based TW-COTDR technique. In Proceedings of the 6th Asia Pacific Optical Sensors Conference, Shanghai, China, 11–14 October 2016.

24. Farhoud, R.; Bertrand, J.; Buschaert, S.; Delepine-Lesoille, S.; Hermand, G. Full scale in situ monitoring section test in the Andra's Underground Research Laboratory. In Proceedings of the 1st Conference on Technological Innovations in Nuclear Civil Engineering (TINCE), Paris, France, 29–31 October 2013.

25. Farhoud, R.; Gilardi, N.; Delepine-Lesoille, S.; Hermand, G.; Zghondi, J. Tunnel monitoring system qualification for an underground radioactive waste disposal. In Proceedings of the International Conference on Advanced Technology Innovation, Samui, Thailand, 25–28 June 2017.

26. MoDeRn European Project 7th Framework n°232598. Monitoring Developments for Safe Repository Operation and Staged Closure. (Deliverable D3.5.1) "Disposal cell monitoring system installation and testing demonstrator in Bure Underground Research Laboratory". Available online: http://www.modern-fp7.eu/publications/project-reports/index.html (accessed on 7 June 2017).

27. Berghmans, F.; Brichard, B.; Fernandez, A.F.; Gusarov, A.; Van Uffelen, M.; Girard, S. An introduction to radiation effects on optical components and fiber optic sensors. *Opt. Waveguide Sens. Imaging* **2008**, 127–165. [CrossRef]

28. Stone, J. Interactions of hydrogen and deuterium with silica optical fibres: A review. *J. Lightwave Technol.* **1987**, *5*, 712–733. [CrossRef]

29. Leparmentier, S.; Auguste, J.L.; Humbert, G.; Pilorget, G.; Lablonde, L.; Delepine-Lesoille, S. Study of the hydrogen influence on the acoustic velocity of single-mode fibres by Rayleigh and Brillouin backscattering measurements. In Proceedings of the International Optical Fibre Sensors Conference (OFS24), SPIE 9634-118, Curitiba, Brazil, 28 September–2 October 2015.

30. Delepine-Lesoille, S.; Bertrand, J.; Lablonde, L.; Phéron, X. Hydrogen influence on Brillouin and Rayleigh distributed temperature or strain sensors. In Proceedings of the 22nd International Conference on Optical Fibre Sensors (OFS22), Beijing, China, 15–19 October 2012.

31. Benabdesselam, M.; Mady, F.; Girard, S.; Mebrouk, Y.; Duchez, J.B.; Gaillardin, M.; Paillet, P. Performance of Ge-doped Optical Fibre as a Thermoluminescent, Performance of Ge-doped Optical Fibre as a Thermoluminescent Dosimeter. *IEEE Trans. Nucl. Sci.* **2013**, *60*, 4251–4256. [CrossRef]

32. Capoen, B.; el Hamzaoui, H.; Bouazaoui, M.; Ouerdane, Y.; Boukenter, A.; Girard, S.; Marcandella, C.; Duhamel, O. Sol-gel derived copper-doped silica glass as a sensitive material for X-ray beam dosimetry. *Opt. Mater.* **2016**, *51*, 104–109. [CrossRef]

33. Girard, S.; Capoen, B.; el Hamzaoui, H.; Bouazaoui, M.; Bouwmans, G.; Morana, A.; di Francesca, D.; Boukenter, A.; Duhamel, O.; Paillet, P.; et al. Potential of Copper- and Cerium-doped Optical Fibre Materials for Proton Beam Monitoring. *IEEE Trans. Nucl. Sci.* **2016**. [CrossRef]

34. O'Keefe, A.; Fitzpatrick, C.; Lewis, E.; Al-Shamma'a, A.I. A review of optical fibre radiation dosimeters. *Sens. Rev.* **2008**, *28*, 136–142. [CrossRef]

35. Girard, S.; Ouerdane, Y.; Marcandella, C.; Boukenter, A.; Quesnard, S.; Authier, N. Feasibility of radiation dosimetry with phosphorus-doped optical fibres in the ultraviolet and visible domain. *J. Non-Cryst. Solids* **2011**, *357*, 1871–1874. [CrossRef]

36. Yang, M.; Yang, Z.; Dai, J.; Zhang, D. Fibre optic hydrogen sensors with sol-gel WO3 coatings. *Sens. Actuators B* **2012**, *166–167*, 632–636. [CrossRef]

37. Huiberts, J.N.; Griessen, R.; Rector, J.H.; Wijngaarden, R.J.; Dekker, J.P.; de Groot, D.G.; Koeman, N.J. Yttrium and lanthanum hybride films with switchable optical properties. *Nature* **1996**, *380*, 231–234. [CrossRef]

38. Goddard, L.; Wong, K.Y.; Garg, A.; Behymer, E.; Cole, G.; Bond, T. Measurements of the complex refractive index of Pd and Pt films in air and upon adsorption of H$_2$ gas. In Proceedings of the IEEE Lasers and Electro-Optics Society (LEOS 2008), Newport Beach, CA, USA, 9–13 November 2008; pp. 569–570.

39. Butler, M.A. Optical fibre hydrogen sensor. *Appl. Phys. Lett.* **1984**, *45*, 1007–1009. [CrossRef]

40. Bevenot, X.; Trouillet, A.; Veillas, C.; Gagnaire, H.; Clement, M. Hydrogen leak detection using an optical fibre sensor for aerospace applications. *Sens. Actuators B Chem.* **2000**, *67*, 57–67. [CrossRef]

41. Monzon-Hernandez, D.; Luna-Moreno, D.; Martinez-Escobar, D. Fast response fibre optic hydrogen sensor based on palladium and gold nano-layers. *Sens. Actuators B* **2009**, *136*, 562–566. [CrossRef]

42. Azar, M.T.; Sutapun, B.; Petrick, R.; Kazemi, A. Highly sensitive hydrogen sensors using palladium coated fibre optics with exposed cores and evanescent field interactions. *Sens. Actuators B* **1999**, *56*, 158–163. [CrossRef]

43. Villatoro, J.; Monzon-Hernandez, D. Fast detection of hydrogen with nanofibre tapers coated with ultra-thin Pd layers. *Opt. Express* **2005**, *13*, 5087–5093. [CrossRef] [PubMed]

44. Greco, F.; Ventrelli, L.; Dario, P.; Mazzolai, B.; Mattoli, V. Micro-wrinckled palladium surface for hydrogen sensing and switched detection of lower flammability limit. *Int. J. Hydrogen Energy* **2012**, *37*, 17529–17539. [CrossRef]

45. Delepine-Lesoille, S.; Bertrand, J.; Lablonde, L.; Phéron, X. "Distributed hydrogen sensing with Brillouin scattering in optical fibres. *Photonics Technol. Lett.* **2012**, *24*, 1475–1477. [CrossRef]

46. Kishida, K.; Li, C. H.; Nishiguchi, K. I.; Yamauchi, Y.; Guzik, A.; Tsuda, T. Hybrid Brillouin-Rayleigh distributed sensing system. In Proceedings of the 2012 22nd Optical Fiber Sensors Conference (OFS-22), Beijing, China, 15–19 October 2012.

47. Metrodecom EMRP European Project Number ENV54, Dedicated to Metrology for Decommissioning Nuclear Facilities, 2014–2017. Available online: https://www.euramet.org/research-innovation/search-research-projects/details/?eurametCtcp_project_show%5Bproject%5D=1221&eurametCtcp_project%5Bback%5D=67&cHash=93346f2e6f28e44285655e34debaa039 (accessed on 7 June 2017).

48. Failleau, G.; Beaumont, O.; Razouk, R.; Delepine-Lesoille, S.; Bertrand, J.; Courthial, B.; Martinot, F.; Hay, B. A metrological support dedicated to Raman-based distributed temperature sensing techniques applied for the thermal structural monitoring. Presented at the International Metrology Congress (CIM), Paris, France, 19–21 September 2017.

49. Delepine-Lesoille, S.; Bertrand, J.; Pilorget, G.; Farhoud, R.; Buschaert, S.; Vogt, T.; Müller, H.; Verstricht, J. Monitoring natural and engineered clay in geological repository with optical fibre sensors—Lessons learnt from large scale experiments. In Proceedings of the 6th International Clay Conference, Brussels, Belgium, 23–26 March 2015.

50. Bertrand, J.; Morice, R.; Beaumont, O.; Dubois, J.P. Field calibration device for Raman Backscatter based Fibre Optic Distributed Temperature System (DTS) Technology. In Proceedings of the 22nd International Conference on Optical Fibre Sensor Conference (OFS2012), Beijing, China, 15–19 October 2012.

51. Cangialosi, C. Effets des Radiations sur les Fibres Optiques: Impact sur les Capteurs à Fibres Optiques Réparties Raman et Brillouin. Ph.D. Thesis, Université de Saint Etienne, Saint Etienne, France, 2016.

52. Cangialosi, C.; Ouerdane, Y.; Girard, S.; Boukenter, A.; Cannas, M.; Delepine-Lesoille, S.; Bertrand, J.; Paillet, P. Hydrogen and radiation induced effects on performances of Raman fiber-based temperature sensors. *IEEE J. Light. Tech.* **2015**, *33*, 2432–2438. [CrossRef]

53. Cangialosi, C.; Girard, S.; Cannas, M.; Boukenter, A.; Marin, E.; Agnello, S.; Delepine-Lesoille, S.; Marcandella, C.; Paillet, P.; Ouerdane, Y. On-line characterization of gamma radiation vulnerability of Raman based distributed fibre optic sensor. *IEEE Trans. Nucl. Sci.* **2016**, *63*, 2051–2057. [CrossRef]

54. Fernandez, A.F.; Rodeghiero, P.; Brichard, B.; Berghmans, F.; Hartog, A.H.; Hughes, P.; Williams, K.; Leach, A.P. Radiation-tolerant Raman distributed temperature monitoring system for large nuclear infrastructures. *IEEE Trans. Nucl. Sci.* **2005**, *52*, 2689–2694. [CrossRef]

55. Di Francesca, D.; Girard, S.; Planes, I.; Cebollada, A.; Alessi, A.; Reghioua, I.; Cangialosi, C.; Ladaci, A.; Rizzolo, S.; Lecoeuche, V.; et al. Radiation Hardened Architecture of a Single-Ended Raman-Based Distributed Temperature Sensor. In Proceedings of the IEEE Nuclear and Space Radiation Effects Conference (NSREC), Portland, OR, USA, 11–15 July 2016.

56. Cangialosi, C.; Ouerdane, Y.; Girard, S.; Boukenter, A.; Delepine-Lesoille, S.; Bertrand, J.; Marcandella, C.; Paillet, P.; Cannas, M. Development of a Temperature Distributed Monitoring System Based On Raman Scattering in Harsh Environment. *IEEE Trans. Nucl. Sci.* **2014**, *61*, 3315–3322. [CrossRef]

57. Planes, S.D.I.; Landolt, M.; Hermand, G.; Perrochon, O. Compared performances of Rayleigh, Raman, and Brillouin distributed temperature measurements during concrete container fire test. In Proceedings of the 25th International Conference on Optical Fibre Sensors (OFS-25), Jeju, Korea, 24–28 April 2017.

58. Sikali-Mamdem, Y. Séparation des Influences de la Température et des Deformations dans les Capteurs à Fibre Optique Fondés sur la Rétrodiffusion Brillouin. Ph.D. Thesis, Telecom ParisTech, Paris, France, 2012.

59. Mamdem, Y.S.; Burov, E.; De Montmorillon, L.A.; Jaouën, Y.; Moreau, G.; Gabet, R.; Taillade, F. Importance of residual stresses in the Brillouin gain spectrum of single mode optical fibers. *Opt. Express* **2012**, *20*, 1790–1797. [CrossRef] [PubMed]

60. Henault, J.M.; Quiertant, M.; Delepine-Lesoille, S.; Salin, J.; Moreau, G.; Taillade, F.; Benzarti, K. Quantitative strain measurement and crack detection in RC structures using a truly distributed fibre optic sensing system. *Constr. Build. Mater.* **2012**, *37*, 916–923. [CrossRef]

61. Hénault, J.-M. Approche Méthodologique Pour L'Evaluation des Performances et de la Durabilité des Systèmes de Mesure Répartie de Déformation. Application à un Câble à Fibre Optique Noyé Dans le Béton. Ph.D. Thesis, University Paris East, Marne-la-vallée, France, 2013.

62. Buchoud, E.; Henault, J.-M.; D'Urso, G.; Girard, A.; Blairon, S.; Mars, J.; Vrabie, V. Development of an automatic algorithm to analyze the cracks evolution in a reinforced concrete structure from strain measurements performed by an Optical Backscatter Reflectometer. In Proceedings of the 4th Workshop on Civil Structural Health Monitoring, Berlin, Germany, 23 November 2012.

63. Billon, A.; Henault, J.M.; Quiertant, M.; Taillade, F.; Khadour, A.; Martin, R.P.; Benzarti, K. Quantitative Strain Measurement with Distributed Fibre Optic Systems: Qualification of a Sensing Cable Bonded to the Surface of a Concrete Structure. In Proceedings of the EWSHM-7th European Workshop on Structural Health Monitoring, Nantes, France, 8–11 July 2014.

64. Artieres, O.; Beck, Y.-L.; Guidoux, C. Scour Erosion Detection with a Fibre Optic Sensor-Enabled Geotextile. In Proceedings of the ICSE6, Paris, France, 27–31 August 2012.

65. Clauzon, T.; Martinot, F.; Moreau, G.; Henault, J.-M.; Courtois, A. EPR Flamanville 3: A technical showcase for innovation in the EDF Group. In Proceedings of the 23rd SMiRT Conference (Conference on Structural Mechanics in Reactor Technology), Manchester, UK, 10–14 August 2015.

66. ITN FINESSE European Union's Horizon 2020 Research and Innovation Programm under the Marie Slodowska–Curie Grant Agreement n° 722509, FIbre NErvous Sensing SystEms 2016–2020. Available online: http://itn-finesse.eu/ (accessed on 7 June 2017).

67. Delepine-Lesoille, S.; Pilorget, G.; Lablonde, L.; Bertrand, J. Influence of doping element in distributed hydrogen optical fiber sensors with Brillouin scattering. In Proceedings of the International Conference on Hydrogen Safety (ICHS), Brussels, Belgium, 9–11 September 2013.

68. Girard, S.; Kuhnhenn, J.; Gusarov, A.; Brichard, B.; van Uffelen, M.; Ouerdane, Y.; Boukenter, A.; Marcandella, C. Radiation Effects on Silica-based Optical Fibres: Recent Advances and Future Challenges. *IEEE Trans. Nucl. Sci.* **2013**, *60*, 2015–2036. [CrossRef]

69. Phéron, X.; Girard, S.; Boukenter, A.; Brichard, B.; Delepine-Lesoille, S.; Bertrand, J.; Ouerdane, Y. High γ-ray dose radiation effects on the performances of Brillouin scattering based optical fibre sensors. *Opt. Express* **2012**, *20*, 26978–26985. [CrossRef] [PubMed]

70. Phéron, X. Durabilité de Capteur à Fibre Optique par Diffusion Brillouin en Environnement Radiatif. Ph.D. Thesis, University of Saint Etienne, Saint Etienne, France, 4 November 2013.

71. Cangialosi, C.; Girard, S.; Boukenter, A.; Marin, E.; Cannas, M.; Delepine-Lesoille, S.; Marcandella, C.; Paillet, P.; Ouerdane, Y. Steady Sate y-radiation effects on Brillouin fiber sensors. In Proceedings of the 24th International Conference on Optical Fiber Sensor, Curitiba, Brazil, 28 September–2 October 2015.

72. Dianov, E.M.; Golant, K.M.; Khrapko, R.R.; Tomashuk, A.L. Low-loss nitrogen-doped silica fibers: The prospects for applications in radiation environment. In Proceedings of the 1996 Optical Fiber Communications (OFC'96), San Jose, CA, USA, 25 February–1 March 1996.

73. Rizzolo, S.; Boukenter, A.; Marin, E.; Cannas, M.; Perisse, J.; Bauer, S.; Mace, J.-R.; Ouerdane, Y.; Girard, S. Vulnerability of OFDR-based distributed sensors to high γ-ray doses. *Opt. Express* **2015**, *23*, 18997–19009. [CrossRef] [PubMed]

74. Rizzolo, S.; Marin, E.; Cannas, M.; Boukenter, A.; Ouerdane, Y.; Périsse, J.; Macé, J.R.; Bauer, S.; Marcandella, C.; Paillet, P.; Girard, S. Radiation effects on OFDR-based sensors. *Opt. Lett.* **2015**, *40*, 4571–4574. [CrossRef] [PubMed]

75. Rizzolo, S.; Marin, E.; Boukenter, A.; Ouerdane, Y.; Cannas, M.; Périsse, J.; Bauer, S.; Mace, J.R.; Marcandella, C.; Paillet, P.; Girard, S. Radiation hardened optical frequency domain reflectometry distributed temperature fibre-based sensors. *IEEE Trans. Nucl. Sci.* **2016**, *62*, 171–176.

76. Leparmentier, S.; Auguste, J.-L.; Humbert, G.; Delaizir, G.; Delepine-Lesoille, S. Fabrication of optical fibres with palladium metallic particles embedded into the silica cladding. *Opt. Mater. Express* **2015**, *5*, 2578–2586. [CrossRef]

77. Beauvois, G.; Caussanel, M.; Lupi, J.-F.; Ude, M.; Trzesien, S.; Dussardier, B.; Duval, H.; Grieu, S. Projet DROÏD: Développement d'un dosimètre distribué à fibre optique. In Proceedings of the 7ème Journée sur les Fibres Optiques en Milieu Radiatif (FMR2016), Chatenay Malabry, France, 12–13 December 2016.

78. Toccafondo, Y.; Marin, E.; Guillermain, E.; Kuhnhenn, J.; Mekki, J.; Brugger, M.; di Pasquale, F. Distributed Optical Fiber Radiation Sensing in a Mixed-Field Radiation Environment at CERN. *J. Lightwave Technol.* **2016**. [CrossRef]

79. Di Francesca, D.; Toccafondo, I.; Calderini, S.; Vecchi, G.L.; Girard, S.; Alessi, A.; Ferraro, R.; Danzeca, S.; Kadi, Y.; Brugger, M. Distributed Optical Fiber Radiation Sensing in the Proton Synchrotron Booster at CERN. In Proceedings of the RADECS 2017 Conference, Geneva, Switzerland, 2–6 October 2017.

![sensors logo] *sensors*

MDPI

Review

A Versatile Electronic Tongue Based on Surface Plasmon Resonance Imaging and Cross-Reactive Sensor Arrays—A Mini-Review

Laurie-Amandine Garçon [1,2,3,4], Maria Genua [1,2,3], Yanjie Hou [1,2,3], Arnaud Buhot [1,2,3], Roberto Calemczuk [1,2,3], Thierry Livache [1,2,3], Martial Billon [1,2,3], Christine Le Narvor [5], David Bonnaffé [5], Hugues Lortat-Jacob [6,7,8] and Yanxia Hou [1,2,3,*]

[1] Institut Nanosciences et Cryogénie, University of Grenoble Alpes, INAC-SyMMES,
    F-38000 Grenoble, France; laurie-amandine.garcon@hotmail.fr (L.-A.G.); mariagenua@hotmail.com (M.G.);
    yanjiehou@yahoo.com (Y.H.); arnaud.buhot@cea.fr (A.B.); rcalemczuk@yahoo.com (R.C.);
    thierry.livache@aryballe.com (T.L.); martial.billon@cea.fr (M.B.)
[2] Centre National de la Recherche Scientifique, SyMMES UMR 5819, F-38000 Grenoble, France
[3] Commissariat à l'Energie Atomique et aux Energies Alternatives (CEA), INAC-SyMMES,
    F-38000 Grenoble, France
[4] Institut Néel, F-38000 Grenoble, France
[5] ICMMO/G2M/LCOM, UMR 8182 (CNRS-UPS), LabEx LERMIT, Université Paris-Sud 11,
    91405 Orsay CEDEX, France; christine.le-narvor@u-psud.fr (C.L.N.); david.bonnaffe@u-psud.fr (D.B.)
[6] Institut de Biologie Structurale, University of Grenoble Alpes, UMR 5075, 38027 Grenoble, France;
    hugues.lortat-jacob@ibs.fr
[7] Centre National de la Recherche Scientifique, Institut de Biologie Structurale, UMR 5075,
    38027 Grenoble, France
[8] Commissariat à l'Energie Atomique et aux Energies Alternatives (CEA), Institut de Biologie Structurale,
    UMR 5075, 38027 Grenoble, France
* Correspondence: yanxia.hou-broutin@cea.fr; Tel.: +33-4-38-78-94-78; Fax: +33-4-38-78-51-45

Academic Editors: Nicole Jaffrezic-Renault and Gaelle Lissorgues
Received: 20 March 2017; Accepted: 2 May 2017; Published: 6 May 2017

Abstract: Nowadays, there is a strong demand for the development of new analytical devices with novel performances to improve the quality of our daily lives. In this context, multisensor systems such as electronic tongues (eTs) have emerged as promising alternatives. Recently, we have developed a new versatile eT system by coupling surface plasmon resonance imaging (SPRi) with cross-reactive sensor arrays. In order to largely simplify the preparation of sensing materials with a great diversity, an innovative combinatorial approach was proposed by combining and mixing a small number of easily accessible molecules displaying different physicochemical properties. The obtained eT was able to generate 2D continuous evolution profile (CEP) and 3D continuous evolution landscape (CEL), which is also called 3D image, with valuable kinetic information, for the discrimination and classification of samples. Here, diverse applications of such a versatile eT have been summarized. It is not only effective for pure protein analysis, capable of differentiating protein isoforms such as chemokines CXCL12α and CXCL12γ, but can also be generalized for the analysis of complex mixtures, such as milk samples, with promising potential for monitoring the deterioration of milk.

Keywords: electronic tongues; cross-reactive sensor array; surface plasmon resonance imaging; pattern recognition; continuous evolution landscape; protein; beverages; milk

---

1. Introduction

At present, the development of sensors with improved performances and new capabilities is driven by the ever-expanding monitoring needs of a variety of gases and liquid species in diverse domains, including environment monitoring such as air/water quality control, the detection of pollution or leaks of hazardous materials for personal and public safety, food safety and quality control, and non-invasive medical diagnostics. In this context, the electronic noses (eNs) and electronic tongues (eTs) have emerged as promising alternatives. They are engineered to mimic the mammalian olfactory system, consisting of an array of low-selective sensors with cross-sensitivity to different species in complex mixtures and using advanced mathematical procedures for signal processing based on pattern recognition and/or multivariate analysis. Herein, eNs refer to cross-reactive sensor arrays dedicated for the analysis of gas samples and eTs for liquid samples. Unlike gas or liquid chromatography, these devices do not provide information on sample composition in detail, but rather give a characteristic fingerprint through pattern recognition, thus allowing the identification of the sample as a whole.

The last three decades have witnessed great progress in the domain of eNs/eTs thanks to the improvement of sensor technology combined with the artificial intelligence approach. When considering the development of eT systems for analysis of liquid samples, the most common technique is based on electrochemical [1,2], potentiometric [3,4], voltammetric [5], amperometric [6], and impedimetric [7–9]) sensors. The potential of these systems for applications in the food and beverages industries as well as the pharmaceutical industry have been reported in several reviews [4,10–13]. In general, most eT systems developed in Europe are based on electrochemical sensors. For example, the leading company in the eN/eTs domain, Alpha MOS (Toulouse, France), has developed an eT based on the ChemFET (Chemical modified Field Effect Transistor) sensor technology using potentiometric measurements. Remarkably, in the past decade, new eT systems have emerged based on mass sensors (multichannel quartz crystal microbalance (QCM) [14,15] and surface acoustic wave (SAW) [16–18]) and optical sensors. In particular, in the US, several groups have worked on cross-reactive sensor arrays using optical sensing approaches and novel sensing materials for the analysis of proteins and complex mixtures. For example, colorimetric sensor arrays were developed by two groups: Suslick's group, based on a series of functional dyes (solvatochromic dyes, pH indicators, and metalloporphyrins) for analysis of different beverages [19–21], and Anslyn's group, by using an array of 29 boronic acid-containing oligopeptide functionalized resin beads for discriminating proteins [22]. Other groups have focused on fluorescence sensor arrays. Hamilton and co-workers designed and prepared synthetic tetra-phenylporphyrin derivatives as differential receptors for protein sensing [23,24]. More recently, Rotello's group developed a more sensitive system based on nanoparticle-fluorophore complexes. They carried out the identification of proteins, in serum, at physiologically relevant concentrations [25], which is promising for future application of eTs in the diagnosis of disease states. However, until now, most of the eTs' application domains remain limited. Thus, there is an identified and current need for the development of new measurement methods and for the search of novel sensing materials to promote a more versatile eT system.

Very recently, our laboratories developed a novel eT by combining surface plasmon resonance imaging (SPRi) with cross-reactive sensor arrays [26]. It is well known that SPRi has been widely applied for "lock-and-key"-based biochips to investigate and quantify biomolecules and interactions [27,28]. To the best of our knowledge, our laboratory was the first to use it for nonspecific and cross-reactive sensor arrays. Indeed, the ability to immobilize many receptors (up to hundreds) on the same surface and to monitor the interactions simultaneously with kinetic information, in a real-time and label-free microarray format, is particularly interesting for eT development. In order to largely simplify the preparation of a diverse variety of sensing materials, an innovative combinatorial approach was proposed. The idea was to use a small number of molecules as building blocks (BBs). BBs were defined as small and easily accessible molecules displaying various physicochemical properties; these BBs are hydrophilic, hydrophobic, negatively charged, positively charged, etc. They can be mixed in varying and controlled proportions, and the different mixtures are arrayed to give combinatorial cross-reactive receptors (CoCRRs) with an exceptional potential for rapid growth in diversity. For example, 11 combinations can be obtained using only two BBs mixed in concentrations varying from 0 to 100% in 10% increments, while 66 can theoretically be accessed by adding a third

BB. Moreover, such growth can be generalized to n BBs and i % concentration increments, leading to [(100/i) + n − 1]!/[(n − 1)!(100/i)!] potentially different CoCRRs. The obtained eT is capable of generating 3D images as fingerprints for samples, providing useful kinetic information for discrimination and identification purposes. This mini-review will concentrate on diverse applications we carried out with such a versatile eT [26,29–31].

2. Electronic Tongue Based on SPRi and Cross-Reactive Sensor Array

For the proof-of-concept study, two small molecules with different physicochemical properties were used as BBs, such as disaccharides lactose (BB1, hydrophilic and neutral) and sulfated lactose (BB2, hydrophilic and negatively charged). A CoCRR array was then constructed, containing 11 cross-reactive receptors made of pure and mixed solutions of BB1 and BB2 at different ratios, as shown in Figure 1.

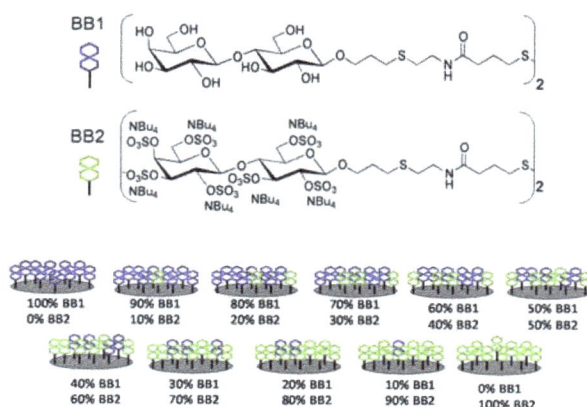

Figure 1. Schematic illustration of the CoCRR array prepared with only 2 building blocks (BBs) such as lactose (BB1) and sulfated lactose (BB2). Reprinted from [30] with permission from JoVE.

In practice, at a constant concentration of 20 μM for [BB1 + BB2], 11 pure and mixed solutions with [BB1]/([BB1] + [BB2]) ratios of 0, 10, 20, 30, 40, 50, 60, 70, 80, 90, and 100% were prepared. They were deposited by an automated micro-spotter on a prism surface covered with a thin gold layer. A quadruplicate was deposited for each cross-reactive receptors. The immobilization of the BBs was carried out by the interaction between the disulfide bond and gold surface *via* the formation of self-assembled monolayers (SAMs). Thanks to SAMs, the physicochemical properties of the CoCRRs can be easily modulated by simply varying BB proportions in the mixtures.

The SPRi apparatus was placed in an incubator at 25 °C. It was connected to a microfluidic system composed of a syringe pump, a degassing system, a 10 μL PEEK flow cell in a hexagonal configuration, and a 6-port injection valve (for more details, see [30]). The chip containing the CoCRR array was mounted on SPRi device. Upon sample injection on the CoCRR array, molecular binding gave rise to a light-up of spots on the prism with different intensities, as shown in the SPR image (Figure 2a). The signal was then converted to variations of reflectivity (expressed as R%) versus time, yielding sensorgrams composed of kinetic binding curves for all the spots (Figure 2b). Afterwards, based on the sensorgrams, a classical pattern was generated in the form of a histogram by combining the signals obtained with all of the receptors at equilibrium. However, thanks to our combinatorial approach and array design, the composition of each sensing receptor was linked closely to that of its neighbors, giving each receptor a signal that is correlated to that of its neighbor. Thus, all signals generated by the CoCRR array can be considered continuous. In this way, for each sample, we obtained a 2D continuous evolution profile (CEP) by plotting the variation of reflectivity (R%) at equilibrium versus BB1% evolution (Figure 2c). In addition, "time" was added as the third dimension, since SPRi is capable of monitoring the binding events in real time, thereby generating a 3D continuous evolution landscape (CEL), also called a 3D image, as shown in Figure 2d. The 2D CEP and 3D CEL can be used as fingerprints for the discrimination and identification of samples. Moreover, principal component analysis (PCA) was utilized for the classification of samples. Finally, it is important to mention that the obtained eT is reusable after regenerating with an appropriate solution.

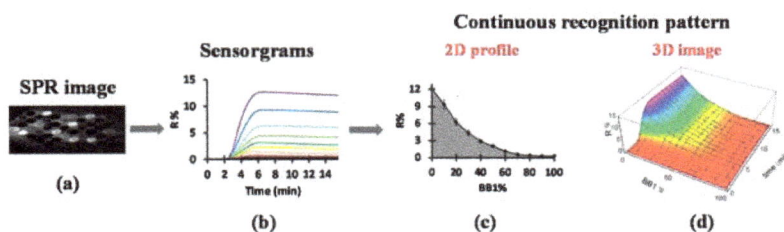

Figure 2. Data treatment for the generation of continuous recognition patterns. (**a**) SPR image recorded by a CCD camera; (**b**) Sensorgrams for all the spots; (**c**) a 2D continuous evolution profile (CEP) and (**d**) a 3D continuous evolution landscape (CEL), also called a 3D image, generated by the electronic tongue (eT). Reprinted from [30] with permission from JoVE.

3. Diverse Applications of the eTs

3.1. Analysis of Pure Proteins

Given that the CoCRR array is composed of receptors with different charge densities, we have assumed that such an array can be effective for common protein analysis. For this purpose, a preliminary test was carried out using three proteins, *Arachis hypogaea* lectin (AHL) (isoelectric point (pI) 6.0), myoglobin (pI 7.2), and lysozyme (pI 11). They have different charges under experimental conditions in HEPES (pH 7.4). Satisfyingly, as shown in Figure 3, the eT generated unique response patterns (2D profile and 3D image) for each protein. Based on them, the three proteins were easily distinguishable. For AHL, though slightly negatively charged under such experimental conditions, it has a stronger interaction with the CoCRRs rich in negatively charged BB2, which is possibly due to the interaction between the positively charged domain of the protein and negatively charged CoCRRs. For neutral myoglobin, there is not much difference between its interactions with all of the CoCRRs. As for positively charged lysozyme, as expected, the maximum signal was observed for the CoCRR of pure BB2 with a much higher signal intensity. These preliminary results are very promising, showing that the eT is sufficiently sensitive and selective for the discrimination and the identification of proteins.

It is important to mention that this model array was initially designed by taking inspiration from the way that cell surface heparan sulfates (HSs) recognize HS binding proteins (HSbps). HS are negatively charged polysaccharides with different negatively charged topologies in accordance with cell type and activation state so as to promote selective interactions with HSbps. We hypothesized that the CoCRR array may be more sensitive to HSbps and may promote differential binding for different HSbps, such as chemokines CXCL12α and CXCL12γ. From a structural point of view, they both have the same first part of 68 amino acids, which are folded in a similar manner with a HS binding site (K24-K27-R41) located in a highly structured domain. In addition, CXCL12γ has a second HS binding site composed of 30 amino acids in the unfolded C-terminal extension with good flexibility. For this study, a third protein *Erythrina cristagalli* lectin (ECL) was added. It is a non-HSbp and thus used as a negative control for its binding to CoCRRs rich in BB2. Meanwhile, ECL is known to bind lactose and could thus play the role of a positive reference for the CoCRRs with a high ratio of BB1.

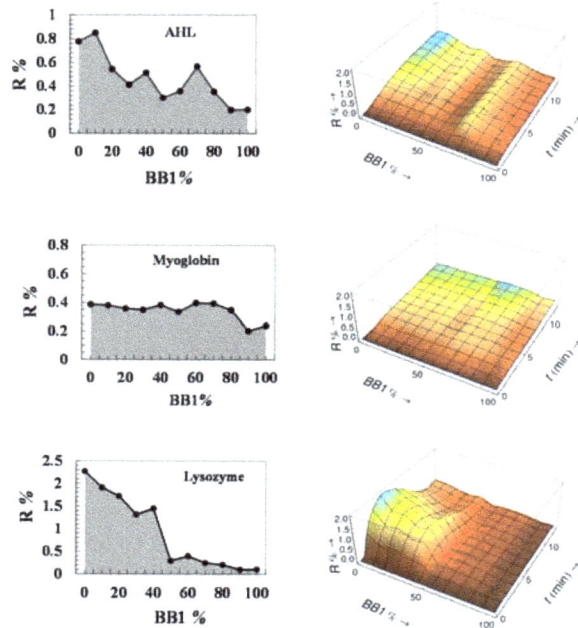

Figure 3. Analysis of common proteins by the eT: CEPs and CELs of *Arachis hypogaea* lectin (AHL) (500 nM), myoglobin (1 μM), and lysozyme (500 nM). Adapted from [30] with permission from JoVE.

Upon the injection of these proteins, it was confirmed that the eT was much more sensitive to HSbps and was able to detect CXCL12α and CXCL12γ at low nanomolar concentrations. In Figure 4, the CEPs of the three proteins are given. For ECL, it clearly has a higher affinity for BB1-rich CoCRRs with a maximum for the CoCRR containing 70% of BB1. In contrast, CXCL12α and CXCL12γ have a higher affinity for BB2-rich CoCRRs, displaying maximal signals at 10%. More importantly, the combinatorial surfaces do deliver new and supplementary information compared to the pure ones since the signals of all receptors are non-linear. For instance, it is not possible to differentiate CXCL12α and CXCL12γ based on the signals obtained on the two pure receptors containing 100% of BB1 and 100% of BB2. However, their 2D continuous profiles obtained with all the CoCRRs are clearly distinct from each other. Moreover, for those CoCRRs with the BB1 content at 50% or higher, the obtained reflectivity for CXCL12α is nearly zero. In contrast, CXCL12γ binds significantly on the CoCRRs containing up to 70% of BB1. This can be explained by the difference in global HS binding site(s) topologies and/or rigidity for those two isoforms (see above): the two distant HS binding domains in CXCL12γ can bind in a cooperative way, thus enhancing the affinity for low charge density CoCRRs. Moreover, the second HS binding domains in CXCL12γ is located in an unfolded part of the protein with flexibility. Thus, it could maximize contact points with their ligands through conformational fluctuations.

Thus, from these results, it is evident that the CEPs are correlated with the structures of the proteins so that they are characteristic of the proteins. In consequence, the obtained eT is also efficient for protein identification. Importantly, the continuous evolution of patterns are advantageous when compared with uncorrelated discrete data sets obtained using traditional eT systems, since a defective sensing receptor providing an abnormal signal can be easily identified. Thus, it provides unique "auto-corrective" behavior to the eT.

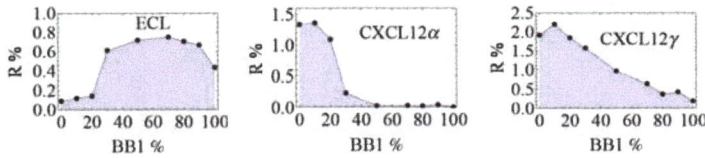

Figure 4. Discrimination of heparan sulfate binding proteins (HSbps) such as CXCL12α and CXCL12γ by the eT. 2D CEPs of ECL (200 nM) used as a non-HSbp for control, CXCL12α and CXCL12γ (both at 100 nM). Adapted from [26].

3.2. Analysis of Protein Mixtures

Furthermore, we have investigated the capacity of the eT for the analysis of protein mixtures. Some preliminary tests were carried out using simple protein mixtures. As an example, CEP of the Mix1 containing 200 nM ECL and 100 nM CXCL12α was shown in Figure 5a together with the CEP of ECL and CXCL12α at the same concentration for comparison. Gratifyingly, the CEP of the Mix1 is distinct from that of the individual pure proteins. Further observations revealed that the CEP of the Mix1 was the combination of the signals obtained from the pure proteins. An identical response was also observed with some other protein mixtures, confirming the potency of the eT for the identification of components in this kind of mixtures by a simple linear decomposition of the CEP into the pure analytes [26].

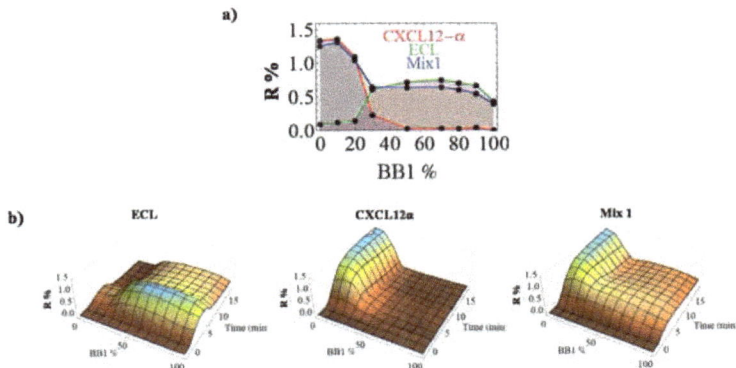

Figure 5. Analysis of simple protein mixtures by the eT. (**a**) CEPs of Mix1 (ECL + CXCL12α) compared to the ones of pure *Erythrina cristagalli* lectin (ECL) and CXCL12α; (**b**) CELs of ECL, CXCL12α, and their mixture Mix1. Adapted from [26].

In addition, as mentioned before, one of the main advantages of SPRi is the ability to monitor the binding events in real time with useful information on adsorption and desorption kinetics, which could serve as a supplementary parameter for sample discrimination. For example, in Figure 5b, the 3D recognition patterns for ECL and CXCL12α are displayed. It is evident that the association phase and dissociation phase of the two proteins on the CoCRR array are different. In particular, the desorption of ECL is much faster than that of CXCL12α. Consequently, discrimination between the two proteins based on the CEL 3D patterns is more straightforward compared to 2D CEP. Thus, such 3D images clearly demonstrate the added value of SPRi for eT development. Furthermore, when compared with the CEL of the Mix1, satisfyingly, it follows the additive behavior found for the CEP. We anticipate that such supplementary discrimination information should facilitate the future identification of analytes in a mixture, since two samples with the same relative affinity for the eT may still differ in their interaction kinetics.

3.3. Analysis of Complex Mixtures

Beyond the detection and identification of pure samples, the main goal of the eT technology is to analyze complex mixtures in diverse domains for exploring their potential applications. Quality control of food and beverages is an important issue for both industrial and personal concerns. In the last two decades, eTs have been developed as promising tools in these fields. We thus decided to challenge the capacity of the eT for the study of complex mixtures, such as beverage samples including wines, beers, and milk. Our main concern was whether the 11 sensing receptors, prepared by mixing only two small disaccharides, were able to respond differently to these complex mixtures so as to give good selectivity. Thus, for this study, we attempted to see, on the one hand, whether the eT was able to differentiate between different kinds of beverages and, on the other hand, whether it was able to discriminate different brands of the same species.

To our delight, the eT responded very differently to these samples with quite good sensitivity, as we can see from their CEPs and CELs, given in Figure 6. The sample of red wine (Bourgogne) had a maximum signal for the CoCRR with 100% BB1. Its CEP was clearly distinct from those of beer and milk. Notably, the milk sample had very low signal intensity on the CoCRRs rich in lactose BB1, which is most likely due to the competition of abundant lactose present in milk. In contrast, its signal intensity on the CoCRRs rich in sulfated lactose BB2 was much higher even using a highly diluted samples, most likely due to the high protein content in the milk sample. These results demonstrated that both their CELs and CEPs can be used as "fingerprints" for differentiation and identification of each sample.

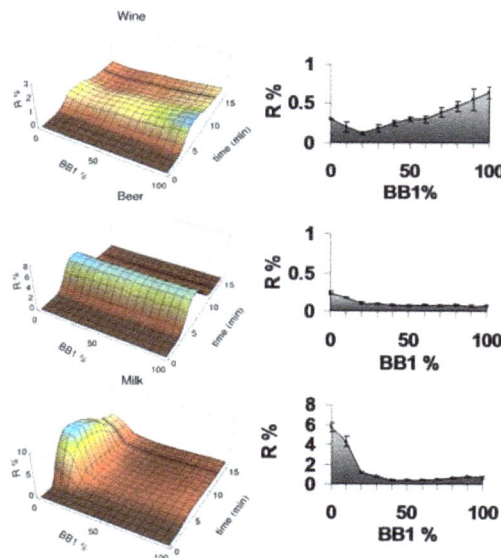

Figure 6. Differentiation of complex mixtures such as beverages by the eT. CEPs and CELs of red wine (Côtes du Rhône), beer (Leffe), and milk (UHT demi-écrémé). Reprinted from [29].

Moreover, the eT was used for the classification of a larger number of samples by principal component analysis (PCA). For this study, three different brands of red wines (Côtes du Rhône, Bordeaux and Bourgogne) and beers (Stella Artois, Leffe and Pelforth-dark), as well as UHT milk were analyzed. As shown in Figure 7, the clusters of the three species were well separated using the two principle component axes corresponding to 97% of the variance. However, the eT was not able to differentiate between different brands for wine and beer, which was probably limited by the low signal intensity. This would require the design and introduction of more appropriate BBs.

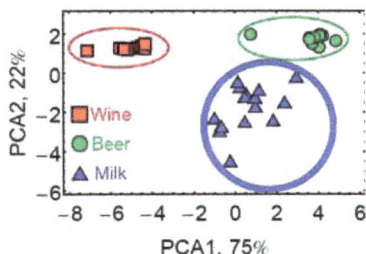

Figure 7. Classification of complex mixtures by the eT based on principal component analysis (PCA) with the two principal components representing 97% of the variance. Reprinted from [29].

On the other hand, our eT gave a very good signal for samples rich in proteins like milk. In this regard, the milk sample could be a good model for further study. We then analyzed five different milk samples either animal-based or plant-based, including UHT pasteurized cow milk, unpasteurized cow milk, soy milk, soy milk with chocolate, and rice milk. As demonstrated in Figure 8, their 3D CELs show evident differences. Therefore, these results demonstrated that the eT is very efficient for the analysis of complex mixtures such as milk samples.

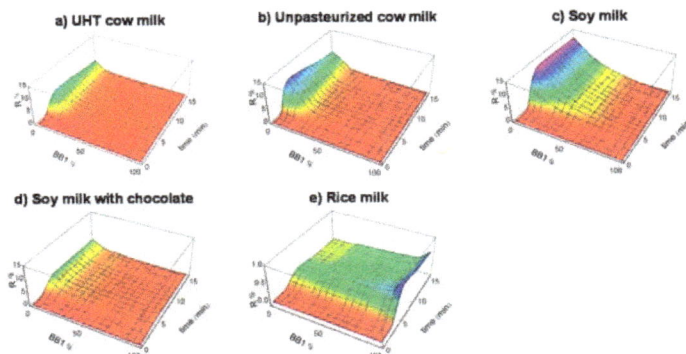

Figure 8. Analysis of protein-rich complex mixtures by the eT. CELs of various milk samples. Adapted from [31].

To further evaluate the discrimination capacity of the eT, PCA was performed to classify all these milk samples. For this, initially, PCA was performed using the data of 2D CEPs for all 20 milk sample injections with parameters in 11 dimensions corresponding to the 11 CoCRRs. As shown in Figure 9a, there was distinct separation between different milk clusters, except for some overlap between the UHT cow milk and the soy milk with chocolate. However, from their CELs, we can see that the kinetics of cow milk and soy milk are quite different, particularly during the dissociation phase. Thus, we decided to take into account all the kinetic information, including the association/dissociation phase and performed PCA based on 3D CELs. To do so, parameters in more than 300 dimensions were analyzed for each sample using 30 cross sections of CEL, taken every 30 s after the beginning of sample injection for 15 min. Notably, a much better discrimination for all the milk samples was achieved, as shown in Figure 9b. This confirms that 3D CEL-based PCA is more efficient and reliable. Consequently, the kinetic information obtained thanks to SPR imaging is also very important for eT development.

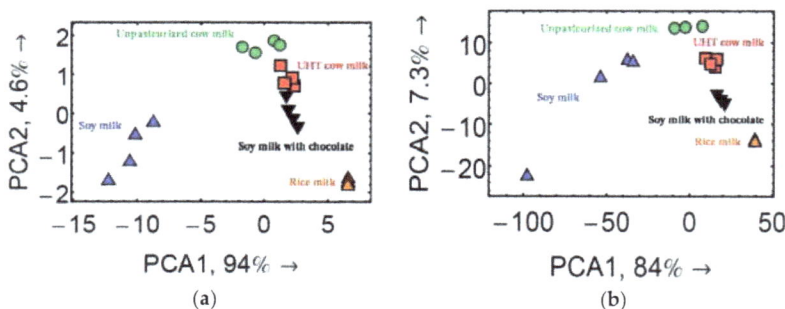

Figure 9. Classification of different milk samples by the eTs with PCA based on 2D CEPs (**a**) and 3D CELs (**b**). For each sample quadruplicate measurements were performed. Reprinted from [31] Copyright@American Scientific Publishers.

3.4. Monitoring Spoilage of Milk

During the experiments for the analysis of milk samples, we observed a quite large distribution of data points in the milk cluster in contrast to those of beers and wines, see Figure 7. This is most likely due to the age difference between the 15 milk samples. In fact, all of them were from the same bottle. So their freshness at the moment they were analyzed was not exactly the same; some were used immediately after opening the bottle, and some others were used after 24 h storage at 4 °C. Thus, it is very probably that the eT is sensitive to the minor changes in such complex mixtures, showing potential for quality control applications. To verify this, we conducted a systematic study to distinguish fresh milk from spoiled milk and to follow the spoilage of milk at room temperature. In practice, immediately after opening, undiluted aliquots of milk samples were stored at 25 °C in an open tube and measured by the eT 1, 24, 48, and 72 h after exposing the sample to air.

As shown in Figure 10, the CEL of the milk sample 1 h after opening was difficult to distinguish from that obtained after 24 h. There was no major difference on the pattern profile. Nevertheless, it was observed that the signal intensity especially for the CoCRRs rich in BB2 continuously increased over storage time. Therefore, it is most likely that the large distribution of data points in the milk cluster in PCA is due to such signal intensity variation. Satisfyingly, CELs of the 48th and 72nd hours were distinct from those collected in the 1st and 24th hours since there were major modification in the samples in this stage, as confirmed by the PCA plot. According to these preliminary results, the eT is sensitive to the changes associated with the milk spoilage and thus has a potential for quality control applications.

Finally, it is important to mention that the eT demonstrated good repeatability and stability. In practice, for all the applications, pure protein ECL at 200 nM was used as a reference sample and was systematically analyzed at the very beginning, several times in the middle in a random order, and at the end of each analysis set. For each sample to analyze, at least triplicates were used. As reported in our previous work, for both the reference sample and the complex mixture samples, there was a correlation of >99% between any two full patterns of their replicated CEP [29]. A good correlation of >93% was obtained for batch-to-batch reproducibility [26]. Furthermore, the eTs remained very stable under continuous use for at least 50 sample injection/regeneration cycles and for at least two weeks without any loss in sensitivity or sensibility. It had also good stability upon storage at 4 °C over a period of 5 months with no significant loss of signal.

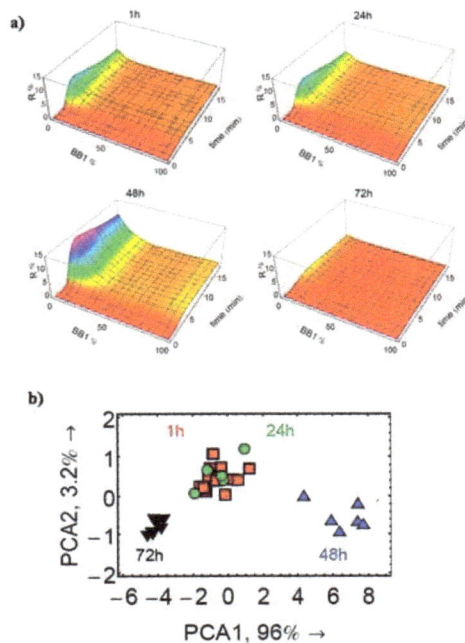

Figure 10. Monitoring spoilage of milk by the eTs: (**a**) 3D CELs of the milk sample in the 1st, 24th, 48th, and 72nd hours after opening; (**b**) PCA score plot derived from the data obtained using these milk samples. Reprinted from [29].

4. Conclusions

In summary, a novel eT was developed by combining an innovative combinatorial approach, which simplifies largely the preparation of sensing materials, with the optical detection technique of SPR imaging. Such an eT approach presents certain advantages over existing eT methods. Firstly, based on the combinatorial approach, a great diversity of sensing receptors can be obtained without increasing the cost of the synthesis of new molecules. Secondly, the continuous evolution of patterns (such as 2D profiles and 3D images) are advantageous compared to uncorrelated discrete data sets obtained using traditional eT systems. Abnormal signals obtained with a defective sensing receptor can be easily identified and excluded if necessary, providing a unique "auto-corrective" behavior to our eT system. Third, thanks to SPR imaging, the eT is able to provide a temporal response with vivid 3D images as "fingerprints" for the samples, giving valuable information on adsorption and desorption kinetics. We have demonstrated that the eT is not only effective for pure protein analysis, especially with the ability to differentiate protein isoforms with good sensitivity and selectivity, but can also be generalized for the analysis of complex mixtures with promising potentials for quality control applications. In the future, new building blocks with complementary physico-chemical properties will be designed and introduced to greatly improve the performances of the device.

Acknowledgments: The authors thank Labex LANEF program (ANR-10-LABX-51-01) and Labex Arcane program (ANR-12-LABX-003), both funded by the French National Research Agency for their support, and the Labex LERMIT ANR (ANR-10-LABX-33, under the program "Investissements d'Avenir" ANR-11-IDEX-0003-01). The authors also thank Programme Transversal Nanosciences CEA.

Author Contributions: Y.X.H., T.L. and A.B. conceived and designed the experiments; L.A.G., M.G. and Y.J.H. performed the experiments; A.B. analyzed the data; C.L.N., D.B., H.L.J., R.C. and M.B. contributed reagents/materials/analysis tools; Y.X.H. wrote the paper.

Conflicts of Interest: The authors declare no conflict of interest.

References

1. Lvova, L.; D'Amico, A.; Pede, A.; Di Natale, C.; Paolesse, R. Metallic Sensors in Multisensor Analysis. In *Multisensor Systems for Chemical Analysis: Materials and Sensors*; Lvova, L., Kirsanov, D., Di Natale, C., Legin, A., Eds.; Pan Stanford Publishing Pte. Ltd.: Singapore, 2014; pp. 69–138.
2. Del Valle, M. Electronic tongues employing electrochemical sensors. *Electroanal* **2010**, *22*, 1539–1555. [CrossRef]
3. Ciosek, P.; Wroblewski, W. Potentiometric electronic tongues for foodstuff and biosample recognition—An overview. *Sensors* **2011**, *11*, 4688–4701. [CrossRef] [PubMed]
4. Escuder-Gilabert, L.; Peris, M. Review: Highlights in recent applications of electronic tongues in food analysis. *Anal. Chim. Acta* **2010**, *665*, 15–25. [CrossRef] [PubMed]
5. Winquist, F. Voltammetric electronic tongues—Basic principles and applications. *Microchim. Acta* **2008**, *163*, 3–10. [CrossRef]
6. Scampicchio, M.; Ballabio, D.; Arecchi, A.; Cosio, S.M.; Mannino, S. Amperometric electronic tongue for food analysis. *Microchim. Acta* **2008**, *163*, 11–21. [CrossRef]
7. Pioggia, G.; Di Francesco, F.; Marchetti, A.; Ferro, A.; Leardi, R.; Ahluwalia, A. A composite sensor array impedentiometric electronic tongue Part II. Discrimination of basic tastes. *Biosens. Bioelectron.* **2007**, *22*, 2624–2628. [CrossRef] [PubMed]
8. Hou, Y.X.; Jaffrezic-Renault, N.; Martelet, C.; Zhang, A.D.; Minic-Vidic, J.; Gorojankina, T.; Persuy, M.A.; Pajot-Augy, E.; Salesse, R.; Akimov, V.; et al. A novel detection strategy for odorant molecules based on controlled bioengineering of rat olfactory receptor 17. *Biosens. Bioelectron.* **2007**, *22*, 1550–1555. [CrossRef] [PubMed]
9. Hou, Y.X.; Jaffrezic-Renault, N.; Martelet, C.; Tlili, C.; Zhang, A.; Pernollet, J.C.; Briand, L.; Gomila, G.; Errachid, A.; Samitier, J.; et al. Study of Langmuir and Langmuir-Blodgett films of odorant-binding protein/amphiphile for odorant biosensors. *Langmuir* **2005**, *21*, 4058–4065. [CrossRef] [PubMed]
10. Rodríguez Méndez, M.L.; Preedy, V.R. *Electronic Noses and Tongues in Food Science*; Academic Press: London, UK, 2016; pp. 1–309.
11. Baldwin, E.A.; Bai, J.H.; Plotto, A.; Dea, S. Electronic noses and tongues: Applications for the food and pharmaceutical industries. *Sensors* **2011**, *11*, 4744–4766. [CrossRef] [PubMed]
12. Smyth, H.; Cozzoino, D. Instrumental methods (spectroscopy, electronic nose, and tongue) as tools to predict taste and aroma in beverages: advantages and limitations. *Chem. Rev.* **2013**, *113*, 1429–1440. [CrossRef] [PubMed]
13. Wadehra, A.; Patil, P.S. Application of electronic tongues in food processing. *Anal. Methods UK* **2016**, *8*, 474–480. [CrossRef]
14. Tuantranont, A.; Wisitsora-at, A.; Sritongkham, P.; Jaruwongrungsee, K. A review of monolithic multichannel quartz crystal microbalance: A review. *Anal. Chim. Acta* **2011**, *687*, 114–128. [CrossRef] [PubMed]
15. Jin, X.X.; Huang, Y.; Mason, A.; Zeng, X.Q. Multichannel monolithic quartz crystal microbalance gas sensor array. *Anal. Chem.* **2009**, *81*, 595–603. [CrossRef] [PubMed]
16. Leonte, I.I.; Sehra, G.; Cole, M.; Hesketh, P.; Gardner, J.W. Taste sensors utilizing high-frequency SH-SAW devices. *Sens. Actuator B Chem.* **2006**, *118*, 349–355. [CrossRef]
17. Sehra, G.; Cole, M.; Gardner, J.W. Miniature taste sensing system based on dual SH-SAW sensor device: An electronic tongue. *Sens. Actuator B Chem.* **2004**, *103*, 233–239. [CrossRef]
18. Cole, M.; Sehra, G.; Gardner, J.W.; Varadan, V.K. Development of smart tongue devices for measurement of liquid properties. *IEEE Sens. J.* **2004**, *4*, 543–550. [CrossRef]
19. Zhang, C.; Suslick, K.S. Colorimetric sensor array for soft drink analysis. *J. Agric. Food Chem.* **2007**, *55*, 237–242. [CrossRef] [PubMed]
20. Zhang, C.; Bailey, D.P.; Suslick, K.S. Colorimetric sensor arrays for the analysis of beers: A feasibility study. *J. Agric. Food Chem.* **2006**, *54*, 4925–4931. [CrossRef] [PubMed]
21. Suslick, B.A.; Feng, L.; Suslick, K.S. Discrimination of Complex Mixtures by a Colorimetric Sensor Array: Coffee Aromas. *Anal. Chem.* **2010**, *82*, 2067–2073. [CrossRef] [PubMed]

22. Wright, A.T.; Griffin, M.J.; Zhong, Z.; McCleskey, S.C.; Anslyn, E.V.; McDevitt, J.T. Differential receptors create patterns that distinguish various proteins. *Angew. Chem. Int. Ed.* **2005**, *44*, 6375–6378. [CrossRef] [PubMed]

23. Baldini, L.; Wilson, A.J.; Hong, J.; Hamilton, A.D. Pattern-based detection of different proteins using an array of fluorescent protein surface receptors. *J. Am. Chem. Soc.* **2004**, *126*, 5656–5657. [CrossRef] [PubMed]

24. Zhou, H.C.; Baldini, L.; Hong, J.; Wilson, A.J.; Hamilton, A.D. Pattern recognition of proteins based on an array of functionalized porphyrins. *J. Am. Chem. Soc.* **2006**, *128*, 2421–2425. [CrossRef] [PubMed]

25. De, M.; Rana, S.; Akpinar, H.; Miranda, O.R.; Arvizo, R.R.; Bunz, U.H.F.; Rotello, V.M. Sensing of proteins in human serum using conjugates of nanoparticles and green fluorescent protein. *Nat. Chem.* **2009**, *1*, 461–465. [CrossRef] [PubMed]

26. Hou, Y.X.; Genua, M.; Batista, D.T.; Calemczuk, R.; Buhot, A.; Fornarelli, P.; Koubachi, J.; Bonnaffe, D.; Saesen, E.; Laguri, C.; et al. Continuous Evolution Profiles for Electronic-Tongue-Based Analysis. *Angew. Chem. Int. Edit* **2012**, *51*, 10394–10398. [CrossRef] [PubMed]

27. Scarano, S.; Mascini, M.; Turner, A.P.F.; Minunni, M. Surface plasmon resonance imaging for affinity-based biosensors. *Biosens. Bioelectron.* **2010**, *25*, 957–966. [CrossRef] [PubMed]

28. Campbell, C.T.; Kim, G. SPR microscopy and its applications to high-throughput analyses of biomolecular binding events and their kinetics. *Biomaterials* **2007**, *28*, 2380–2392. [CrossRef] [PubMed]

29. Genua, M.; Garcon, L.A.; Mounier, V.; Wehry, H.; Buhot, A.; Billon, M.; Calemczuk, R.; Bonnaffe, D.; Hou, Y.X.; Livache, T. SPR imaging based electronic tongue via landscape images for complex mixture analysis. *Talanta* **2014**, *130*, 49–54. [CrossRef] [PubMed]

30. Hou, Y.X.; Genua, M.; Garcon, L.A.; Buhot, A.; Calemczuk, R.; Bonnaffe, D.; Lortat-Jacob, H.; Livache, T. Electronic Tongue Generating Continuous Recognition Patterns for Protein Analysis. *JoVE J. Vis. Exp.* **2014**, *91*, 51901. [CrossRef] [PubMed]

31. Garçon, L.A.; Hou, Y.J.; Genua, M.; Buhot, A.; Calemczuk, R.; Bonnaffé, D.; Hou, Y.; Livache, T. Landscapes of taste by a novel electronic tongue for the analysis of complex mixtures. *Sens. Lett.* **2014**, *12*, 1059–1064. [CrossRef]

![sensors logo] *sensors*

MDPI

Review

Synergetic Effects of Combined Nanomaterials for Biosensing Applications

Michael Holzinger *, Alan Le Goff and Serge Cosnier

Department of Molecular Chemistry (DCM) UMR 5250, University Grenoble Alpes—CNRS,
F-38000 Grenoble, France; alan.le-goff@univ-grenoble-alpes.fr (A.L.G.);
Serge.Cosnier@univ-grenoble-alpes.fr (S.C.)
* Correspondence: michael.holzinger@univ-grenoble-alpes; Tel.: +33-456-520-811

Academic Editors: Nicole Jaffrezic-Renault and Gaelle Lissorgues
Received: 31 March 2017; Accepted: 27 April 2017; Published: 3 May 2017

Abstract: Nanomaterials have become essential components for the development of biosensors since such nanosized compounds were shown to clearly increase the analytical performance. The improvements are mainly related to an increased surface area, thus providing an enhanced accessibility for the analyte, the compound to be detected, to the receptor unit, the sensing element. Nanomaterials can also add value to biosensor devices due to their intrinsic physical or chemical properties and can even act as transducers for the signal capture. Among the vast amount of examples where nanomaterials demonstrate their superiority to bulk materials, the combination of different nano-objects with different characteristics can create phenomena which contribute to new or improved signal capture setups. These phenomena and their utility in biosensor devices are summarized in a non-exhaustive way where the principles behind these synergetic effects are emphasized.

Keywords: nanomaterials; biosensors; hybrids; carbon; metals; semiconductors; energy transfer

1. Introduction

The particularity of biosensors, compared to classic sensors, is that the sensing element, also called the receptor unit, is a biological entity or a bioinspired compound which confers an excellent selectivity towards the analyte to be detected. The unique specificity of such bioreceptors represents the main advantage within all sensor devices and the development of biosensors has become a huge research topic since highly complex solutions like blood can be analyzed for one specific target [1–3]. Biosensors are mainly used for the monitoring of diseases and are based on the recognition event of immune systems, viruses, bacteria, or cells, but also find utility for the detection of chemicals like blood sugar or pollutants [4,5]. One challenge is the signal capture during the biological recognition event [6,7].

Voltammetric biosensors rely on a redox process where the involved electron transfers are proportional to the analyte concentration [8]. For instance, the enzyme glucose oxidase (GOx) recognizes very specifically β-D-glucose, which is oxidized to gluconolactone. The reduced enzyme generally regenerates itself by reducing oxygen to hydrogen peroxide [9], a electroactive molecule which can be detected by the electrode. For immunosensing and the detection of DNA, more sophisticated setups are needed since an immune reaction or a hybridization of DNAs does not produce an electrochemical signal. For these cases, labeled secondary antibodies or DNA strands have to be involved after the recognition event where these labels will give the electrochemical signal. To avoid such supplemental time consuming preparation steps, electrochemical impedance spectroscopy (EIS) represents a very appropriate tool for immune and DNA sensors. EIS works with alternating currents (ACs) of small amplitude within a wide range of frequencies. The biorecognition event changes the sensing capacitance and interfacial electron transfer resistance of the electrode leading to a highly sensitive signal capture down to the femtomolar range [10,11].

Gravimetric biosensors are mostly piezoelectric devices where the detection of biological targets provokes a change of the resonance frequency related to the mass of the analyte [12,13]. One famous example is quartz crystal microbalance (QCM) but also micro- (or nano-) mechanical cantilever setups [14] are promising candidates for highly sensitive label free transduction techniques.

Most optical biosensors are based on a change in fluorescence or color during or after the recognition event [15]. As for electrochemical biosensors, some techniques need the use of supplemental labelling steps to introduce a photosensitive probe. Label-free optical detection can be achieved using surface plasmon resonance (SPR), which is a highly sensitive and quantitative transduction technique. The principle is based on the change of light-induced electron oscillations (surface plasmons) in the conduction band of metallic coatings (usually gold) when the dielectric constant of its environment changes [16]. This is the case, among others, for immune reactions or DNA hybridization where the recognition event changes the oscillation frequency which results in an angle change of the reflected light, its change of intensity, refractive index, or its phase [17,18].

The use of nanomaterials clearly already enhances the signal capture of all these transduction techniques used for biosensing thanks to their enhanced specific surface which allows the immobilization of an enhanced amount of bioreceptor units with an improved accessibility for the analytes. The advantages of different nanomaterials for biosensors are summarized in many review articles [19–27]. Here, we want to present some selected examples of synergetic effects achievable by combining different nanomaterials, thus enabling new or original transduction of biorecognition events.

2. Nanoparticles

Nanoparticles have become important components in biosensing devices since almost every material can be shaped into nanosized structures, thus conferring specific properties to the sensing element [28]. For instance, noble metal particles like silver and gold are famous for their localized resonant surface plasmons tremendously enhancing SPR or Raman signals [29–31]. These and other materials like platinum nanoparticles [32], or metal oxide nanoparticles [33] also provide improvements in catalysis and conductivity in electrochemical biosensors, while original setups were developed using magnetic nanoparticles [34]. Since many of these materials shows synergetic effects with other nano-objects, several examples will be described in more detail in the following sections. The most exploited synergetic effect between nanostructured materials is based on non-radiative energy transfer using upconverting nanoparticles (UCNPs) and quantum dots (QDs).

2.1. Upconverting Nanoparticles

UCNPs have the capacity to absorb several photons in the infrared range and to convert this absorbed energy into an emission in the visible range via a nonlinear optical process [35]. Contrary to common multiphoton absorption materials, these nanoparticles do not need high excitation densities for efficient anti-Stokes type emission. The phenomenon of high wavelength absorption and low wavelength emission strongly depends on the ion-ion distance of a dopant (mostly lanthanides) in a host material (generally Na^+ or Ca^{2+} fluorides). The confinement of the lanthanides in the matrix also determines the color of emitted light [36,37]. UCNPs became promising alternatives to other fluorescent labels for biosensing applications since they have very low background emissions and the high excitation wavelength does not provoke luminescence or absorption effects with other components of the biosensor [38,39]. UCNPs can be used as simple labels but more interesting are transduction principles based on non-radiative resonance energy transfer (RET) from the exited UCNPs to an acceptor where it should be emphasized that lanthanides are luminescent and not fluorescent and the RET is called luminescent resonant energy transfer (LRET) contrary to fluorescent (or Förster) resonant energy transfer (FRET) using fluorescent dyes [40]. The acceptor also plays a crucial role for this type of transduction because RET can only be achieved at corresponding quantum yields and cross sections, donor-acceptor distances, and their spectral emission and absorption overlap [41]. Furthermore, when the acceptor is a fluorescent probe, it can be excited by the UCNP leading to a change of the emission spectrum or simply to a change of the color of emitted light, or, when the acceptor is not a fluorophore, this results in the quenching of the emission of UCNPs [42] as illustrated in Figure 1.

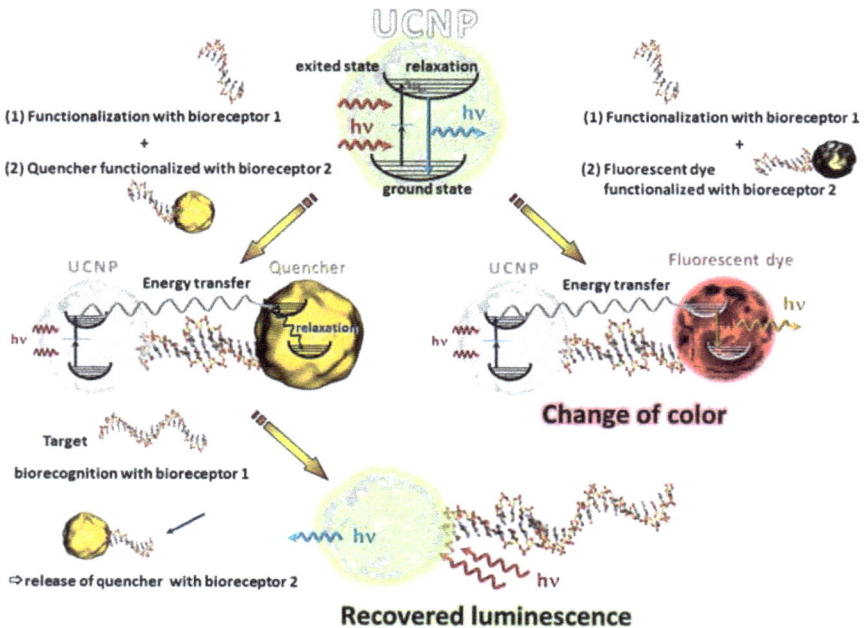

Figure 1. Schematic presentation of an UCNP and its anti-stokes type emission (top) and their functioning as bioanalytical transducer using a nanosized quencher (left) or a fluorescent dye (right).

Many approaches have been proposed relying on a quenching effect. For instance, Wang et al. demonstrated LRET between biotin-functionalized UCNPs and biotin-functionalized gold nanoparticles in the presence of avidin, which served here as the analyte which brings the two nano-objects in close contact leading to a linear reduction of the intensity of emitted light as a function of the avidin concentration [42]. A more sophisticated strategy was applied for the detection of thrombin in human plasma [43]. The specific thrombin aptamer was attached to the UCNPs while its luminescence is quenched in presence of carbon nanoparticles which form weak interactions with the aptamer. When the analyte is added, the carbon nanoparticles are released due to the stronger interaction between the aptamer and thrombin leading to a linear luminescence increase. A similar transduction principle was chosen by Zhang et al. who modified UCNPs with concanavalin A which interacts with saccharides. As quencher, chitosan-labeled graphene oxide was chosen and, due to the concanavalin A-chitosan interaction, the two components are assembled in close contact leading to the extinction of light. Then, glucose was used as analyte which forms stronger interaction with concanavalin A than chitosan leading to a glucose concentration dependent increase of emitted light [44].

The possibility to transfer upconverted energy to fluorophores thus changing the emission band after a biorecognition event has been extensively studied by Mattsson et al. [45]. For the proof of concept the biotin-streptavidin binding event was also used here as model recognition system. Streptavidin-modified UCNPs were attached to biotin-modified quantum dots leading to a change of the emission wavelength (the one for the quantum dots). In the presence of free biotin, mixed emissions or only the UCNPs' luminescence could be observed. The possibility to calibrate the intensities of different wavelengths might represent a promising platform for multiplex biosensing.

However, one drawback of UCNPs for biosensing applications has to be noted. All mentioned examples need an excitation wavelength of 980 nm which is right in the absorption band of water and heats the sample. This inconvenience can be overcome by doping UCNPs with neodymium ions lowering the excitation wavelength to 808 nm. Sample heating can thus be avoided which is of particular importance for in vivo bioimaging [46] and this approach also shows advantages in the monitoring of enzymatic reactions [47].

2.2. Quantum Dots

QDs have become almost the nanomaterial of choice for fluorescence-based transduction in bioanalytics. QDs are luminescent semiconducting nanocrystals principally based on cadmium chalcogenides [48–50]. Most of them are available as core-shell particles coated with ZnS or CdS for enhanced quantum yields and photostability [51,52]. They can absorb in a large wavelength range but have a narrow emission spectrum which is dependent on the particle size [53] (Figure 2).

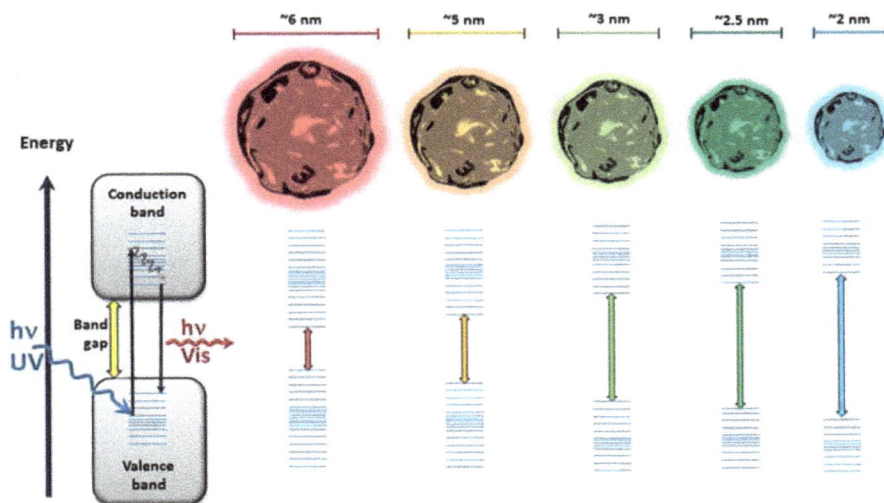

Figure 2. Illustration of QDs with different sizes and the related band gaps leading to different emission wavelengths after excitation with UV light.

The availability of QDs with different emission wavelengths has made them promising candidates for multiplexed analysis [54–56], as depicted in Figure 3. Furthermore, a final coating of QDs allows efficient functionalization with bioreceptor units and can overcome possible toxicity issues [57].

QDs are, as UCNPs, excellent optical transducers in combination with other nanomaterials. The principle is mostly based on the release of a quencher after the recognition event and the recovery of fluorescence. This strategy is particularly efficient for aptamer- and DNA sensors [58,59]. As a representative example, the assembly of a QD-labeled receptor DNA with a shorter corresponding DNA tagged with a gold nanoparticle is cut by the (longer) analyte DNA due to its higher hybridization kinetics. The gold nanoparticle is released and the QDs start to emit light again where the intensity is proportional to the analyte concentration [60,61] (Figure 3).

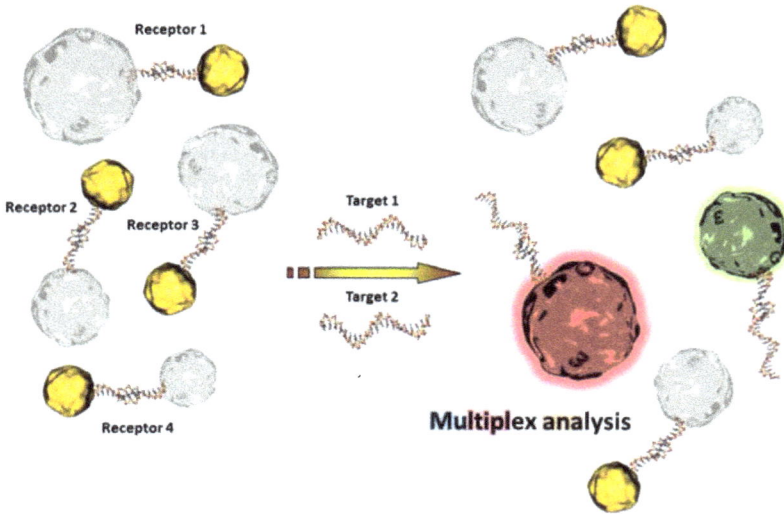

Figure 3. Scheme of a multiplex sensing principle using QDs and quenchers.

Gold nanoparticles are not only used as non-radiative quenchers, but can also act as antennas for increased fluorescence of QDs due to their high plasmonic behavior. When gold nanoparticles are localized at around 30 nm to the QD surface, the gold nanoparticle provokes an increase of the excitation rates of the QDs and hence the intensity of the fluorescence [62]. Further non-radiative energy transfer leading to QD fluorescence can be achieved using emitting protein labels which eliminate the need of external excitation light source [63]. There are also charge transfer quenching and chemiluminescence resonance energy transfer phenomena [64] to complete the most common applied principles of FRET-based biosensing using QDs [65–67].

QDs were also combined with magnetic nanoparticles for improved biodetection [68]. The magnetic nanoparticles are used for the separation of biological analytes in complex media like blood or any type of body fluids. In detail, receptor unit-modified magnetic nanoparticles are introduced in the analyte solution and interact specifically with the target molecule. The particles then migrate in a magnetic field until settling to form a deposit. The remaining solution can then be removed and the deposit can even be rinsed to eliminate any trapped species. QDs functionalized with a secondary receptor unit interact with the analyte on the magnetic particles and can quantify the detection via the intensity of the QD emission. A more sophisticated setup was proposed by Kurt et al. [69]. QDs and UCNPs, functionalized with different aptamers for different targets, served as recognition and transduction element in combination with magnetic nanoparticles modified with corresponding short DNA strands. The principle is based on the affinity interaction between aptamers and DNA linking weakly the magnetic nanoparticles with QDs and UCNPs which can then be separated from the solution in a magnetic field. In presence of the analytes (here the pathogens *Salmonella typhimurium* and *Staphylococcus aureus*), the DNA-aptamer link is broken by the competitive interaction with the target and the luminescent particles remain in solution after applying a magnetic field. The *Salmonella typhimurium*-UCNP and *Staphylococcus aureus*-QD assemblies can then be removed by washing thus leading to reduced intensities of emitted light. The authors observed a linear decrease of luminescence intensity with the analyte concentration. For the multiplex sensing setup, the remaining *Staphylococcus aureus* specific QDs and *Salmonella typhimurium* specific UCNPs are exited at 325 nm and 980 nm. UCNP excitation at 980 nm cannot excite the QDs since the energy of photons is below the band gap of the QDs. The authors observed negligible excitation of the QDs by the emission of UCNPs at 470 nm and could also exclude FRET. This setup might be a promising strategy for facilitated multiplex analysis but will need materials with narrower excitation and emission lines to prevent overlap or crosstalk effects.

Besides FRET with other nanomaterials, QDs can also interact with propagating surface plasmons of gold surfaces leading to light emission of the QDs or the light induced excited state of QDs are transferred to the surface plasmons [70] as illustrated in Figure 4. The second effect led to clear signal enhancements in SPR setups where a 25-fold increase was observed for ss-DNA and a 50-fold increase could be obtained with prostate-specific antigens compared to bare gold surfaces. [71].

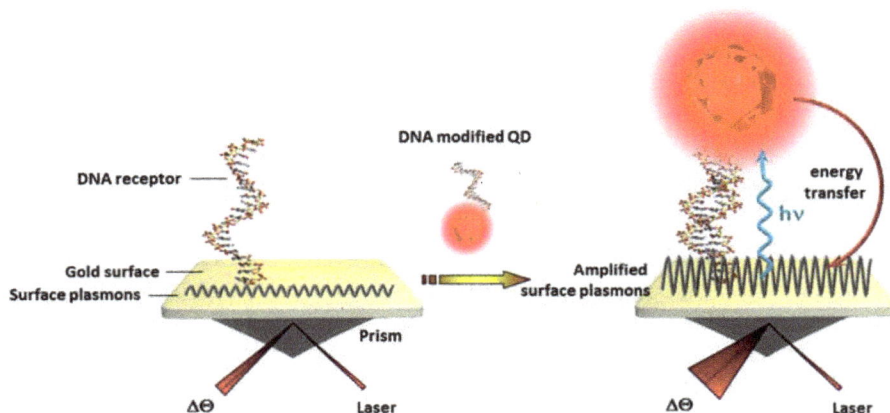

Figure 4. Schematic illustration of SPR signal amplification after the biorecognition event using QD labeled biomarkers.

QDs also show remarkable properties in electrochemical biosensing devices in combination with CNTs [72]. The distribution of these semiconductors within a CNT composite matrix forms domains with altered conductivities behaving like a microelectrode array. This phenomenon results in a clear reduction of the double layer capacitance and thus to an improved noise signal ratio. This setup was validated for the detection of hydrogen peroxide and ascorbic acid.

3. Carbon Nanomaterials

Carbon is a privileged material for biosensing applications, especially for electrochemical transduction due to its excellent conductivity and biocompatibility [73]. Carbon appears in many different allotropes based on graphite (sp^2), diamond (sp^3) and intermittent sp^2-sp^3 hybridized macroscopic structures generally called amorphous carbon, from which vast amounts of substructures can be synthesized [74]. For electrochemical biosensing, glassy carbon, doped diamond, and graphite are standard materials for electrodes [75]. Their nanostructured part in the form of carbon nanotubes [76], fullerenes [77], or graphene [78] partly became the material of choice for improved performances of bioanalytical devices [79]. More recently, fluorescent carbon nanodots have attracted attention as non-toxic alternatives to quantum dots for optical biosensing and bioimaging [80]. Efficient functionalization techniques were established for carbon nanomaterials which allow the formation of bioassemblies and to combine the beneficial properties with those of other nanosized materials [81]. This also allows reproducible processing and shaping to obtain the desired properties. An elegant way to assemble different materials is the formation of composites. As an example for electrochemical transduction, carbon paste electrodes provide unlimited possibilities to combine any type of carbon material with (nanosized) fillers conferring improved performances to the biosensor device. Selected examples and procedures of customized carbon paste-based biosensors were summarized by Muñoz et al. [82]. The following section presents some further examples of successful combinations of nanostructured carbon allotropes with other materials with synergetic effects for enhanced biosensor performance.

3.1. Graphene

Graphene has become a fashionable material for biosensing because it is considered less toxic than CNTs [83,84]. Even though graphene is per definition not a nanomaterial [85], it is worth summarizing some examples of its synergetic effects with other nano-objects since it belongs to the rich carbon allotrope family.

For electrochemical transduction, graphite-based layered materials are used in bulk form but are also often called graphene or graphene-like 2D materials [86]. Obtained after mechanic exfoliation [87,88], chemical oxidation of graphite [89] and/or subsequent reduction [90], these carbon materials are represented in many biosensor application examples [91] such as electrochemical immunosensors [92] or enzymatic biosensors [93]. In terms of synergetic hybrid materials, and similar to CNT hybrids, many different metal nanoparticles like gold [94], platinum [95–99], or palladium [100], or metal oxide nanoparticles [101] clearly improved the sensing performance when combined with graphene and graphene-like 2D materials.

For optical transduction, graphene materials can also act as non-radiant energy acceptors in FRET-based biosensors using organic dye- [102] or quantum dot [103] -labeled bioreceptors like DNAs, aptamers, or proteins [104]. Graphene oxide itself shows photoluminescence and can act as both, energy donor and energy acceptor [105] with excellent quenching efficiency [106].

In particular, graphene can interact with DNA or oligonucleotide receptors in a non-covalent and reversible manner, contrary to CNTs, and these dye-labeled receptors desorb after the recognition, recovering the fluorescence of the labels. This principle could even be applied for a multiplexed colorimetric DNA sensor [107]. Weak interactions with graphene oxide can also be obtained with antibodies labeled with QDs, which were exploited for the detection of the model pathogen *E. coli* [108]. The fluorescence of a corresponding antibody-modified QD is quenched in the presence of graphene oxide and recovered after the recognition event with the bacteria. This setup was successfully applied on nanocellulose-based papers for fluorescence biosensing using QDs and UCNPs [109]. Furthermore, such papers also provide an excellent platform for colorimetric sensing of biorelevant chemicals when functionalized with silver or gold nanoparticles.

Furthermore, real monolayer graphene provides impressive beneficial properties in resonant plasmon transduction techniques [110]. Firstly predicted by theoretical models, these plasmonic properties of a single layer of graphene can interact with the surface plasmons of gold surfaces thus significantly amplifying the optical sensitivity of surface plasmon resonance (SPR) sensors [111]. By excitation in the visible light range [112] the propagation constant of surface plasmon polaritons (SPPs) is changed and the refractive index response in particular is amplified [113]. An almost two- fold increase of the SPR signal could be obtained just in presence of a graphene monolayer on gold which was validated in a highly sensitive anti-cholera toxin SPR sensor [114]. This phenomenon can in theory be further optimized using intermittent MoS_2 layers [115]. Due to the improved optical absorption efficiency, the graphene-MoS_2 layer can transfer this energy to the underlying gold layer thus further exciting and amplifying the resonant surface plasmons (Figure 5).

The authors calculated an up to 500-fold increase of phase sensitivity of the SPR signal with theoretic models when a biomolecule is adsorbed on the graphene layer via π-π stacking interactions. The authors unfortunately did not precise which biomolecule these calculations were based on.

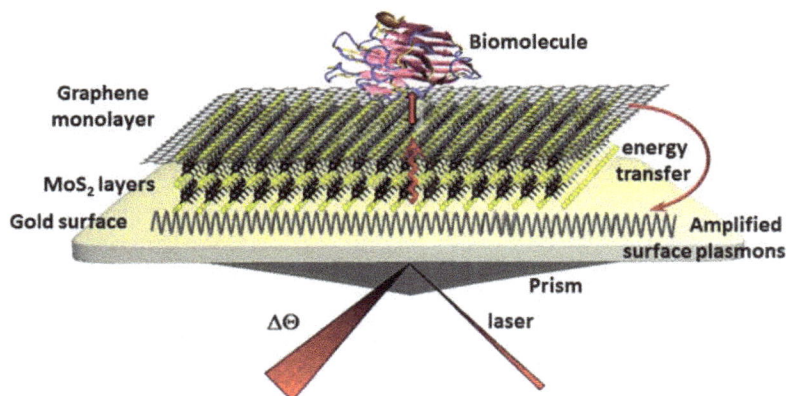

Figure 5. Principle of improved SPR signals after adsorption of a biomolecules using MoS$_2$ as intermittent layer between monolayer graphene and the gold surface.

3.2. Carbon Nanotubes

Carbon nanotubes (CNTs) can be seen as seamlessly rolled up graphene with one to up to hundreds of concentric wall layers and provide excellent 1D conductivity and high aspect ratios which form entangled porous structures in bulk, drastically increasing the accessible surface area of electrodes [116–118]. Furthermore, efficient and reliable functionalization methods were developed for the immobilization of bioreceptor units on CNTs without altering the biological activity [119].

CNTs were confined with Pt nanoparticles in a Nafion matrix for improved DNA sensing using daunomycin, a redox active compound which intercalates hybridized DNAs [120]. Single stranded receptor DNA was immobilized on this composite and was exposed to different concentrations of the analyte, the corresponding ssDNA. Since daunomycin only intercalates after the recognition event, the differential pulse voltammetric signal increased for the electrocatalytic reduction of the electrochemical probe. The combination of the enhanced specific surface area of CNTs and the catalytic properties of Pt led to clearly improved performances compared to setups using the individual compounds.

CNT-gold nanoparticle (AuNPs) assemblies clearly improve electrochemical transduction due to enhanced electron transfer rates between an enzymatically generated substrate and the AuNPs-CNT composites. A highly sensitive choline biosensor was developed based on choline oxidase modified AuNPs-CNT electrodes [121]. After the enzyme-catalyzed oxidation of choline, hydrogen peroxide is released which is finally oxidized on the electrocatalytic nanocomposite electrode. Beside the beneficial effect of AuNP-CNT assemblies for electrochemical biosensors [122,123], other metal or metal oxide nanoparticles showed improved biosensing performances when combined with carbon nanotubes. Most examples describe the improved electrocatalytic oxidation of enzymatically generated H$_2$O$_2$ using cobalt hexacyanoferrate nanoparticles [124], Pt nanoparticles [125], or ZnO nanoparticles [126] while for this example CNT-graphene hybrids were used. There are many further examples of using different nanomaterials in combination with CNTs which are summarized in reference [127].

3.3. C$_{60}$ Fullerenes and Carbon Dots

C$_{60}$ is the first fully characterized carbon nano-object and is classified as a 0D material. Its molecular structure is composed of 12 five-membered rings surrounded by a total of 20 six-membered rings and it obeys perfectly Euler's rule [128]. The particular electrochemical properties of C$_{60}$ [129] evoked much attention for its possible application as a redox mediator in enzymatic biosensors [130]. C$_{60}$ also showed remarkably enhancements of the specific surface of electrodes and was used as a building block for original nanoscaffolds [131]. An electrochemical aptasensor was reported where an electrode surface is modified with onion-like mesoporous graphene sheets, gold nanoparticles, and a first aptamer receptor. Prussian Blue-modified gold nanoparticles were adsorbed on amine-functionalized C_{60} together with a second aptamer receptor and alkaline phosphatase as label. After the recognition event of the model target platelet-derived growth factor B-chain, and the formation of the sandwich structure, the immobilized enzyme label hydrolyzes ascorbic acid phosphate to ascorbic acid which is then oxidized on the Prussian Blue/gold nanoparticles/C_{60} electrode. Further examples describe the combination of C_{60} with mostly gold or platinum nanoparticles for improved electrochemical [132] or gravimetric [133] immunosensors, or an electrochemiluminescent aptasensor [134] but all of them rely on the electrochemical behavior of C_{60} or the capability to enhance the surface area and not on synergetic phenomena between these nano-objects.

Carbon dots or carbon quantum dots, sometimes also called graphene quantum dots, can be considered further 0D carbon nanomaterials. Accidently discovered as a side product during arc discharge synthesis of single walled CNTs [135], these carbon QDs are promising candidates to replace heavy metal semiconductor QDs since they are still considered as a non-toxic carbon material with very satisfying quantum yields where even upconverting nanoparticles could be isolated and studied [136,137]. The fluorescence phenomenon is related to isolated domains of conjugated sp^2 carbon surrounded by diamond like sp^3 carbon [138] as depicted in Figure 6.

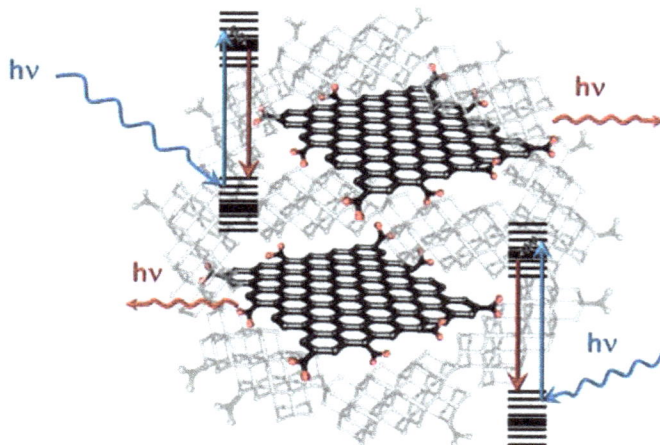

Figure 6. Sketch of a carbon QD with its defined sp^2 domains isolated and surrounded with diamond-like carbon which is highly oxidized on the surface of the particle.

The fluorescence is also influenced by the mostly carboxylated surface which confers carbon QDs excellent solubility, but also strong pH-dependent fluorescence emission [139]. Even when great progress was achieved in the synthesis and isolation of carbon QDs with specific properties, the controlled synthesis of defined domain distribution and surface functionalities leading to distinguished absorption and emission spectra, as it is the case for semiconductor QDs, remains a challenge [140,141]. In terms of biosensing applications, carbon QDs show similar performances as semiconductor QDs concerning FRET-based biosensing and as fluorescence labels [80]. Efficient FRET between gold nanoparticles and carbon QDs could be achieved when each nano-object is modified with a corresponding antibody-antigen system [142]. In the presence of the analyte, in this example an organic pollutant, these assemblies are broken, leading to the recovery of fluorescence. Based on the same principle, a DNA sensor was proposed using assemblies of carbon QDs and fluorescence dye quenching each other whereby fluorescence reappears after the recognition event [143]. It would be interesting to study the efficiency and performances of carbon QDs and semiconductor QDs under identical condition to gain insight into the real potential of carbon QDs. It might be assumed that such studies will be reported in the near future.

4. Conclusions

Synergetic effects of different nanoparticles became promising tools for highly sensitive biodetection applications where the FRET effect is at the moment the most promising example where QDs were shown to be particularly versatile when combined with nanosized acceptors. Carbon QDs or UCNPs are promising candidates with lower toxicity issues to one day replace semiconductor QDs. In regards of the steady growing availability of different nanomaterials with different properties revealing new phenomena when in contact, other original electronic, electrochemical, or magnetic transduction methods can be developed. There remains one famous example of a nano-object with synergetic properties to which a section was not dedicated in this review: gold nanoparticles. This is simply due to the fact that nanosized gold was mentioned with almost all discussed materials and to avoid repetition, a separate discussion about gold-hybrids was intentionally omitted. However, for more information, examples of the beneficial properties of gold nanoparticle hybrids for biosensing and diagnostics are summarized in reference [144].

Conflicts of Interest: The authors declare no conflict of interest.

References

1. De Corcuera, J.I.R.; Cavalieri, R.P. Biosensors. In *Encyclopedia of Agricultural, Food, and Biological Engineering*; Taylor & Francis: Abingdon, UK, 2007; pp. 119–123.
2. Tothill, I.E.; Turner, A.P.F. Biosensors. In *Encyclopedia of Food Sciences and Nutrition*, 2nd ed.; Caballero, B., Ed.; Academic Press: Oxford, UK, 2003; pp. 489–499.
3. Cosnier, S. *Electrochemical Biosensors*; Pan Stanford Publishing: Singapore, 2015.
4. Mehrotra, P. Biosensors and their applications: A review. *J. Oral Biol. Craniofacial Res.* **2016**, *6*, 153–159. [CrossRef] [PubMed]
5. Wang, X.; Lu, X.; Chen, J. Development of biosensor technologies for analysis of environmental contaminants. *Trends Environ. Anal. Chem.* **2014**, *2*, 25–32. [CrossRef]
6. Mohanty, S.P.; Kougianos, E. Biosensors: A tutorial review. *IEEE Potentials* **2006**, *25*, 35–40. [CrossRef]
7. Perumal, V.; Hashim, U. Advances in biosensors: Principle, architecture and applications. *J. Appl. Biomed.* **2014**, *12*, 1–15. [CrossRef]
8. Ronkainen, N.J.; Halsall, H.B.; Heineman, W.R. Electrochemical biosensors. *Chem. Soc. Rev.* **2010**, *39*, 1747–1763. [CrossRef] [PubMed]
9. Wang, H.-C.; Lee, A.-R. Recent developments in blood glucose sensors. *J. Food Drug Anal.* **2015**, *23*, 191–200. [CrossRef]
10. Baur, J.; Gondran, C.; Holzinger, M.; Defrancq, E.; Perrot, H.; Cosnier, S. Label-free femtomolar detection of target DNA by impedimetric DNA sensor based on poly(pyrrole-nitrilotriacetic acid) film. *Anal. Chem.* **2010**, *82*, 1066–1072. [CrossRef] [PubMed]
11. Giroud, F.; Gorgy, K.; Gondran, C.; Cosnier, S.; Pinacho, D.G.; Marco, M.-P.; Sánchez-Baeza, F.J. Impedimetric Immunosensor Based on a Polypyrrole-Antibiotic Model Film for the Label-Free Picomolar Detection of Ciprofloxacin. *Anal. Chem.* **2009**, *81*, 8405–8409. [CrossRef] [PubMed]
12. Skládal, P. Piezoelectric biosensors. *TrAC Trends Anal. Chem.* **2016**, *79*, 127–133. [CrossRef]
13. Johannsmann, D. Gravimetric Sensing. In *The Quartz Crystal Microbalance in Soft Matter Research: Fundamentals and Modeling*; Springer: Cham, Switzerland, 2015; pp. 191–204.
14. Meisam, O.; Malakoutian, M.A.; Mohammadmehdi, C.; Oroojalian, F.; Haghiralsadat, F.; Yazdian, F. A Label-Free Detection of Biomolecules Using Micromechanical Biosensors. *Chin. Phys. Lett.* **2013**, *30*, 068701.
15. Damborský, P.; Švitel, J.; Katrlík, J. Optical biosensors. *Essays Biochem.* **2016**, *60*, 91. [CrossRef] [PubMed]
16. Kelly, K.L.; Coronado, E.; Zhao, L.L.; Schatz, G.C. The Optical Properties of Metal Nanoparticles: The Influence of Size, Shape, and Dielectric Environment. *J. Phys. Chem. B* **2002**, *107*, 668–677. [CrossRef]

17. Wijaya, E.; Lenaerts, C.; Maricot, S.; Hastanin, J.; Habraken, S.; Vilcot, J.-P.; Boukherroub, R.; Szunerits, S. Surface plasmon resonance-based biosensors: From the development of different SPR structures to novel surface functionalization strategies. *Curr. Opin. Solid State Mater. Sci.* **2011**, *15*, 208–224. [CrossRef]

18. Guo, X. Surface plasmon resonance based biosensor technique: A review. *J. Biophotonics* **2012**, *5*, 483–501. [CrossRef] [PubMed]

19. Hou, S.; Zhang, A.; Su, M. Nanomaterials for Biosensing Applications. *Nanomaterials* **2016**, *6*, 58. [CrossRef] [PubMed]

20. Holzinger, M.; Le Goff, A.; Cosnier, S. Nanomaterials for biosensing applications: A review. *Front. Chem.* **2014**, *2*, 63. [CrossRef] [PubMed]

21. Aragay, G.; Pino, F.; Merkoçi, A. Nanomaterials for Sensing and Destroying Pesticides. *Chem. Rev.* **2012**, *112*, 5317–5338. [CrossRef] [PubMed]

22. Ge, X.; Asiri, A.M.; Du, D.; Wen, W.; Wang, S.; Lin, Y. Nanomaterial-enhanced paper-based biosensors. *TrAC Trends Anal. Chem.* **2014**, *58*, 31–39. [CrossRef]

23. Song, Y.; Luo, Y.; Zhu, C.; Li, H.; Du, D.; Lin, Y. Recent advances in electrochemical biosensors based on graphene two-dimensional nanomaterials. *Biosens. Bioelectron.* **2016**, *76*, 195–212. [CrossRef] [PubMed]

24. Oliveira, S.F.; Bisker, G.; Bakh, N.A.; Gibbs, S.L.; Landry, M.P.; Strano, M.S. Protein functionalized carbon nanomaterials for biomedical applications. *Carbon* **2015**, *95*, 767–779. [CrossRef]

25. Chimene, D.; Alge, D.L.; Gaharwar, A.K. Two-Dimensional Nanomaterials for Biomedical Applications: Emerging Trends and Future Prospects. *Adv. Mater.* **2015**, *27*, 7261–7284. [CrossRef] [PubMed]

26. Ju, H.; Zhang, X.; Wang, J. Nanomaterials for Immunosensors and Immunoassays. In *NanoBiosensing*; Springer: New York, NY, USA, 2011; pp. 425–452.

27. Lei, J.; Ju, H. Signal amplification using functional nanomaterials for biosensing. *Chem. Soc. Rev.* **2012**, *41*, 2122–2134. [CrossRef] [PubMed]

28. El-Ansary, A.; Faddah, L.M. Nanoparticles as biochemical sensors. *Nanotechnol. Sci. Appl.* **2010**, *3*, 65–76. [CrossRef] [PubMed]

29. Doria, G.; Conde, J.; Veigas, B.; Giestas, L.; Almeida, C.; Assunção, M.; Rosa, J.; Baptista, P.V. Noble Metal Nanoparticles for Biosensing Applications. *Sensors* **2012**, *12*, 1657–1687. [CrossRef] [PubMed]

30. Liao, H.; Nehl, C.L.; Hafner, J.H. Biomedical applications of plasmon resonant metal nanoparticles. *Nanomedicine* **2006**, *1*, 201–208. [CrossRef] [PubMed]

31. Anker, J.N.; Hall, W.P.; Lyandres, O.; Shah, N.C.; Zhao, J.; Van Duyne, R.P. Biosensing with plasmonic nanosensors. *Nat. Mater.* **2008**, *7*, 442–453. [CrossRef] [PubMed]

32. Rioux, R.M.; Song, H.; Grass, M.; Habas, S.; Niesz, K.; Hoefelmeyer, J.D.; Yang, P.; Somorjai, G.A. Monodisperse platinum nanoparticles of well-defined shape: Synthesis, characterization, catalytic properties and future prospects. *Top. Catal.* **2006**, *39*, 167–174. [CrossRef]

33. Taurino, I.; Sanzò, G.; Antiochia, R.; Tortolini, C.; Mazzei, F.; Favero, G.; De Micheli, G.; Carrara, S. Recent advances in Third Generation Biosensors based on Au and Pt Nanostructured Electrodes. *TrAC Trends Anal. Chem.* **2016**, *79*, 151–159. [CrossRef]

34. Li, X.; Wei, J.; Aifantis, K.E.; Fan, Y.; Feng, Q.; Cui, F.-Z.; Watari, F. Current investigations into magnetic nanoparticles for biomedical applications. *J. Biomed. Mater. Res. Part A* **2016**, *104*, 1285–1296. [CrossRef] [PubMed]

35. Haase, M.; Schäfer, H. Upconverting Nanoparticles. *Angew. Chem. Int. Ed. Engl.* **2011**, *50*, 5808–5829. [CrossRef] [PubMed]

36. Heer, S.; Kömpe, K.; Güdel, H.U.; Haase, M. Highly Efficient Multicolour Upconversion Emission in Transparent Colloids of Lanthanide-Doped NaYF4 Nanocrystals. *Adv. Mater.* **2004**, *16*, 2102–2105. [CrossRef]

37. Wang, F.; Deng, R.; Wang, J.; Wang, Q.; Han, Y.; Zhu, H.; Chen, X.; Liu, X. Tuning upconversion through energy migration in core–shell nanoparticles. *Nat. Mater.* **2011**, *10*, 968–973. [CrossRef] [PubMed]

38. Zhang, F. Upconversion Nanoparticles for Biosensing. In *Photon Upconversion Nanomaterials*; Zhang, F., Ed.; Springer: Berlin/Heidelberg, Germany, 2015; pp. 255–284.

39. Achatz, D.E.; Ali, R.; Wolfbeis, O.S. Luminescent Chemical Sensing, Biosensing, and Screening Using Upconverting Nanoparticles. In *Luminescence Applied in Sensor Science*; Prodi, L., Montalti, M., Zaccheroni, N., Eds.; Springer: Berlin/Heidelberg, Germany, 2011; pp. 29–50.

40. Su, Q.; Feng, W.; Yang, D.; Li, F. Resonance Energy Transfer in Upconversion Nanoplatforms for Selective Biodetection. *Acc. Chem. Res.* **2017**, *50*, 32–40. [CrossRef] [PubMed]

41. Lakowicz, J.R. Principles of Fluorescence Spectroscopy, Third Edition. *J. Biomed. Opt.* **2008**, *13*, 029901. [CrossRef]

42. Wang, L.; Yan, R.; Huo, Z.; Wang, L.; Zeng, J.; Bao, J.; Wang, X.; Peng, Q.; Li, Y. Fluorescence Resonant Energy Transfer Biosensor Based on Upconversion-Luminescent Nanoparticles. *Angew. Chem. Int. Ed. Engl.* **2005**, *44*, 6054–6057. [CrossRef] [PubMed]

43. Wang, Y.; Bao, L.; Liu, Z.; Pang, D.-W. Aptamer Biosensor Based on Fluorescence Resonance Energy Transfer from Upconverting Phosphors to Carbon Nanoparticles for Thrombin Detection in Human Plasma. *Anal. Chem.* **2011**, *83*, 8130–8137. [CrossRef] [PubMed]

44. Zhang, C.; Yuan, Y.; Zhang, S.; Wang, Y.; Liu, Z. Biosensing Platform Based on Fluorescence Resonance Energy Transfer from Upconverting Nanocrystals to Graphene Oxide. *Angew. Chem. Int. Ed. Engl.* **2011**, *50*, 6851–6854. [CrossRef] [PubMed]

45. Mattsson, L.; Wegner, K.D.; Hildebrandt, N.; Soukka, T. Upconverting nanoparticle to quantum dot FRET for homogeneous double-nano biosensors. *RSC Adv.* **2015**, *5*, 13270–13277. [CrossRef]

46. Wang, Y.-F.; Liu, G.-Y.; Sun, L.-D.; Xiao, J.-W.; Zhou, J.-C.; Yan, C.-H. Nd^{3+}-Sensitized Upconversion Nanophosphors: Efficient In Vivo Bioimaging Probes with Minimized Heating Effect. *ACS Nano* **2013**, *7*, 7200–7206. [CrossRef] [PubMed]

47. Himmelstoß, S.F.; Wiesholler, L.M.; Buchner, M.; Muhr, V.; Märkl, S.; Baeumner, A.J.; Hirsch, T. 980 nm and 808 nm Excitable Upconversion Nanoparticles for the Detection of Enzyme Related Reactions. *Proc. SPIE* **2017**, *10077*. [CrossRef]

48. Murray, C.B.; Norris, D.J.; Bawendi, M.G. Synthesis and characterization of nearly monodisperse CdE (E = sulfur, selenium, tellurium) semiconductor nanocrystallites. *J. Am. Chem. Soc.* **1993**, *115*, 8706–8715. [CrossRef]

49. Park, J.; Joo, J.; Kwon, S.G.; Jang, Y.; Hyeon, T. Synthesis of Monodisperse Spherical Nanocrystals. *Angew. Chem. Int. Ed. Engl.* **2007**, *46*, 4630–4660. [CrossRef] [PubMed]

50. Reiss, P.; Protière, M.; Li, L. Core/Shell Semiconductor Nanocrystals. *Small* **2009**, *5*, 154–168. [CrossRef] [PubMed]

51. Dabbousi, B.O.; Rodriguez-Viejo, J.; Mikulec, F.V.; Heine, J.R.; Mattoussi, H.; Ober, R.; Jensen, K.F.; Bawendi, M.G. (CdSe)ZnS Core−Shell Quantum Dots: Synthesis and Characterization of a Size Series of Highly Luminescent Nanocrystallites. *J. Phys. Chem. B* **1997**, *101*, 9463–9475. [CrossRef]

52. Jaiswal, J.K.; Mattoussi, H.; Mauro, J.M.; Simon, S.M. Long-term multiple color imaging of live cells using quantum dot bioconjugates. *Nat. Biotech.* **2003**, *21*, 47–51. [CrossRef] [PubMed]

53. Weller, H. Colloidal Semiconductor Q-Particles: Chemistry in the Transition Region between Solid State and Molecules. *Angew. Chem. Int. Ed. Engl.* **1993**, *32*, 41–53. [CrossRef]

54. Geißler, D.; Charbonnière, L.J.; Ziessel, R.F.; Butlin, N.G.; Löhmannsröben, H.-G.; Hildebrandt, N. Quantum Dot Biosensors for Ultrasensitive Multiplexed Diagnostics. *Angew. Chem. Int. Ed. Engl.* **2010**, *49*, 1396–1401. [CrossRef] [PubMed]

55. Petryayeva, E.; Algar, W.R. Multiplexed Homogeneous Assays of Proteolytic Activity Using a Smartphone and Quantum Dots. *Anal. Chem.* **2014**, *86*, 3195–3202. [CrossRef] [PubMed]

56. Algar, W.R.; Khachatrian, A.; Melinger, J.S.; Huston, A.L.; Stewart, M.H.; Susumu, K.; Blanco-Canosa, J.B.; Oh, E.; Dawson, P.E.; Medintz, I.L. Concurrent Modulation of Quantum Dot Photoluminescence Using a Combination of Charge Transfer and Förster Resonance Energy Transfer: Competitive Quenching and Multiplexed Biosensing Modality. *J. Am. Chem. Soc.* **2017**, *139*, 363–372. [CrossRef] [PubMed]

57. Biju, V.; Itoh, T.; Ishikawa, M. Delivering quantum dots to cells: Bioconjugated quantum dots for targeted and nonspecific extracellular and intracellular imaging. *Chem. Soc. Rev.* **2010**, *39*, 3031–3056. [CrossRef] [PubMed]

58. Zhang, C.-Y.; Yeh, H.-C.; Kuroki, M.T.; Wang, T.-H. Single-quantum-dot-based DNA nanosensor. *Nat. Mater.* **2005**, *4*, 826–831. [CrossRef] [PubMed]

59. Freeman, R.; Girsh, J.; Willner, I. Nucleic Acid/Quantum Dots (QDs) Hybrid Systems for Optical and Photoelectrochemical Sensing. *ACS Appl. Mater. Interfaces* **2013**, *5*, 2815–2834. [CrossRef] [PubMed]

60. Dyadyusha, L.; Yin, H.; Jaiswal, S.; Brown, T.; Baumberg, J.J.; Booy, F.P.; Melvin, T. Quenching of CdSe quantum dot emission, a new approach for biosensing. *Chem. Commun.* **2005**, 3201–3203. [CrossRef] [PubMed]

61. Dai, Z.; Zhang, J.; Dong, Q.; Guo, N.; Xu, S.; Sun, B.; Bu, Y. Adaption of Au Nanoparticles and CdTe Quantum Dots in DNA Detection. *Chin. J. Chem. Eng.* **2007**, *15*, 791–794. [CrossRef]

62. Maye, M.M.; Gang, O.; Cotlet, M. Photoluminescence enhancement in CdSe/ZnS-DNA linked-Au nanoparticle heterodimers probed by single molecule spectroscopy. *Chem. Commun.* **2010**, *46*, 6111–6113. [CrossRef] [PubMed]

63. So, M.-K.; Xu, C.; Loening, A.M.; Gambhir, S.S.; Rao, J. Self-illuminating quantum dot conjugates for in vivo imaging. *Nat. Biotech.* **2006**, *24*, 339–343. [CrossRef] [PubMed]

64. Huang, X.; Li, L.; Qian, H.; Dong, C.; Ren, J. A Resonance Energy Transfer between Chemiluminescent Donors and Luminescent Quantum-Dots as Acceptors (CRET). *Angew. Chem. Int. Ed. Engl.* **2006**, *118*, 5264–5267. [CrossRef]

65. Algar, W.R.; Tavares, A.J.; Krull, U.J. Beyond labels: A review of the application of quantum dots as integrated components of assays, bioprobes, and biosensors utilizing optical transduction. *Anal. Chim. Acta* **2010**, *673*, 1–25. [CrossRef] [PubMed]

66. Frasco, M.; Chaniotakis, N. Semiconductor Quantum Dots in Chemical Sensors and Biosensors. *Sensors* **2009**, *9*, 7266–7286. [CrossRef] [PubMed]

67. Petryayeva, E.; Algar, W.R.; Medintz, I.L. Quantum Dots in Bioanalysis: A Review of Applications Across Various Platforms for Fluorescence Spectroscopy and Imaging. *Appl. Spectrosc.* **2013**, *67*, 215–252. [CrossRef] [PubMed]

68. Moro, L.; Turemis, M.; Marini, B.; Ippodrino, R.; Giardi, M.T. Better together: Strategies based on magnetic particles and quantum dots for improved biosensing. *Biotechnol. Adv.* **2017**, *35*, 51–63. [CrossRef] [PubMed]

69. Kurt, H.; Yüce, M.; Hussain, B.; Budak, H. Dual-excitation upconverting nanoparticle and quantum dot aptasensor for multiplexed food pathogen detection. *Biosens. Bioelectron.* **2016**, *81*, 280–286. [CrossRef] [PubMed]

70. Wei, H.; Ratchford, D.; Li, X.; Xu, H.; Shih, C.-K. Propagating Surface Plasmon Induced Photon Emission from Quantum Dots. *Nano Lett.* **2009**, *9*, 4168–4171. [CrossRef] [PubMed]

71. Malic, L.; Sandros, M.G.; Tabrizian, M. Designed Biointerface Using Near-Infrared Quantum Dots for Ultrasensitive Surface Plasmon Resonance Imaging Biosensors. *Anal. Chem.* **2011**, *83*, 5222–5229. [CrossRef] [PubMed]

72. Muñoz, J.; Bastos-Arrieta, J.; Muñoz, M.; Muraviev, D.; Céspedes, F.; Baeza, M. CdS quantum dots as a scattering nanomaterial of carbon nanotubes in polymeric nanocomposite sensors for microelectrode array behavior. *J. Mater. Sci.* **2016**, *51*, 1610–1619. [CrossRef]

73. Săndulescu, R.; Tertiş, M.; Cristea, C.; Bodoki, E. New Materials for the Construction of Electrochemical Biosensors. In *Biosensors—Micro and Nanoscale Applications*; Rinken, T., Ed.; InTech: Rijeka, Croatia, 2015; pp. 1–36.

74. Zahra, K.; Majid, M.; Mircea, V.D. Main Allotropes of Carbon: A Brief Review. In *Sustainable Nanosystems Development, Properties, and Applications*; Mihai, V.P., Marius Constantin, M., Eds.; IGI Global: Hershey, PA, USA, 2017; pp. 185–213.

75. Uslu, B.; Ozkan, S.A. Electroanalytical Application of Carbon Based Electrodes to the Pharmaceuticals. *Anal. Lett.* **2007**, *40*, 817–853. [CrossRef]

76. Tîlmaciu, C.-M.; Morris, M.C. Carbon nanotube biosensors. *Front. Chem.* **2015**, *3*, 59. [CrossRef] [PubMed]

77. Pilehvar, S.; De Wael, K. Recent Advances in Electrochemical Biosensors Based on Fullerene-C_{60} Nano-Structured Platforms. *Biosensors* **2015**, *5*, 712–735. [CrossRef] [PubMed]

78. Celik, N.; Balachandran, W.; Manivannan, N. Graphene-based biosensors: Methods, analysis and future perspectives. In *IET Circuits, Devices and Systems*; Institution of Engineering and Technology: Stevenage, UK, 2015; Volume 9, pp. 434–445.

79. Yáñez-Sedeño, P.; Campuzano, S.; Pingarrón, J. Carbon Nanostructures for Tagging in Electrochemical Biosensing: A Review. *J. Carbon Res.* **2017**, *3*, 3. [CrossRef]

80. Shi, H.; Wei, J.; Qiang, L.; Chen, X.; Meng, X. Fluorescent Carbon Dots for Bioimaging and Biosensing Applications. *J. Biomed. Nanotechnol.* **2014**, *10*, 2677–2699. [CrossRef] [PubMed]

81. Bardhan, N.M. 30 years of advances in functionalization of carbon nanomaterials for biomedical applications: A practical review. *J. Mater. Res.* **2016**, *32*, 107–127. [CrossRef]

82. Muñoz, J.; Baeza, M. Customized Bio-Functionalization of Nanocomposite Carbon Paste Electrodes for Electrochemical Sensing: A Mini Review. *Electroanalysis* **2017**. [CrossRef]

83. Yang, W.; Ratinac, K.R.; Ringer, S.P.; Thordarson, P.; Gooding, J.J.; Braet, F. Carbon Nanomaterials in Biosensors: Should You Use Nanotubes or Graphene? *Angew. Chem. Int. Ed. Engl.* **2010**, *49*, 2114–2138. [CrossRef] [PubMed]

84. Pumera, M. Graphene in biosensing. *Mater. Today* **2011**, *14*, 308–315. [CrossRef]

85. Potočnik, J. Commission Recommendation of 18 October 2011 on the definition of nanomaterial. *Off. J. Eur. Union* **2011**, *54*, 38–40.

86. Bonaccorso, F.; Lombardo, A.; Hasan, T.; Sun, Z.; Colombo, L.; Ferrari, A.C. Production and processing of graphene and 2D crystals. *Mater. Today* **2012**, *15*, 564–589. [CrossRef]

87. Paton, K.R.; Varrla, E.; Backes, C.; Smith, R.J.; Khan, U.; O'Neill, A.; Boland, C.; Lotya, M.; Istrate, O.M.; King, P.; et al. Scalable production of large quantities of defect-free few-layer graphene by shear exfoliation in liquids. *Nat. Mater.* **2014**, *13*, 624–630. [CrossRef] [PubMed]

88. Coleman, J.N. Liquid Exfoliation of Defect-Free Graphene. *Acc. Chem. Res.* **2012**, *46*, 14–22. [CrossRef] [PubMed]

89. Hummers, W.S.; Offeman, R.E. Preparation of Graphitic Oxide. *J. Am. Chem. Soc.* **1958**, *80*, 1339. [CrossRef]

90. Kuila, T.; Mishra, A.K.; Khanra, P.; Kim, N.H.; Lee, J.H. Recent advances in the efficient reduction of graphene oxide and its application as energy storage electrode materials. *Nanoscale* **2013**, *5*, 52–71. [CrossRef] [PubMed]

91. Morales-Narváez, E.; Baptista-Pires, L.; Zamora-Gálvez, A.; Merkoçi, A. Graphene-Based Biosensors: Going Simple. *Adv. Mater.* **2017**, *29*. [CrossRef] [PubMed]

92. Campuzano, S.; Pedrero, M.; Nikoleli, G.P.; Pingarrón, J.M.; Nikolelis, D.P. Hybrid 2D-nanomaterials-based electrochemical immunosensing strategies for clinical biomarkers determination. *Biosens. Bioelectron.* **2017**, *89*, 269–279. [CrossRef] [PubMed]

93. Pavlidis, I.V.; Patila, M.; Bornscheuer, U.T.; Gournis, D.; Stamatis, H. Graphene-based nanobiocatalytic systems: Recent advances and future prospects. *Trends Biotechnol.* **2014**, *32*, 312–320. [CrossRef] [PubMed]

94. Hong, W.; Bai, H.; Xu, Y.; Yao, Z.; Gu, Z.; Shi, G. Preparation of Gold Nanoparticle/Graphene Composites with Controlled Weight Contents and Their Application in Biosensors. *J. Phys. Chem. C* **2010**, *114*, 1822–1826. [CrossRef]

95. Dey, R.S.; Raj, C.R. Development of an Amperometric Cholesterol Biosensor Based on Graphene−Pt Nanoparticle Hybrid Material. *J. Phys. Chem. C* **2010**, *114*, 21427–21433. [CrossRef]

96. Claussen, J.C.; Kumar, A.; Jaroch, D.B.; Khawaja, M.H.; Hibbard, A.B.; Porterfield, D.M.; Fisher, T.S. Nanostructuring Platinum Nanoparticles on Multilayered Graphene Petal Nanosheets for Electrochemical Biosensing. *Adv. Funct. Mater.* **2012**, *22*, 3399–3405. [CrossRef]

97. Borisova, B.; Sánchez, A.; Jiménez-Falcao, S.; Martín, M.; Salazar, P.; Parrado, C.; Pingarrón, J.M.; Villalonga, R. Reduced graphene oxide-carboxymethylcellulose layered with platinum nanoparticles/PAMAM dendrimer/magnetic nanoparticles hybrids. Application to the preparation of enzyme electrochemical biosensors. *Sens. Actuators B Chem.* **2016**, *232*, 84–90. [CrossRef]

98. Loaiza, O.A.; Lamas-Ardisana, P.J.; Añorga, L.; Jubete, E.; Ruiz, V.; Borghei, M.; Cabañero, G.; Grande, H.J. Graphitized carbon nanofiber–Pt nanoparticle hybrids as sensitive tool for preparation of screen printing biosensors. Detection of lactate in wines and ciders. *Bioelectrochemistry* **2015**, *101*, 58–65. [CrossRef] [PubMed]

99. Vanegas, D.C.; Taguchi, M.; Chaturvedi, P.; Burrs, S.; Tan, M.; Yamaguchi, H.; McLamore, E.S. A comparative study of carbon-platinum hybrid nanostructure architecture for amperometric biosensing. *Analyst* **2014**, *139*, 660–667. [CrossRef] [PubMed]

100. Zeng, Q.; Cheng, J.-S.; Liu, X.-F.; Bai, H.-T.; Jiang, J.-H. Palladium nanoparticle/chitosan-grafted graphene nanocomposites for construction of a glucose biosensor. *Biosens. Bioelectron.* **2011**, *26*, 3456–3463. [CrossRef] [PubMed]

101. Halder, A.; Zhang, M.; Chi, Q. Electrocatalytic Applications of Graphene–Metal Oxide Nanohybrid Materials. In *Advanced Catalytic Materials—Photocatalysis and Other Current Trends*; Norena, L.E., Wang, J.-A., Eds.; InTech: Rijeka, Croatia, 2016. [CrossRef]

102. Li, F.; Huang, Y.; Yang, Q.; Zhong, Z.; Li, D.; Wang, L.; Song, S.; Fan, C. A graphene-enhanced molecular beacon for homogeneous DNA detection. *Nanoscale* **2010**, *2*, 1021–1026. [CrossRef] [PubMed]

103. Dong, H.; Gao, W.; Yan, F.; Ji, H.; Ju, H. Fluorescence Resonance Energy Transfer between Quantum Dots and Graphene Oxide for Sensing Biomolecules. *Anal. Chem.* **2010**, *82*, 5511–5517. [CrossRef] [PubMed]

104. Ma, H.; Wu, D.; Cui, Z.; Li, Y.; Zhang, Y.; Du, B.; Wei, Q. Graphene-Based Optical and Electrochemical Biosensors: A Review. *Anal. Lett.* **2012**, *46*, 1–17. [CrossRef]

105. Morales-Narváez, E.; Merkoçi, A. Graphene Oxide as an Optical Biosensing Platform. *Adv. Mater.* **2012**, *24*, 3298–3308. [CrossRef] [PubMed]

106. Morales-Narvaez, E.; Perez-Lopez, B.; Pires, L.B.; Merkoci, A. Simple Forster resonance energy transfer evidence for the ultrahigh quantum dot quenching efficiency by graphene oxide compared to other carbon structures. *Carbon* **2012**, *50*, 2987–2993. [CrossRef]

107. He, S.; Song, B.; Li, D.; Zhu, C.; Qi, W.; Wen, Y.; Wang, L.; Song, S.; Fang, H.; Fan, C. A Graphene Nanoprobe for Rapid, Sensitive, and Multicolor Fluorescent DNA Analysis. *Adv. Funct. Mater.* **2010**, *20*, 453–459. [CrossRef]

108. Morales-Narváez, E.; Hassan, A.-R.; Merkoçi, A. Graphene Oxide as a Pathogen-Revealing Agent: Sensing with a Digital-Like Response. *Angew. Chem. Int. Ed. Engl.* **2013**, *52*, 13779–13783. [CrossRef] [PubMed]

109. Morales-Narváez, E.; Golmohammadi, H.; Naghdi, T.; Yousefi, H.; Kostiv, U.; Horák, D.; Pourreza, N.; Merkoçi, A. Nanopaper as an Optical Sensing Platform. *ACS Nano* **2015**, *9*, 7296–7305. [CrossRef] [PubMed]

110. Islam, M.S.; Kouzani, A.Z. Variable Incidence Angle Localized Surface Plasmon Resonance Graphene Biosensor. In Proceedings of the 2011 IEEE/ICME International Conference on Complex Medical Engineering, Harbin, China, 22–25 May 2011; pp. 58–63.

111. Wu, L.; Chu, H.S.; Koh, W.S.; Li, E.P. Highly sensitive graphene biosensors based on surface plasmon resonance. *Opt. Express* **2010**, *18*, 14395–14400. [CrossRef] [PubMed]

112. Bruna, M.; Borini, S. Optical constants of graphene layers in the visible range. *Appl. Phys. Lett.* **2009**, *94*, 031901. [CrossRef]

113. Jacek, G.; Dawn, T.H.T. Graphene-based waveguide integrated dielectric-loaded plasmonic electro-absorption modulators. *Nanotechnology* **2013**, *24*, 185202.

114. Singh, M.; Holzinger, M.; Tabrizian, M.; Winters, S.; Berner, N.C.; Cosnier, S.; Duesberg, G.S. Non-covalently functionalized monolayer graphene for sensitivity enhancement of surface plasmon resonance immunosensors. *J. Am. Chem. Soc.* **2015**, *137*, 2800–2803. [CrossRef] [PubMed]

115. Zeng, S.; Hu, S.; Xia, J.; Anderson, T.; Dinh, X.-Q.; Meng, X.-M.; Coquet, P.; Yong, K.-T. Graphene–MoS₂ hybrid nanostructures enhanced surface plasmon resonance biosensors. *Sens. Actuators B Chem.* **2015**, *207*, 801–810. [CrossRef]

116. Battigelli, A.; Ménard-Moyon, C.; Da Ros, T.; Prato, M.; Bianco, A. Endowing carbon nanotubes with biological and biomedical properties by chemical modifications. *Adv. Drug Deliv. Rev.* **2013**, *65*, 1899–1920. [CrossRef] [PubMed]

117. Le Goff, A.; Holzinger, M.; Cosnier, S. Enzymatic biosensors based on SWCNT-conducting polymer electrodes. *Analyst* **2011**, *136*, 1279–1287. [CrossRef] [PubMed]

118. Wang, J. Carbon-Nanotube Based Electrochemical Biosensors: A Review. *Electroanalysis* **2005**, *17*, 7–14. [CrossRef]

119. Ménard-Moyon, C.; Kostarelos, K.; Prato, M.; Bianco, A. Functionalized Carbon Nanotubes for Probing and Modulating Molecular Functions. *Chem. Biol.* **2010**, *17*, 107–115. [CrossRef] [PubMed]

120. Zhu, N.; Chang, Z.; He, P.; Fang, Y. Electrochemical DNA biosensors based on platinum nanoparticles combined carbon nanotubes. *Anal. Chim. Acta* **2005**, *545*, 21–26. [CrossRef]

121. Wu, B.; Ou, Z.; Ju, X.; Hou, S. Carbon Nanotubes/Gold Nanoparticles Composite Film for the Construction of a Novel Amperometric Choline Biosensor. *J. Nanomater.* **2011**, *2011*, 464919. [CrossRef]

122. Zhang, H.; Meng, Z.; Wang, Q.; Zheng, J. A novel glucose biosensor based on direct electrochemistry of glucose oxidase incorporated in biomediated gold nanoparticles–carbon nanotubes composite film. *Sens. Actuators B Chem.* **2011**, *158*, 23–27. [CrossRef]

123. Wu, B.-Y.; Hou, S.-H.; Yin, F.; Zhao, Z.-X.; Wang, Y.-Y.; Wang, X.-S.; Chen, Q. Amperometric glucose biosensor based on multilayer films via layer-by-layer self-assembly of multi-wall carbon nanotubes, gold nanoparticles and glucose oxidase on the Pt electrode. *Biosens. Bioelectron.* **2007**, *22*, 2854–2860. [CrossRef] [PubMed]

124. Yang, M.; Jiang, J.; Yang, Y.; Chen, X.; Shen, G.; Yu, R. Carbon nanotube/cobalt hexacyanoferrate nanoparticle-biopolymer system for the fabrication of biosensors. *Biosens. Bioelectron.* **2006**, *21*, 1791–1797. [CrossRef] [PubMed]

125. Yang, M.; Yang, Y.; Liu, Y.; Shen, G.; Yu, R. Platinum nanoparticles-doped sol–gel/carbon nanotubes composite electrochemical sensors and biosensors. *Biosens. Bioelectron.* **2006**, *21*, 1125–1131. [CrossRef] [PubMed]

126. Hwa, K.-Y.; Subramani, B. Synthesis of zinc oxide nanoparticles on graphene–carbon nanotube hybrid for glucose biosensor applications. *Biosens. Bioelectron.* **2014**, *62*, 127–133. [CrossRef] [PubMed]

127. Kumar, S.; Ahlawat, W.; Kumar, R.; Dilbaghi, N. Graphene, carbon nanotubes, zinc oxide and gold as elite nanomaterials for fabrication of biosensors for healthcare. *Biosens. Bioelectron.* **2015**, *70*, 498–503. [CrossRef] [PubMed]

128. Kroto, H.W.; Heath, J.R.; O'Brien, S.C.; Curl, R.F.; Smalley, R.E. C$_{60}$: Buckminsterfullerene. *Nature* **1985**, *318*, 162–163. [CrossRef]

129. Chlistunoff, J.; Cliffel, D.; Bard, A.J. Electrochemistry of fullerene films. *Thin Solid Films* **1995**, *257*, 166–184. [CrossRef]

130. Afreen, S.; Muthoosamy, K.; Manickam, S.; Hashim, U. Functionalized fullerene (C$_{60}$) as a potential nanomediator in the fabrication of highly sensitive biosensors. *Biosens. Bioelectron.* **2015**, *63*, 354–364. [CrossRef] [PubMed]

131. Han, J.; Zhuo, Y.; Chai, Y.; Yuan, R.; Xiang, Y.; Zhu, Q.; Liao, N. Multi-labeled functionalized C$_{60}$ nanohybrid as tracing tag for ultrasensitive electrochemical aptasensing. *Biosens. Bioelectron.* **2013**, *46*, 74–79. [CrossRef] [PubMed]

132. Li, Y.; Fang, L.; Cheng, P.; Deng, J.; Jiang, L.; Huang, H.; Zheng, J. An electrochemical immunosensor for sensitive detection of *Escherichia coli* O157:H7 using C$_{60}$ based biocompatible platform and enzyme functionalized Pt nanochains tracing tag. *Biosens. Bioelectron.* **2013**, *49*, 485–491. [CrossRef] [PubMed]

133. Pan, N.-Y.; Shih, J.-S. Piezoelectric crystal immunosensors based on immobilized fullerene C$_{60}$-antibodies. *Sens. Actuators B Chem.* **2004**, *98*, 180–187. [CrossRef]

134. Zhuo, Y.; Ma, M.-N.; Chai, Y.-Q.; Zhao, M.; Yuan, R. Amplified electrochemiluminescent aptasensor using mimicking bi-enzyme nanocomplexes as signal enhancement. *Anal. Chim. Acta* **2014**, *809*, 47–53. [CrossRef] [PubMed]

135. Xu, X.; Ray, R.; Gu, Y.; Ploehn, H.J.; Gearheart, L.; Raker, K.; Scrivens, W.A. Electrophoretic Analysis and Purification of Fluorescent Single-Walled Carbon Nanotube Fragments. *J. Am. Chem. Soc.* **2004**, *126*, 12736–12737. [CrossRef] [PubMed]

136. Xiao, A.; Wang, C.; Chen, J.; Guo, R.; Yan, Z.; Chen, J. Carbon and Metal Quantum Dots toxicity on the microalgae *Chlorella pyrenoidosa*. *Ecotoxicol. Environ. Saf.* **2016**, *133*, 211–217. [CrossRef] [PubMed]

137. Zheng, X.T.; Ananthanarayanan, A.; Luo, K.Q.; Chen, P. Glowing Graphene Quantum Dots and Carbon Dots: Properties, Syntheses, and Biological Applications. *Small* **2015**, *11*, 1620–1636. [CrossRef] [PubMed]

138. Baker, S.N.; Baker, G.A. Luminescent Carbon Nanodots: Emergent Nanolights. *Angew. Chem. Int. Ed. Engl.* **2010**, *49*, 6726–6744. [CrossRef] [PubMed]

139. Liu, H.; Ye, T.; Mao, C. Fluorescent Carbon Nanoparticles Derived from Candle Soot. *Angew. Chem. Int. Ed. Engl.* **2007**, *46*, 6473–6475. [CrossRef] [PubMed]

140. Wang, Y.; Hu, A. Carbon quantum dots: Synthesis, properties and applications. *J. Mater. Chem. C* **2014**, *2*, 6921–6939. [CrossRef]

141. Lim, S.Y.; Shen, W.; Gao, Z. Carbon quantum dots and their applications. *Chem. Soc. Rev.* **2015**, *44*, 362–381. [CrossRef] [PubMed]

142. Bu, D.; Zhuang, H.; Yang, G.; Ping, X. An immunosensor designed for polybrominated biphenyl detection based on fluorescence resonance energy transfer (FRET) between carbon dots and gold nanoparticles. *Sens. Actuators B Chem.* **2014**, *195*, 540–548. [CrossRef]

143. Li, H.; Zhang, Y.; Wang, L.; Tian, J.; Sun, X. Nucleic acid detection using carbon nanoparticles as a fluorescent sensing platform. *Chem. Commun.* **2011**, *47*, 961–963. [CrossRef] [PubMed]

144. Kim, J.E.; Choi, J.H.; Colas, M.; Kim, D.H.; Lee, H. Gold-based hybrid nanomaterials for biosensing and molecular diagnostic applications. *Biosens. Bioelectron.* **2016**, *80*, 543–559. [CrossRef] [PubMed]

MDPI AG

St. Alban-Anlage 66

4052 Basel, Switzerland

Tel. +41 61 683 77 34

Fax +41 61 302 89 18

http://www.mdpi.com

Sensors Editorial Office

E-mail: sensors@mdpi.com

http://www.mdpi.com/journal/sensors

www.ingramcontent.com/pod-product-compliance
Lightning Source LLC
Chambersburg PA
CBHW051844210326
41597CB00033B/5771